Springer Geophysics

The Springer Geophysics series seeks to publish a broad portfolio of scientific books, aiming at researchers, students, and everyone interested in geophysics. The series includes peer-reviewed monographs, edited volumes, textbooks, and conference proceedings. It covers the entire research area including, but not limited to, applied geophysics, computational geophysics, electrical and electromagnetic geophysics, geodesy, geodynamics, geomagnetism, gravity, lithosphere research, paleomagnetism, planetology, tectonophysics, thermal geophysics, and seismology.

More information about this series at http://www.springer.com/series/10173

Arkoprovo Biswas
Editor

Self-Potential Method: Theoretical Modeling and Applications in Geosciences

 Springer

Editor
Arkoprovo Biswas
Department of Geology
Centre of Advanced Study
Institute of Science
Banaras Hindu University
Varanasi, India

ISSN 2364-9119   ISSN 2364-9127 (electronic)
Springer Geophysics
ISBN 978-3-030-79335-7   ISBN 978-3-030-79333-3   (eBook)
https://doi.org/10.1007/978-3-030-79333-3

This Springer imprint is published by the registered company Springer Nature Switzerland AG
The registered company address is: Gewerbestrasse 11, 6330 Cham, Switzerland

Preface

Self-Potential (SP) is a passive method, as it employs measurements of naturally occurring potential differences (PDs) due to electrochemical, electrokinetic, and thermoelectric fields in the earth's subsurface. The method was first used for understanding mineralization potential primarily associated with sulfide ore bodies, which was explained by electrochemical mechanism and oxidation potential. Self-potentials are generated by electrochemical processes involving reductions and oxidations that occur when ores or buried metals come into contact with rocks, groundwater, or rock fluids. Different kinds of electrical potentials are produced in the ground or within the subsurface. Natural potentials can occur between dissimilar materials, near varying concentrations of electrolytic solutions, and due to the flow of fluids through porous rocks. In geophysical prospecting, the spontaneous potential method takes advantage of the natural electrical potential phenomenon in the subsurface. If two electrodes are driven into the ground and connected to a sensitive voltmeter, an electric voltage is present between them. Such PDs may range from a few millivolts to one volt or more. SP prospecting is the simplest in terms of operation and is inexpensive compared to other geophysical methods. The controlling factor in all cases is underground water. Sulfide ore bodies have been detected by the SP generated by ore bodies acting as batteries. These potentials are associated with various occurrences such as weathering of sulfide mineral bodies, variation in rock properties (mineral content) at geological contacts, bioelectric activity of organic material, corrosion of metals, thermal and pressure gradients in underground fluids, and other phenomena of a similar nature such as nuclear blasts, thunderstorms, and charge clouds.

Mineralization potentials are of interest in mineral exploration because they are associated predominantly with massive sulfide ore bodies and explained using an electrochemical mechanism or related to oxidation potential. The amplitude of SP anomalies encountered over mineralized targets can be in the range of hundreds of millivolts (mV) such that other naturally occurring potentials are typically ignored and considered noise. A number of these "background" potentials are signals of interest in other applications of the method. The superposition of multiple source contributions can explain the anomaly. These source mechanisms may be described using the coupled flow theory. Thermoelectric, electrochemical, and electrokinetic source mechanisms give rise to electrical conduction current flow that can be characterized using the SP method.

The fundamental nature of SP anomalies is identified with the half-redox reactions where both electron contributors and acceptors are cooperating with an electronic conductor permitting the long-range transport of electrons. Also, fault, vein, and thin sheet-like mineralization show a solid self-potential anomaly that have demonstrated the mechanism of generation of SP anomaly in a tank experiment. Also, aside from SP anomalies identified with mineralization, it has likewise been explained by the action of plant roots, i.e., *bioelectric potential*. Moreover, stable SP anomalies of comparable magnitude may also be registered due to the presence of metallic cultural noise such as pipelines, steel well casings, metallic fences, and utility boxes. This type of SP signal is believed to be caused by the movement of electrons in response to spatial variations in the redox potential at heterogeneous metal–electrolyte interfaces. Since its discovery, the SP method has been widely used in many geophysical applications, such as sulfide and graphite exploration, groundwater exploration, geothermal exploration, landslide studies, seepage control/dam seepage, cavity detection, earthquake prediction from fluid flow, volcanic eruptions, buried palaeochannels, hydraulic fracturing in rocks for hydrocarbon recovery, archaeology, glaciology/glacial geomorphology, tracing shear zones in the continental crust, coal fire mapping, and engineering and environmental applications. Moreover, SP logging methods have been widely applied in oil exploration. Mineral exploration and very recently in the problems of saline water intrusion.

This edited book provides stimulating, theoretical modeling and its advancement in SP data and its applications in various investigations. Some chapters on the data analysis and inverse theory are provided, and case studies amply illustrate chapters. This is an essential edited book for advanced undergraduate and graduate students in geophysics and a treasured reference for enthusiastic geophysicists, geologists, hydrologists, archaeologists, civil and geotechnical engineers, and others who use geophysics and its application in their professional research and teaching. The book will also serve as a valuable reference for geoscientists, engineers, and others engaged in academic, government, or industrial pursuits that call for SP investigation. The accessible techniques are characterized by different penetration and resolution capabilities. Theoretical advancements, modeling, inversion, and case studies on the application of SP are also illustrated.

Varanasi, India   Arkoprovo Biswas

Acknowledgments

The present work has developed from a prolonged series of energetic communication with my colleagues, seniors, and juniors both in India and abroad, especially during the last few years, and also my past ten year experiences working on the topic *Self-Potential Method: Theoretical Modeling and Applications in Geosciences.* The present book will showcase some advancements in mathematical modeling, inversion on self-potential methods, and its application in actual field data. This work also stresses the significance of SP in various studies such as exploration, contamination, and environmental problems. The book has a broad literature survey, and all pains have taken to take care of proper citation at the requisite places. I would personally like to thank them on behalf of the authors of other chapters and me, respectively. Any inadvertent error/omission in this regard is regretted. Also, we thank the "Authors" of respective chapters, who have contributed to the same with their valuable time, effort, and expertise in the separate area of research/study, as provided in each chapter, and also completed their chapter on time even though the current pandemic situation. I would also like to present my sincere gratitude to my mentor **Prof. S. P. Sharma**, IIT Kharagpur, who introduced me to work in SP and its application in exploration to contamination and other studies. I would also like to thank my University for supporting me in the successful development of this book.

Arkoprovo Biswas

Contents

Editor and Contributors

About the Editor

Dr. Arkoprovo Biswas is an Assistant Professor at the Department of Geology, Institute of Science, Banaras Hindu University (BHU), Varanasi. He received his B.Sc. (2002) in Geology from Presidency College, University of Calcutta, M.Sc. (2004) in Geological Science, M.Tech. (2006) in Earth and Environmental Science from IIT Kharagpur, P. G. Diploma (2009) in Petroleum Exploration from Annamalai University. He joined Geostar Surveys India Pvt. Ltd. as a Geophysicist in 2006 and later joined WesternGeco Electromagnetics, Schlumberger as an On-Board Data Processing Field Engineer/Geophysicist in 2007 and served there till 2008. In 2013 he received his Ph.D. in Exploration Geophysics from IIT Kharagpur. Later, he joined the Department of Earth and Environmental Sciences, Indian Institute of Science Education and Research Bhopal as a Visiting Faculty in 2014 and completed his tenure in 2015. He again joined Wadia Institute of Himalayan Geology (WIHG) Dehradun in 2016 as Research Associate and later he joined BHU in October 2017. His main Research Interest includes Near Surface Geophysics, Integrated Electrical and Electromagnetic Methods, Geophysical Inversion, Mineral, and Groundwater Exploration and Subsurface contamination. He has published more than 40 papers on theoretical modeling, inversion, and application in practical geoscience problems in peer-reviewed international, national journals and 6 book chapters. Dr. Biswas received the Prestigious **M. S. Krishnan Medal** of

Indian Geophysical Union (IGU), Hyderabad for the year 2019 and **B. C. Patnaik Memorial Gold Medal** from the Society of Geoscientists and Allied Technologists (SGAT), Bhubaneshwar, Odisha, India in 2019. He is also a Life Member of the Indian Geophysical Union and an Active member of SEG (USA). He is also an Associate Editor of the *Journal of Earth System Sciences, Spatial Information Research*, Springer, and *Results in Geophysical Sciences*, Elsevier. Dr. Biswas is also a member of the Editorial Advisory Board of *Natural Resources Research*, and, *Contributions to Geophysics and Geodesy* Journals. He also published an Edited Book with Springer on *Advances in Modeling and Interpretation in Near Surface Geophysics*.

Contributors

Maha Abdelazeem Lab of Geomagnetism, National Research Institute for Astronomy and Geophysics, NRIAG, Helwan, Egypt

Erna Apriliani Department of Mathematics, Institut Teknologi Sepuluh Nopember (ITS), Surabaya, Indonesia

Laurenţiu Artugyan Department of Geography, West University of Timisoara, Timisoara, Timis, Romania

Rima Chatterjee Department of Applied Geophysics, IIT-ISM, Dhanbad, India

Mahmoud Elhussein Faculty of Science, Geophysics Department, Cairo University, Giza, Egypt

Lev V. Eppelbaum Faculty of Exact Sciences, Department of Geophysics, Tel Aviv University, Tel Aviv, Israel;
Azerbaijan State Oil and Industry University, Baku, Azerbaijan

Khalid S. Essa Faculty of Science, Geophysics Department, Cairo University, Giza, Egypt

Fajriani Department of Physics, Universitas Samudra, Jl. Prof. Syarief Thayeb, Kota Langsa, Aceh, Indonesia

Juan Luis Fernández-Martínez Group of Inverse Problems, Optimization and Machine Learning. Mathematics Department, University of Oviedo, Oviedo, Spain

Zulima Fernández-Muñiz Group of Inverse Problems, Optimization and Machine Learning. Mathematics Department, University of Oviedo, Oviedo, Spain

Mohamed Gobashy Faculty of Science, Department of Geophysics, Cairo University, Giza, Egypt

Dewashish Kumar CSIR-National Geophysical Research Institute, Hyderabad, India

Madhvi Department of Geophysics, Banaras Hindu University, Varanasi, India

Mangal Maurya Department of Geophysics, Banaras Hindu University, Varanasi, India

Neha Rai Department of Geophysics, Banaras Hindu University, Varanasi, India

Saifuddin Department of Physics, Institut Teknologi Sepuluh Nopember (ITS), Surabaya, Indonesia

Coşkun Sari Engineering Faculty, Department of Geophysics, Dokuz Eylül University, Buca, İzmir, Turkey

Uma Shankar Department of Geophysics, Banaras Hindu University, Varanasi, India

Petek Sindirgi Faculty of Engineering, Department of Geophysical Engineering, Dokuz Eylül University, İzmir, Turkey

Dip Kumar Singha Department of Geophysics, Banaras Hindu University, Varanasi, India

Wahyu Srigutomo Physics of Earth and Complex System, Department of Physics, Faculty of Mathematics and Natural Sciences, Institut Teknologi Bandung (ITB), Bandung, Indonesia

Y. Srinivas Department of GeoTechnology, Manonmaniam Sundaranar University, Tirunelveli, India

N. Sundararajan Department of Earth Sciences, Sultan Qaboos University, Muscat, Oman

Sungkono Department of Physics, Institut Teknologi Sepuluh Nopember (ITS), Surabaya, Indonesia

Emre Timur Engineering Faculty, Department of Geophysics, Dokuz Eylül University, Buca, İzmir, Turkey

Petru Urdea Department of Geography, West University of Timisoara, Timisoara, Timis, Romania

Şenol Özyalin Faculty of Engineering, Department of Geophysical Engineering, Dokuz Eylül University, İzmir, Turkey

Chapter 1
Analytical Methods in the Interpretation of Self-Potential Anomalies—A Comprehensive Review

N. Sundararajan and Y. Srinivas

Abstract The self potential (SP) method is one of the simple, elegant and economical geophysical tools that has many diverse applications in the subsurface exploration for minerals, ground water etc. It is a passive method based on the natural occurrence of electrical field ranging from less than a milli volt to a maximum of one or two volts on/under the surface of earth. The surface measurements of SP caused by subsurface targets can be estimated in terms of its size, dimension, depth, width, inclination etc. often by qualitative methods directly from the shape of the anomalies. In this, the profile shape, amplitude, polarity and contour pattern are also considered. In addition, the target/source is assumed to lie directly below the minimum of the anomaly. On the other hand, a large number of quantitative methods that are in vogue approximate the subsurface targets as regular geometrical shapes such as cylinder, sphere etc. These quantitative methods are mostly based on mathematical transforms such as Fourier, Hartley, Hilbert transforms and some of their modifications in addition to a few soft computing tools like artificial neural network (ANN) etc. In this chapter, some of these methods are illustrated with theoretical examples and also exemplified with real field data. A simplified mathematical treatment of these techniques, the merits of the methods are also included.

Keywords Self-Potential · Analytical methods · Subsurface structures · Mineral exploration

1.1 General Introduction

The self-potential (SP) method is one of the oldest methods that was proposed in the beginning of nineteenth century (1830) by Robert Fox) who carried out SP experiments with mines in UK. The SP method is a passive method similar to gravity and magnetic methods that measures the natural electrical potential observed on/in

N. Sundararajan (✉)
Department of Earth Sciences, Sultan Qaboos University, Muscat, Oman

Y. Srinivas
Department of GeoTechnology, Manonmaniam Sundaranar University, Tirunelveli, India

© The Author(s), under exclusive license to Springer Nature Switzerland AG 2021
A. Biswas (ed.), *Self-Potential Method: Theoretical Modeling and Applications in Geosciences*, Springer Geophysics, https://doi.org/10.1007/978-3-030-79333-3_1

the subsurface due to geological/hydrogeological and geochemical reasons which usually cause potential difference between any two measuring points. The major factor for the origin of SP is ground water and is sensitive to the flow of the ground water and to the chemistry of both the pore water and the pore water mineral interface. The SP is also known as Static potential or Spontaneous potentials that are usually caused by charge separation in clay or other minerals, due to the presence of semi-permeable interface impeding the diffusion of ions through the pore space of rocks, or by natural flow of a conducting fluid through the rocks. In most cases, the SP measured is only related to the electrochemical potential. The potentials are measured in millivolts (mV) relative to a reference point (base station), where the potential is assumed to be zero volts. SP can range from less than a milli volt (mv) to over one volt and the sign of the potential is an important diagnostic factor in the interpretation of SP anomalies.

Further, there exist two different types of SP anomalies namely **mineral potential** and **background potential**. The mineral potential is due to sulphide ore bodies— generally −ve in the range of a few 100 mv and it is the most important in mineral exploration associated with massive sulphide ore bodies. It is characterized by large negative SP anomalies (100–1000 mV) that can be observed particularly over deposits of pyrite, chalcopyrite, pyrrhotite, magnetite,and graphite. The potentials are almost invariably negative over the top of the deposit and are quite stable in time. Also, mineral potential is constant and unidirectional due to electrochemical processes. Background potential is due to geochemical process, bioelecric activity, ground water movement and topography, either + ve or −ve and ranges from less than 300 mv except in the case of topography wherein it varies up to −2 V. Background potentials fluctuate with time caused by different processes ranging from AC currents induced by thunderstorms.

Variations in Earth's magnetic fields, effects of heavy rainfalls etc. also fluctuate background potentials. Generally to measure SP, two non polarisable porous pot electrodes are connected to a high precision multi meter with an impedance $>10^8$ Ω and capable of measuring to at least 1 mv.Each electrode is made up of a copper (cu). Electrodes dipped in a saturated solution of copper sulphate ($CuSO_4$) which can penetrate through the porous base to the pot in order to make electrical contact with the ground. Zinc & silver electrodes and their respective sulphate solutions can also be used.

There are two techniques by which SP can be measured and they are known as **potential gradient and potential amplitude**. In potential gradient method, two electrodes at a fixed separation (5–10 m) between which the potential difference- pd- measured is divided by the electrode separation results potential gradient in mv/m. The point to which this observation applied is the mid point between two electrodes. The electrodes are moved along the traverse and the pd measured every time and recorded. On the other hand, in potential amplitude method, one electrode is fixed at a base station on a mineralized ground and to measure the pd, the second electrode is moved along the traverse. Depth of investigation ranges approximately 30–100 m depending upon the depth to targets as well nature of overburden.

Generally, SP consists of two different components of potentials known as *Static* and *Variable* potential. Static part of SP is called signal and the variable part of SP is termed as noise. The variable part is due to atmospheric effect which is in the range of 5–10 Hz and at best be minimized by repeated measurements along the profile and averaged for further interpretation. Electrical noise may incur if the measurements are made too soon after heavy rain or near running water. Best be avoided by suspending the SP work during rainy days. Topography also causes minor variation and hence require correction.

Further, SP measured over a large area may have a regional trend due to "telluric current" of 100 mv/km. Mineral potential may be superimposed upon this regional gradient. Prior to interpretation of the anomaly which is due to the source/target, it has to be isolated from regional gradient as is done in gravity regional and residual separation (R/R) separation. Topography causes potential variation particularly at the highly elevated locations, the SP anomaly is −ve for which a minor correction need to be applied prior to interpretation.

1.2 Interpretation of SP Anomalies

Usually, interpretation consists of looking for the order of magnitude of anomalies in the range 0–20 mv as normal variation, 20–50 mv possibly of interest, especially if observed over a fairly large area, more than 50 mv as definite anomaly and 400–1000 mv as very large anomalies. Generally, SP anomalies are interpreted by means of "qualitative method", wherein the target/source is assumed to lie directly below the minimum of the anomaly. The anomaly half width provides a rough estimate of depth. The symmetry or asymmetry of anomaly provides the attitude of the body. The presence and type of over burden can have strong effect on the presence or absence of SP anomaly. For example sand has very little effect where as a clay cover may mask the SP anomaly of the subsurface source/target.

In quantitative interpretation of SP anomalies, generally it is assumed that the causative bodies to be regular geometrical shapes that can be described with appropriate analytical formulas. Logarithmic-curve matching (Murty and Haricharan 1984), the method of characteristic points (Rao et al. 1970), and the method of nomograms (Bhattacharya and Roy 1981) all involve many approximations. The method of least squares necessitates a series of trials to minimize the error between the observed and calculated values. Spectral analysis is reliable only for very long profiles. Whereas the ease of use and accuracy of results vary with the specific interpretation technique, they are all subject to many constraints. None of these methods yields a precise location of origin of the source of the anomaly, which is a prerequisite for meaningful interpretation.

Some of the analytical methods of self potential anomaly interpretation based on mathematical tools and techniques employing Hilbert transform, modified Hilbert transform/Sundararajan transform, Fourier transform/Hartley transform, Mellin transform etc.besides soft computing tools such as artificial neural networks (ANN) etc.play a significant role to realize reliable estimation of source parameters of subsurface targets. In this chapter, a brief review of these mathematical tools and the process of interpretation are given as hereunder.

During early 1990s, Sundararajan et al. (1990, 1998), Sundararajan and Chary (1993) suggested the use of Hilbert transforms for the interpretation of SP anomalies because this method is effective in the presence of random noise. This method is based on using the real roots of the anomaly and its Hilbert transform/modified Hilbert transform, horizontal and vertical derivatives, avoids many of the drawbacks listed above. In addition, a precise location of origin is achieved by using the Hilbert transform and its modified version by means of amplitude of analytic signal. The method is simple, elegant, straightforward, and above all is free from any assumptions. It also can be automated easily. Theoretical and field examples illustrate the method in the following section of some selected models and well known field anomalies.

1.3   Hilbert Transforms

The Hilbert transform (HT) and its modified version in geophysical data processing and interpretation has gained importance over the last more than half a century (Nabhigian 1972; Mohan et al. 1982; Sundararajan et al. 1998, 2000; Sundararajan and Srinivas 2010). In these methods, the parameters of the causative bodies are evaluated as functions of some characteristic points of the anomaly and its Hilbert transform. The HT can physically be realized as a 90° (270° in the case of MHT) phase shifter is not only useful in extracting the parameters such as depth, inclination, width etc. of the causative bodies but also plays a significant role in exact spatial location of the subsurface sources. The modified Hilbert transform is also known as 'Sundararajan transform' in literature (Sundararajan et al. 2000). In this section, it is illustrated some basic concepts of HT and how they are applied in the interpretation of self potential anomalies of certain simple geometrical structures such as 2-D horizontal circular cylinder, sphere, inclined sheet etc.

The Hilbert transform HT(x) and modified Hilbert transform MHT(x) of self potential anomalies represented by any real function SP(x) can be defined as

$$HT1(x) = \frac{1}{\pi} \int_0^\infty [Im\ SP(\omega)\cos(\omega x) - ReSP(\omega)\sin(\omega x)]d\omega$$

$$HT2(x) = \frac{1}{\pi} \int_0^\infty [Im\ SP(\omega)\cos(\omega x) + ReSP(\omega)\sin(\omega x)]d\omega$$

where ImSP(ω) and ReSP(ω) are the imaginaray and real components of the Fourier transform of SP(x) implying that the HT can be computed via the Fourier transform.

Alternatively, the HT can also be computed in space domain as convolution of SP(x) with $1/\pi x$ given by

$$HT(x) = (1/\pi x) * SP(x)$$

where * is the convolution operator.

Yet another way by which the Hilbert transform can be conceived is the horizontal and vertical derivatives of SP(x) that form a Hilbert transform pair. That is,

$$SP_x (x)\langle - - - - - \rangle SP_z (x)$$

Here $SP_z (x)$ is the vertical derivative and SPx(x) is the horizontal derivative of the SP anomaly SP(x) which form the Hilbert transform pair.

1.4 Analytic Signal and Amplitude

Locating the origin is of paramount importance in the interpretation of all geophysical anomalies, that can be achieved with utmost accuracy by the amplitude of analytical signal in a couple of ways as discussed hereunder.

The analytic signal of a self potential represented by SP(x) can be expressed as:

$$AS(x) = SP(x) - iHT(x)$$

where HT(x) is the Hilbert transform of SP(x). The amplitude of analytic signal can be deciphered as:

$$A(x) = \sqrt{SP(x)^2 + HT(x)^2}$$

In general, the amplitude A(x) of analytic signal attains its maximum exactly over the subsurface targets/source in structures whose width is less than the depth. On the other hand, for the structures whose width is greater than the depth, A(x) results two peaks flanked by a minimum at the centre. In this case, the minimum corresponds to the centre of the source/target and the distance between the two peaks yields the width of the target. Further, if the modified Hilbert transform HT2(x) is used for the extraction of parameters, then in the above relation, HT(x) can be replaced with its modified version HT2(x). This can be defined as under while using Hilbert transform or modified Hilbert transform (Nabhigian 1972; Sundararajan et al. 1998; Sundararajan and Srinivs 2010)

$$A1(x) = \sqrt{SP(x)^2 + HT1(x)^2}$$

and

$$A2(x) = \sqrt{SP(x)^2 + HT2(x)^2}$$

Alternatively, the intersection of the HT1(x) and HT2(x) also corresponds to the origin (Fig. 1.1). Similarly, the amplitudes as defined above A1(x) and A2(x) do intersect over the origin (Fig. 1.2).

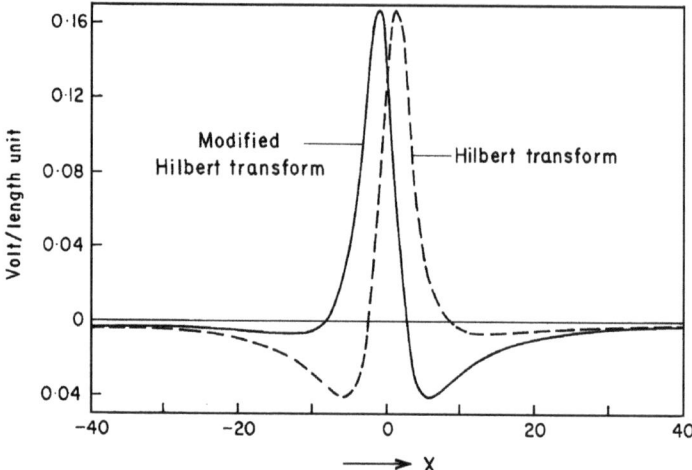

Fig. 1.1 Hilbert transform and its modified version

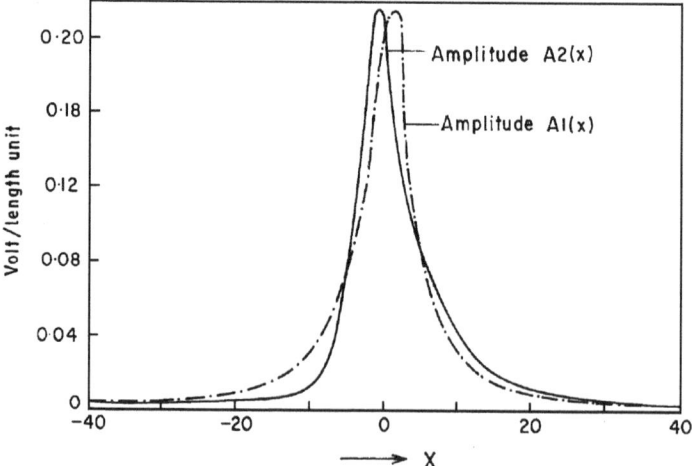

Fig. 1.2 Amplitudes A1(x) and A2(x) of analytic signal

1.5   2-D Horizontal Circular Cylinder

The geometry of obliquely polarized 2-D horizontal circular cylinder with radius is shown in Fig. 1.3. In the Cartesian coordinate system, 'O' is the origin which is on the surface at a point vertically above the center of the cylinder. The axis of the cylinder is parallel to the y-axis. AA′ is the axis of polarization. It makes an angle 'α' with the x-axis. P is the point of observation at a distance 'x' from the origin, 'α' is the angle between the axis of polarization and the line passing through the centre of the sphere and P and Po is the point where the potential is zero. Therefore, the potential at a point P on the surface is given (Sundararajan and Srinivas 1996) as:

The self potential due to such a cylindrical structure can be expressed as (Sundararajan and Srinivas 1996):

$$SP1(x) = \frac{A[\, x \cos(\alpha) - z sin(\alpha)}{\left(x^2 + z^2\right)}$$

where z-is the depth to the centre of the cylinder, 'α' is the angle between the horizontal axis and the axis of polarization and A is a constant comprising the polarization current (I) and the resistivity (ρ).

In this case, the computation of Hilbert transform HT1(x) and its modified version HT2(x) can be realized via the Fourier transform and therefore, the real and imaginary

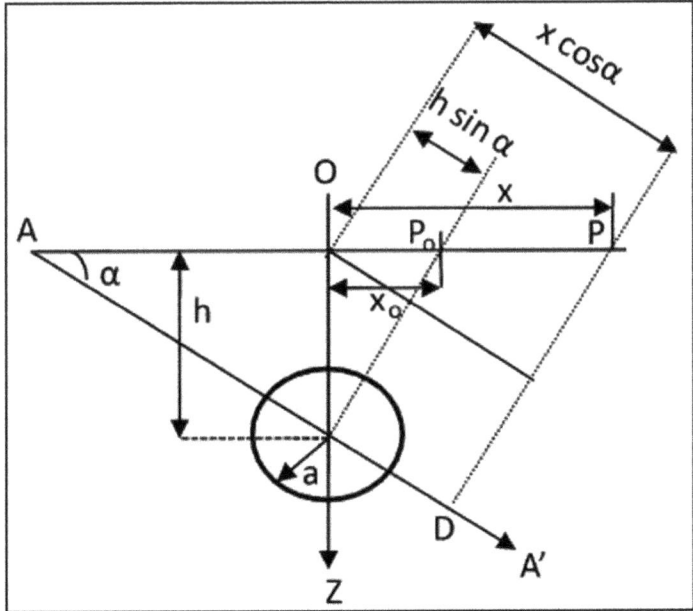

Fig. 1.3 Geometry of the 2-D horizontal circular cylinder

components of the Fourier transform of SP1(x) are derived as:

$$\text{ReSP1}(\omega) = K\pi \, sin(\alpha)e^{-\omega z}$$

$$\text{ImSP1}(\omega) = K\pi \, cos(\alpha)e^{-\omega z}$$

Using these ReSP1 (ω) and ImSP1 (ω) components in the equations of HT and the MHT, the Hilbert transform and its modified version can be obtained as:

$$HT1(x) = \frac{A[\, z\cos(\alpha) + x\, sin(\alpha)}{\left(x^2 + z^2\right)}$$

$$HT2(x) = \frac{A[\, z\cos(\alpha) - x\, sin(\alpha)}{\left(x^2 + z^2\right)}$$

The graphical plots of SP1(x) and HT1(x) or [SP1(x) and HT2(x)] intersect at one point since SP1(x) and HT1(x) or H2(x) are of first degree in x and at this point of intersection say x_1, the following holds good:

$$\text{SP1}(x) = \text{HT1}(x) \text{ at x} = x_1$$

$$\text{SP1}(x) = \text{HT2}(x) \text{ at x} = x_1$$

That is, when

$$\text{SPl}(x) = \text{HT1}(x)$$

$$i, e \quad \frac{A[\, x\cos(\alpha) - z sin(\alpha)}{\left(x^2 + z^2\right)} = \frac{A[\, z\cos(\alpha) + x\, sin(\alpha)}{\left(x^2 + z^2\right)} \text{ at x} = x_1$$

On simplification it results, the depth 'z' as

$$z = x_1 \left[\frac{\sin\alpha - cos\alpha}{\sin\alpha + cos\alpha} \right]$$

This solution for depth 'z' is dependent in 'α' which itself is an unknown to be evaluated and therefore to be ignored. On the other hand, the following results the depth independently as a function of the abscissa as illustrated hereunder.

$$\text{SP1}(x) = \text{HT2}(x)$$

$$i, e \quad \frac{A[\, x\cos(\alpha) - z sin(\alpha)}{\left(x^2 + z^2\right)} = \frac{A[\, z\cos(\alpha) - x\, sin(\alpha)}{\left(x^2 + z^2\right)} \text{ at x} = x_1$$

Further simplification lead to the required solution for depth 'z' as,

$$z = x_1$$

This implies that the depth is directly equal to the point of intersection of the SP anomaly SP1(x) and its modified Hilbert transform HT2(x). Once the depth is evaluated, the angle of polarization 'α' can be determined from the SP anomaly SP1(x) and its modified Hilbert transform HT2(x) as:

$$\alpha = \tan^{-1} \frac{[zSP1(x) - xHT2(x)]}{[xSP1(x) - zHT2(x)]}$$

A more accurate solution for 'α' can be obtained as an average taken over several values of 'x'. Finally, the constant term consisting of I (the polarization current) and ρ (resistivity) can be evaluated at x = 0 from equations SP1(0) and HT2(0) as under:

$$A = \sqrt{SP1(0)^2 + HT2(0)^2}$$

By knowing either current or the resistivity, the other quantity may fairly well be determined. Thus, the depth 'z', the polarization angle 'α' and the constant term can be estimated based on the above analysis.

Theoretical and Field Examples: The interpretation procedure elucidated above is illustrated with a theoretical model and exemplified with a field data of Sulleymonkey anomaly of length 260 m in the Ergani copper district, Turkey. The Hilbert and the modified Hilbert transforms HT1(x) and HT2(x) and the SP anomaly SP1(x) are computed and shown in Fig. 1.4 in the case of theoretical model. The point of

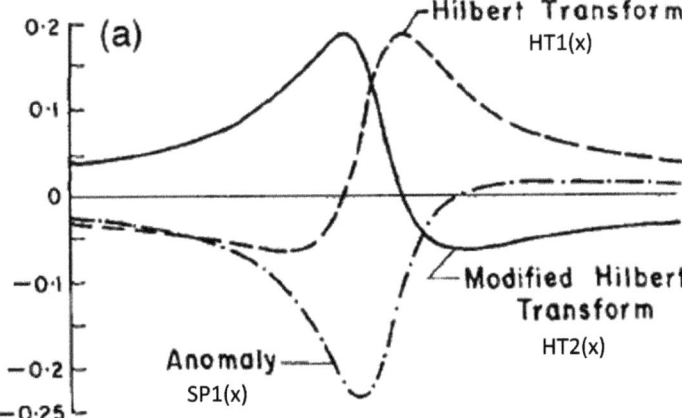

Fig. 1.4 The self potential anomaly due to a 2-D horizontal circular cylinder, the Hilbert transform and the modified Hilbert transform of a theoretical model

intersection of Hilbert transform and the modified Hilbert transform yield precisely the origin, and the abscissa results the depth to the centre of the cylinder. The other parameters are evaluated as discussed in the text. Similarly, the Hilbert transform and its modified versions in addition to the amplitudes of field Sulleymonkey anomaly are computed and shown in Fig. 1.5a–c. The evaluated parameters in both theoretical as well as field examples are presented in Table 1.1 and compared with the other available methods.

1.6 Spherical Structures

The geometry of the obliquely polarized sphere with radius 'a' is considered for the analysis and shown in Fig. 1.6. In the Cartesian coordinate system, 'O' is the origin, on the surface at a point vertically above the centre of the sphere. The axis of the sphere is parallel to the y-axis and AA' is the axis of polarization, 'θ' is included between the polarization and x-axis. P is the point of observation at a distance 'x' from the origin, 'α' is the angle between the axis of polarization and the line passing through the centre of the sphere and P. Q is the point where the potential is zero. The potential at a point P on the surface is given as (Sundararajan and Chary 1993):

$$SP2(x) = C\left[\frac{z\cos(\theta) + x\sin(\theta)}{\left(x^2 + z^2\right)^{1/2}}\right]$$

where 'z' is the depth to the centre of the sphere, 'θ' is the angle of polarization and 'C' is constant comprising the current density (I) and the resistivity (ρ) of the surrounding medium given by $C = I\rho/2\pi$.

As stated earlier, the horizontal and vertical derivatives of SP2(x) are obtained as:

$$SP_x\,2(x) = C\left[\frac{(z^2 - 2x^2)\sin(\theta) - 3xz\cos(\theta)}{\left(x^2 + z^2\right)^{\frac{5}{2}}}\right]$$

$$SP_z\,2(x) = C\left[\frac{(x^2 - 2z^2)\cos(\theta) - 3xz\sin(\theta)}{\left(x^2 + z^2\right)^{\frac{5}{2}}}\right]$$

At x = 0, the horizontal and vertical derivatives reduce to,

$$SP_x\,2(0) = C\sin(\theta)/z^3$$

and

$$SP_z\,2(0) = -2C\cos(\theta)/z^3$$

Field Example

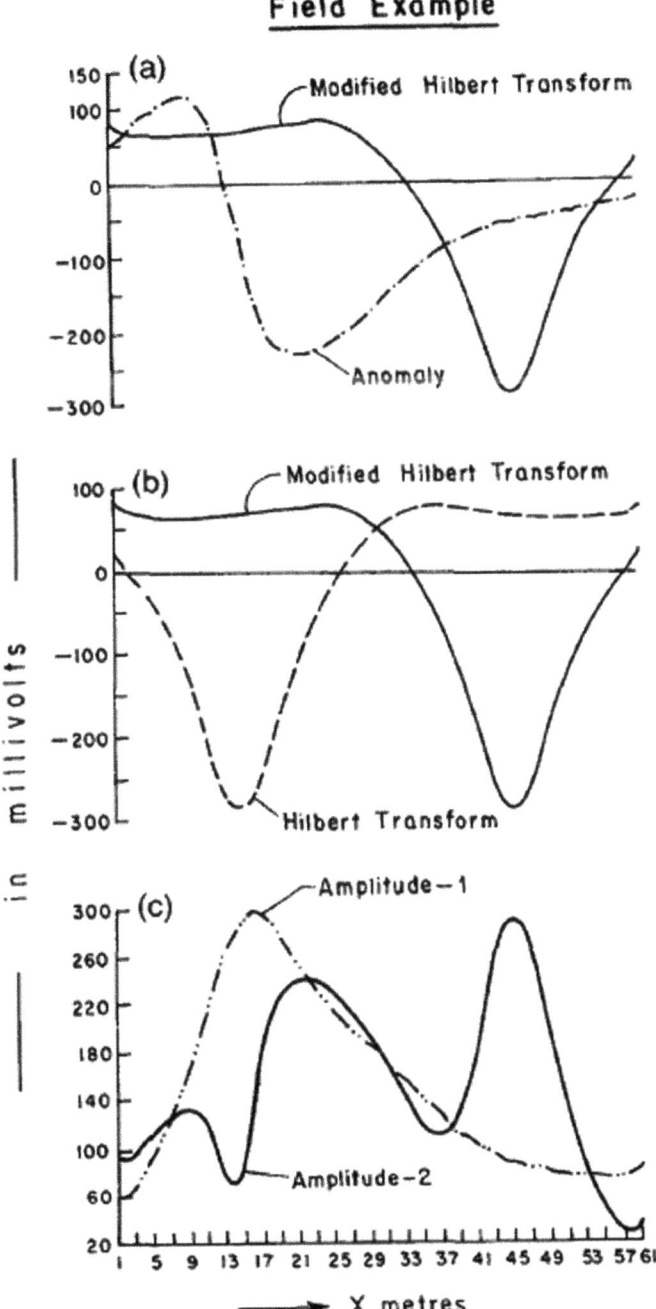

Fig. 1.5 a The self potential field of Sulleymonkey anomaly in the Ergani copper District, Turkey, and the modified Hilbert transform **b** The Hilbert transform of the anomaly and its modified version and **c** The amplitudes A1(x) and A2(x)

Table 1.1 SP interpretation of theoretical model of 2-D horizontal circular cylinder and field self potential Sulleymonkey anomaly, Ergani copper district, Turkey

Parameters	Depth (z)	Polarization angle (α)	A = (I ρ)
Theoretical Example			
Assumed values	4.00 units	60°	1.00
Interpreted values	4.00 units	59°	0.98
Field Example			
Interpreted values by the present method	36.00 m	45°	–
By Youngal (1950)	38.00 m	64°	–
Hartley spectral analysis (Al-Garni and Sundararajan 2011)	35.00 m	48°	–

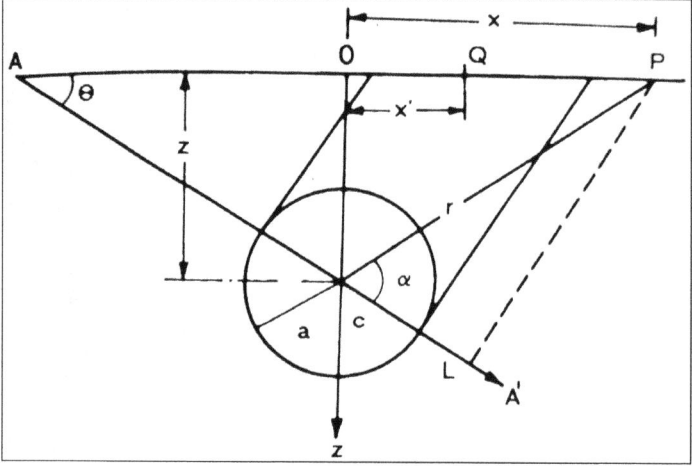

Fig. 1.6 Geometry of the spherical structure

From the above relations, the angle of polarization 'θ' can be determined as:

$$\theta = \tan^{-1}[-2SP_x\,2(0)/SP_z\,2(0)]$$

Further, the derivatives $SP_x\,2(x)$ and $SP_z\,2(x)$ are of second degree in x, they have two real roots say x_1 and x_2 and therefore it can be written as,

$$SP_x\,2(x) \;=\; SP_z\,2(x)\,\text{at x} = x_1 \text{ and } x_2$$

$$i,\,e\;\;C\left[\frac{(z^2 - 2x^2)\sin(\theta) - 3xz\cos(\theta)}{(x^2 + z^2)^{\frac{5}{2}}}\right] = C\left[\frac{(x^2 - 2z^2)\cos(\theta) - 3xz\sin(\theta)}{(x^2 + z^2)^{\frac{5}{2}}}\right] \text{ at x} = x_1 \text{ and } x_2.$$

On simplification, the depth 'z' can be obtained as

$$z = (x_1 + x_2) \left[\frac{(\cos(\theta) + 2\sin(\theta))}{3(\sin(\theta) - \cos(\theta))} \right]$$

As $'\theta'$, is known already, the depth can be evaluated from the above relation. However, it would be worth mentioning here that the depth tends to '∞' at $\theta = 45°$ which is purely a hypothetical in such an analysis and that can be attributed to the fact that $(x_1 + x_2) = 0$. This introduces a catastrophe in the mathematical procedure. That is, the magnitude of the roots of the derivatives are equal and opposite in sign which is seldom encountered in practice in which case the depth further simplified as:

$$z = x_1 = -x_2$$

Finally, the constant term C can be evaluated as by squaring and adding $SP_x\,2(0)$ and $SP_z\,2(0)$ *as*

$$C = \frac{2z^3 \left[SP_x\,2(0)^2 + SP_z\,2(0)^2 \right]^{1/2}}{\left(1 + 3\cos^2(\theta) \right)}$$

Field Example: The procedure detailed above is exemplified by the well known 'Weiss anomaly' of the copper district in eastern Turkey. The anomaly represents the principle profile AA' shown in the contour map (Fig. 1.7). The 'Weiss' anomaly is approximately 1 km north west of the Madam copper mine and is assumed to be due to spherical structure. The assumption is validated by comparing with the computed values and shown in Fig. 1.8. For further clarity, the 'Weiss" anomaly is shown exclusively in Fig. 1.9. The horizontal derivative is obtained by the numerical differentiation and the vertical derivative is computed by the Hilbert transform. The horizontal and vertical derivatives along with amplitude of analytic signal are shown in Fig. 1.10. The depth (z) to the centre of the sphere and the angle of polarization (θ) are evaluated based on the analytical procedure discussed in the text and shown in the following Table 1.2. The depth (z) and the polarization angle (θ) obtained are compared with those of Youngal (1950) and the method of Bhattacharya and Roy (1981).

2-D Inclined Sheets.

The SP field at any given point P on the surface perpendicular to the strike of 2-D inclined sheet of infinite horizontal extent (Fig. 1.11) is given as (Murty and Haricharan 1985)

$$SP3(x) = \frac{I\rho}{2\pi} \ln \left[\frac{(x - a\cos(\alpha))^2 + (h - a\sin(\alpha))^2}{(x + a\cos(\alpha))^2 + (h - a\sin(\alpha))^2} \right]$$

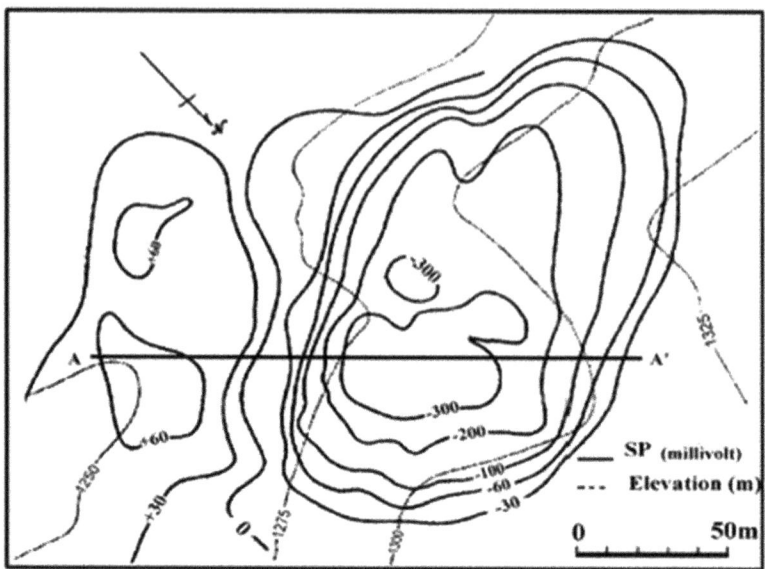

Fig. 1.7 Contour map of the self potential (Weiss) of the Ergani copper district in eastern Turkey with elevation. AA′ is the principle profile

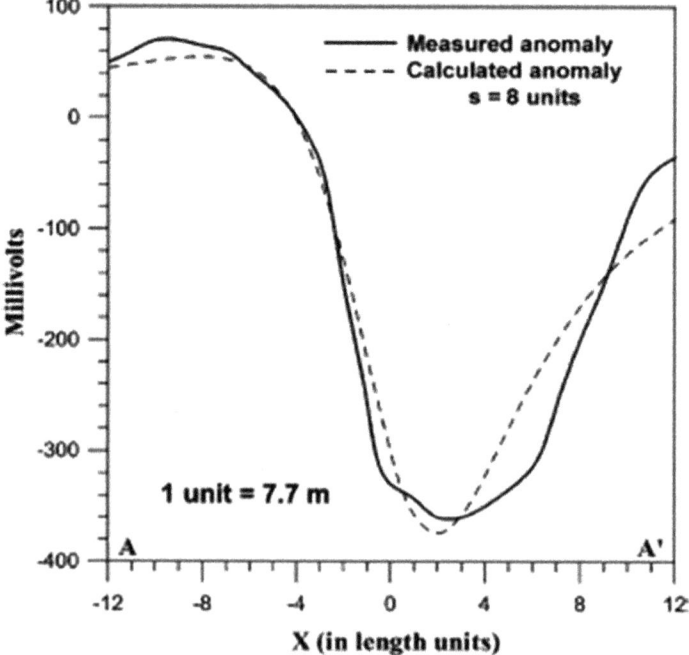

Fig. 1.8 Measured and calculated self potential along AA' over the Weiss anomaly, Ergani, Turkey

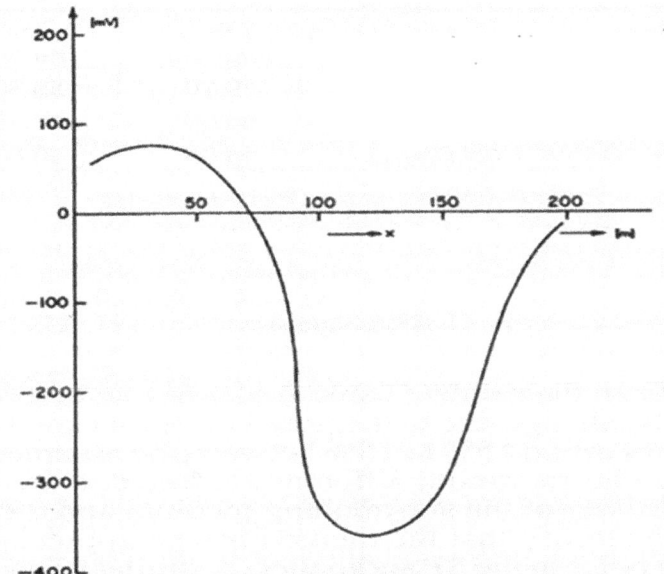

Fig. 1.9 Self potential anomaly (Weiss) of the Ergani copper district in eastern Turkey

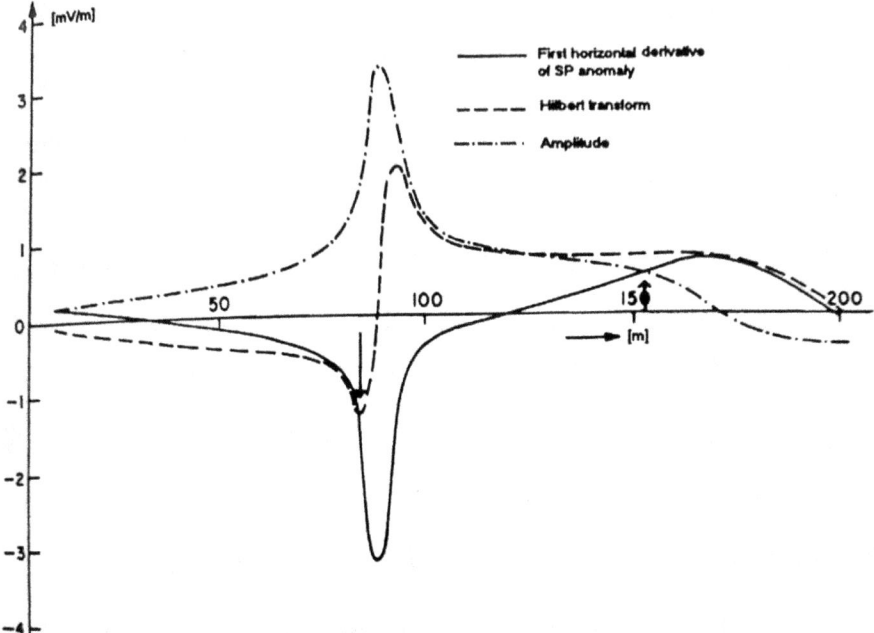

Fig. 1.10 First horizontal derivative of the Weiss SP anomaly, the vertical derivative(the Hilbert transform) and their amplitude

Table 1.2 Interpreted parameters of Weiss SP anomaly, Ergani copper district of eastern Turkey

Parameters	Depth (z) in m	Polarization angle (θ) in degrees
Present method	79.00	52.00
Yungul (1950)	64.00	53.00
Bhattacharya and Roy (1981)	54.00	30.00

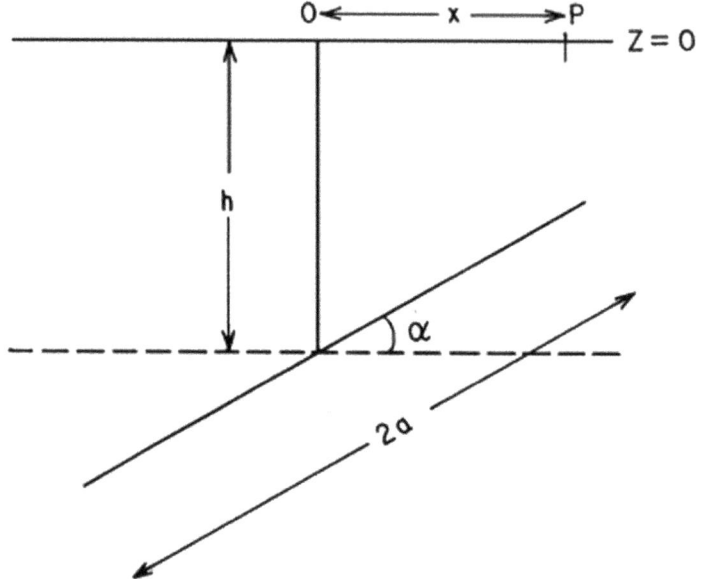

Fig. 1.11 Geometry of the inclined sheet

where 'h' is the depth to the top of the sheet, $'\alpha'$ is the inclination, 'a' is the half width, and 'ρ' is the resistivity and 'I' is the current density of the surrounding medium.

In this case SP3(x) is log function and hence, the partial differentiation with respect to 'x' frees the logarithm and yields the horizontal derivative from which the Hilbert transform and its modified version can be obtained as in the case of 2-D horizontal circular cylinder.

$$SP_x 3(x) = K\left[\frac{(x - a\cos(\alpha))}{(x - a\cos(\alpha))^2 + (h - a\sin(\alpha))^2} + \frac{(x + a\cos(\alpha))}{(x + a\cos(\alpha))^2 + (h + a\sin(\alpha))^2} \right]$$

The real and imaginary components of the Fourier transform of $SP_x 3(x)$ can be obtained as:

$$\mathrm{Re} SP_x\ 3(\omega) = K\,\pi\,\sin(\omega a \cos(\alpha))\big[e^{-\omega(h+a\sin(a))} - e^{-\omega(h-a\sin(\alpha))}\big]$$

$$\mathrm{Im} SP_x\ 3(\omega) = K\,\pi\,\cos(\omega a \sin(\alpha))\big[e^{-\omega(h+a\sin(\alpha))} - e^{-\omega(h-a\sin(\alpha))}\big]$$

Using these components in the equations of HT1(x) and HT2(x), the Hilbert transform and the modified Hilbert transform of $SP_x\ 3(x)$ can be obtained as:

$$HT1(x) = K\left[\frac{(h - a\sin(\alpha))}{(x - a\cos(\alpha))^2 + (h - a\sin(\alpha))^2}\right.$$
$$\left. + \frac{(h + a\sin(\alpha))}{(x + a\cos(\alpha))^2 + (h + a\sin(\alpha))^2}\right]$$

$$HT2(x) = K\left[\frac{(h - a\sin(\alpha))}{(x - a\cos(\alpha))^2 + (h - a\sin(\alpha))^2}\right.$$
$$\left. + \frac{(h + a\sin(\alpha))}{(x - a\cos(\alpha))^2 + (h + a\sin(\alpha))^2}\right]$$

It may be noted that the equation HT1(x) can also be obtained directly as the partial derivative of SP3(x) with respect to 'h'.

Equations of $SP_x\ 3(x)$ and HT1(x) are quadratic in x, hence we can write the following,

$$SP_x\ 3(x) = HT1(x) \quad \text{at } x = x_1 \text{ and } x_2$$

where the roots x_1 and x_2 are nothing but, the abscissa of the points of intersection of the plots of $SP_x\ 3(x)$ and HT1(x).

Further algebraic simplification results,

$$x^2 + 2Q\,ahx + Qa^2 - h^2 = 0$$

Then, sum of the roots of this quadratic equation may be expressed as,

$$x_1 + x_2 = -2Qh$$

where $Q = (\sin(\alpha) - \cos(\alpha))/(\sin(\alpha) + \cos(\alpha))$.

The above relation yields the depth 'h' once the value of 'α' is known.

Further, at x = 0, the equations $SP_x\ 3(x)$ and HT1(x) reduce to,

$$\frac{HT1(0)}{SP_x\ 3(0)} = \left[\frac{(a^2 - h^2)}{(a^2 + h^2)}\right]\tan(\alpha) = S$$

By simplifying the last three equations [quadratic in x, sum of the roots and the S], a cubic equation in $\tan(\alpha)$ is obtained as:

$$A \tan^3(\alpha) + B \tan^2(\alpha) + C \tan(\alpha) + D = 0,$$

where

$$A = 2x_1 \cdot x_2, \quad B = (x_1 + x_2)^2 (S - 1) - A(S + 2)$$

$$C = (x_1 + x_2)^2 (1 - S) - A(2S + 1) \text{ and } D = -SA$$

In this cubic equation in $\tan(\alpha)$, x_1 and x_2 as well as S are known and hence 'α' can be determined.

Subsequently, the depth 'h' and the half width 'a' can also be evaluated as:

$$h = \frac{(x_1 + x_2)(\cos(\alpha) + \sin(\alpha))}{(\cos(\alpha) - \sin(\alpha))}$$

$$a = \sqrt{(x_1 . x_2 + h^2) \frac{(\cos(\alpha) + \sin(\alpha))}{(\cos(\alpha) - \sin(\alpha))}}$$

In evaluating 'h' and 'a', there is a singularity at $\alpha = 45°$ which can be attributed to the fact that $(x_1 + x_2) = 0$ and $(x_1 . x_2 + h^2) = 0$ in the above equations. That is, the roots are equal and opposite. In such a case, the depth 'h' and the half width 'a' are evaluated as

$$h = x_1 = -x_2$$

and

$$a = h = \sqrt{-x_1 \cdot x_2}$$

Finally the constant term 'K' ($K = I \rho$) can be evaluated as:
$K = \frac{NR}{DR}$ where

$$NR = (a^2 + h^2 - 2ah\sin(\alpha))((a^2 + h^2 - 2ah\sin(\alpha))/2a$$

$$DR = \sqrt{\left[\frac{(a^4 + h^4 + 2ah\cos(2\alpha))}{SP_x 3(0)^2 + HT!(0)^2}\right]}$$

where K being the product of polarization current (I) and the resistivity 'ρ', with the knowledge of one of them, the other quantity can be estimated approximately. Thus, all the parameters like depth (h), half width (a), the polarization angle (α) and the

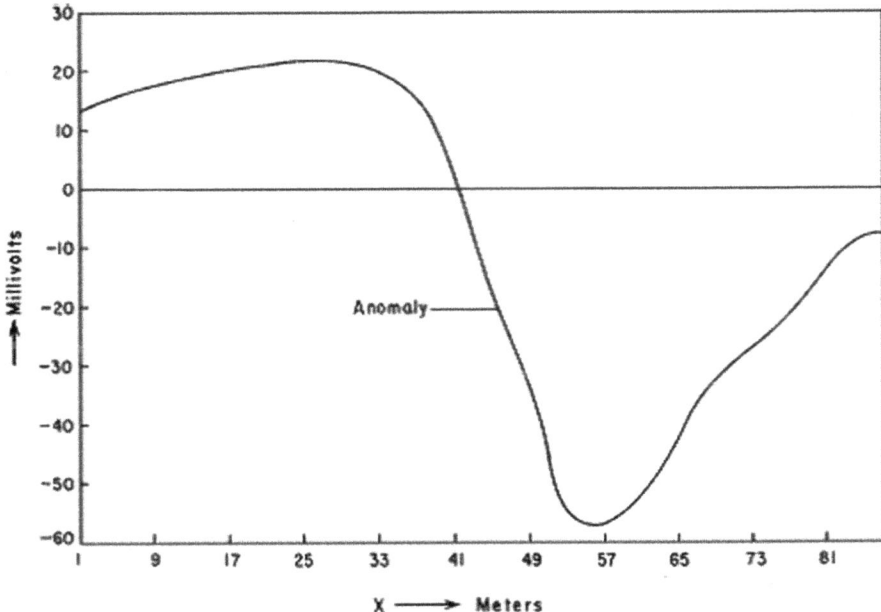

Fig. 1.12 SP field anomaly in Surda area of Rakha mines, Singhbhum copper belt, Bihar, India

constant term (K) can be determined from the above analysis (Sundararajan et al. 1998).

Field Example: The practicability of the method is tested on an SP profile $(E - 19 + 100)$ in the Surda area of the Rakha mines Singhbhum copper belt, Bihar, India. The anomaly (Murthy and Haricharan 1984) of the profile is shown in Fig. 1.12. The first horizontal derivative is computed numerically, and the vertical derivative is obtained by means of the Hilbert transform. The derivatives, along with the amplitude, $A(x)$ are shown in Fig. 1.13. The maximum value of the amplitude corresponds to the origin; the roots, are determined from the points of intersection of the derivatives. The parameters obtained are compared with the results of Paul (1965), Rao et al. (1970), Murthy and Haricharan (1984) and presented in Table 1.3. In the area under discussion, sulfides occur at depths ranging between 12.2 and 30.5 m. The present method yields a depth value of 27.65 m.

1.7 Hartley Spectral Analysis of SP Anomalies

In general, the use of Hartley transform in geophysical data analysis has gained importance since the early 1990s (Saatcilar and Ergintov 1991, Sundararajan 1995, 1997; Sundararajan et al. 2007). The familiarity of the Fourier transforms attracts the

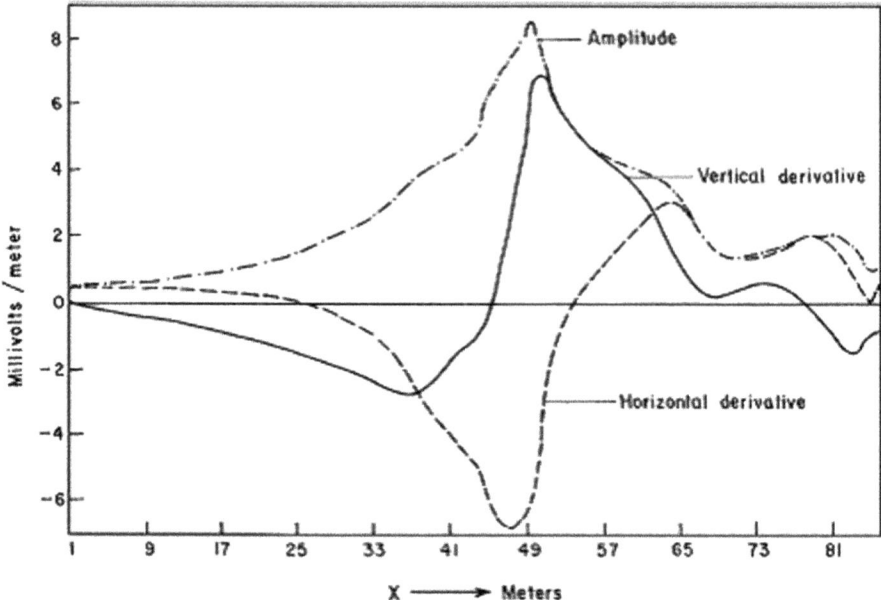

Fig. 1.13 The horizontal and vertical derivatives with their amplitude of the SP anomaly of Rakha mines, Singhbhum copper belt, Bihar, India

Table 1.3 Hilbert transform interpretation of SP anomaly in the Surda area of the Rakha mines, Sighbhum copper belt, Bihar, India

Parameters	Depth (z) in m	Polarization angle (α) in degrees	The half width (a) in m
Present method	27.65	13.20	32.35
Paul (1965)	21.40	20.01	40.20
Rao et al. (1970)	30.48	10.01	34.87
Murthy and Haricharan (1984)	29.50	30.00	29.50

scientists and engineers from the advantages of the Hartley transform. The Hartley and Fourier transforms are fully equivalent; however, Hartley transform differs in phase by 45° from its progenitor—the Fourier transform. The Hartley transform is purely real (Bracewell 1983; and Sundararajan 1995). The physical implication of both transforms is exactly the same and the frequency in both transforms has the same meaning. In this section, the Hartley transform is applied to a theoretical example to illustrate the method and then applied to a field example of the "Sulleymonkey" anomaly in the Ergoni copper district, Turkey to demonstrate the applicability of the method. It may be noted that the Spectral analysis of geophysical data either by

Fourier transform or Hartley transform encores identical results as their amplitude spectra in 1-D are identical.

The Hartley transform H(x) of the real function V(x) is defined by Hartley (1942) as:

$$H(\omega) = \int_{-\infty}^{\infty} V(x) Cas(\omega x) \, dx$$

where

$$Cas(\omega x) = Cos(\omega x) + Sin(\omega x)$$

is considered as the kernel that is 45° phase-shifted sine wave. It takes the harmonics of both cosine and sine functions as real and the frequency (ω) does have the same physical meaning as that of Fourier transform (Bracewell 1983; Sundararajan 1995).

Basically, the Hartley and Fourier transforms can be related using the even and odd components with the real and imaginary components of the Fourier transform (Bracewell 1983; Sundararajan 1995) as

$$H(\omega) = E(\omega) + O(\omega)$$

$$F(\omega) = Re(\omega) - iIm(\omega)$$

where $E(\omega)$ and $O(\omega)$ of Hartley transform $H(\omega)$ are numerically equal to real and imaginary parts $Re(\omega)$ and $Im(\omega)$ of the Fourier transform $F(\omega)$. Thus, the amplitude of the Hartley transform can be expressed as in the case of Fourier amplitude.:

$$A(\omega) = \sqrt{E(\omega)^2 + O(\omega)^2}$$

Alternatively, the amplitude spectrum can also be expressed in terms of the Hartley transform H(w) as:

$$A(\omega) = \sqrt{\frac{[H^2(\omega) + H^2(-\omega)]}{2}}$$

Also, the phase of the Hartley transform can be expressed in the same way as that of Fourier phase as:

$$\emptyset(\omega) = \tan^{-1}\left[-\frac{O(\omega)}{E(\omega)}\right]$$

Alternatively, the phase spectrum can be realized as a function of $H(\omega)$ and $H(-\omega)$

$$\varnothing(\omega) \ = \tan^{-1}\left[\frac{H(-\omega) - H(\omega)}{H(-\omega) + H(\omega)}\right]$$

The self potential due to 2-D horizontal circular cylinder given earlier can be written as:

$$SP4(x) \ = \ \frac{A[\ x \cos(\alpha) - z sin(\alpha)}{(x^2 + z^2)}$$

where 'z', 'α' and 'A' have same meaning as defined earlier.

The even and odd components of the Hartley transform of the SP anomaly due to horizontal circular cylinder given above can be obtained as:

$$E(\omega) \ = \ K\pi sin(\alpha)e^{-\omega z}$$

and

$$O(\omega) \ = \ K\pi \ cos(\alpha)e^{-\omega z}$$

Therefore, the amplitude spectrum of the Hartley transform can be obtained by squaring and adding and taking the square root as:

$$A(\omega) = K\pi \ e^{-\omega z}$$

And the phase also can be obtained as the arctan of odd by even components of the Hartley transform as

$$\varnothing(\omega) \ = \ \alpha \ - \ \pi/4$$

Theoretically, it is feasible to express the amplitude at two different frequencies say ω_i and ω_{i+1} as

$$A(\ \omega_i \ ) = K\pi \ e^{-\omega_i \ z}$$

and

$$A(\ \omega_{i+1} \ ) = K\pi \ e^{-\omega_{i+1}z}$$

By a simple algebraic division of the above equations with $i = 1$ and also taking natural logarithm, the depth 'z' can be obtained as:

$$z = \frac{1}{\omega_{1-} \ \omega_2} \ \ln \frac{A(\ \omega_1 \ )}{A(\ \omega_2 \ )}$$

and the angle of polarization 'α' can be evaluated by dividing the even and odd as

$$\alpha = \tan^{-1}[O(\omega)/E(\omega)]$$

For a complete solution, the constant term $A = I\rho$ can be evaluated by substituting the depth 'z' in the equation of amplitude as:

$$A = \frac{A(\omega)}{\pi} e^{\omega z}$$

Theoretical and Field Examples

The Hartley spectral analysis of geophysical data particularly in SP interpretation is relatively a recent procedure (Al-Garni and Sundararajan 2011) in comparison with the traditional Fourier spectral analysis. It may be emphasized that both are identical in magnitude however differs in phase by 45° and therefore ensure equality in applications. But being a real tool, the computation of Hartley transform is faster than its progenitor the Fourier transform. Therefore in such studies, it makes no difference in either of the transforms. Here it is illustrated with a theoretical model and substantiated with a field data of Sulleymonkey anomaly of length 260 m in the Ergani copper district, Turkey. The even and odd components, the Hartley transform of SP4(x) and amplitude spectrum are computed and shown in Fig. 1.14a–d in the case of theoretical model. The SP field of Sulleymonkey anomaly in Ergani copper district, Turkey is shown in Fig. 1.15. On the other hand, Fig. 1.16a–d illustrate the even, odd components of field SP Sulleymonkey anomaly in addition to the Hartley transform as an algebraic sum of even and odd components and also the amplitude spectrum. All the parameters are evaluated as discussed in the text. The evaluated parameters in both theoretical as well as field examples are presented in Table 1.4 and compared with the other available methods in the literature.

1.8 Artificial Neural Network Analysis

Soft computing tools such as artificial neural network (ANN) has been gaining importance in the recent past in the interpretation of geophysical data particularly self-potential anomalies (Bescoby et al. 2006; Bhagwan Das and Sundararajan 2016). Self-potential anomaly due to a horizontal circular cylinder can be approximated by an artificial neural network, as they are universal approximators. The universal approximation theorem for multilayer perception (MLP) was proved by several authors in the early 1980s although the results depend on how many hidden units are necessary which is yet to be known. In this section, the analysis of self-potential anomalies due to a 2D horizontal circular cylinder (the interpretation of the very same geometrical structure was carried out in the previous sections using modified

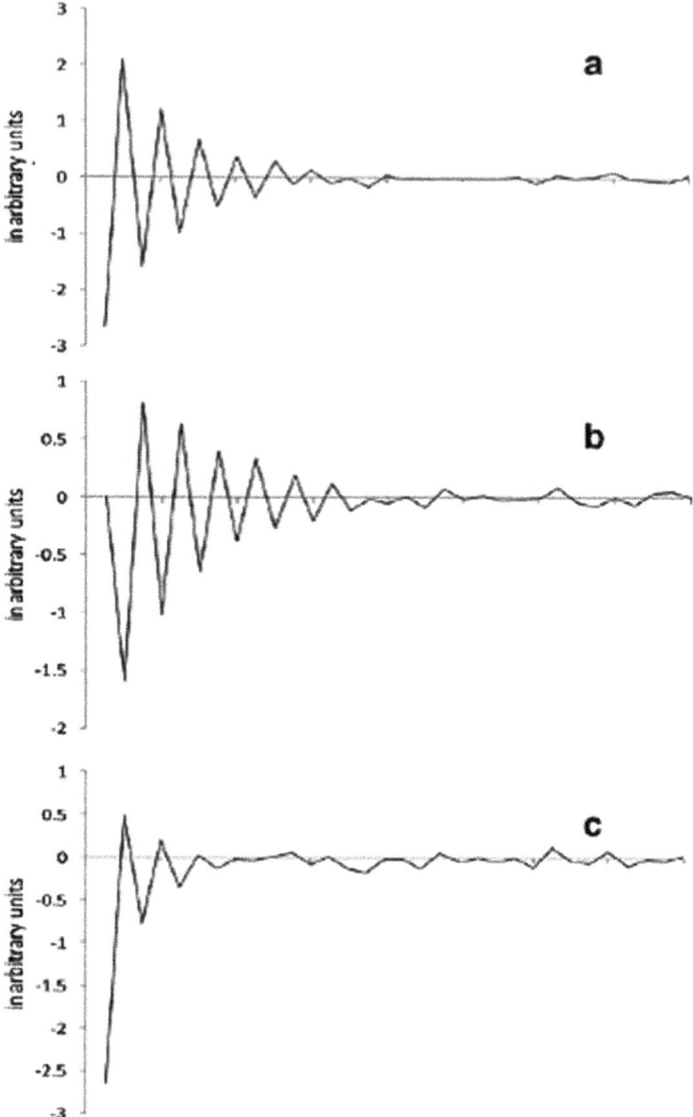

Fig. 1.14 Hartley spectral analysis of theoretical model. **a** the even component, **b** the odd component, **c** the Hartley transform and **d** the amplitude spectrum

Hilbert transform as well as Hartley spectral analysis) is performed using ANN-based committee machine. The soundness of the method is illustrated with the study of theoretical model and a field example.

The salient features of ANNs include that it does not require any prior knowledge about the input/output mapping that is required for model development. The

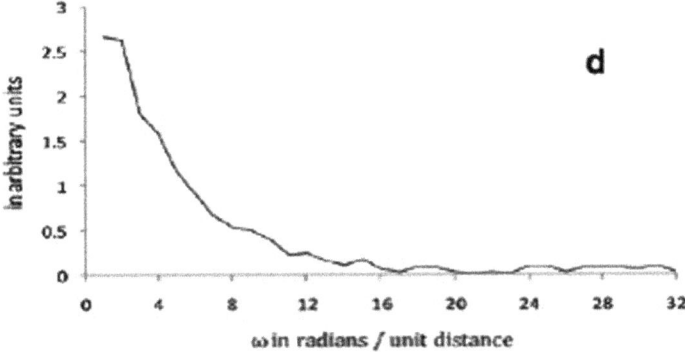

Fig. 1.14 (continued)

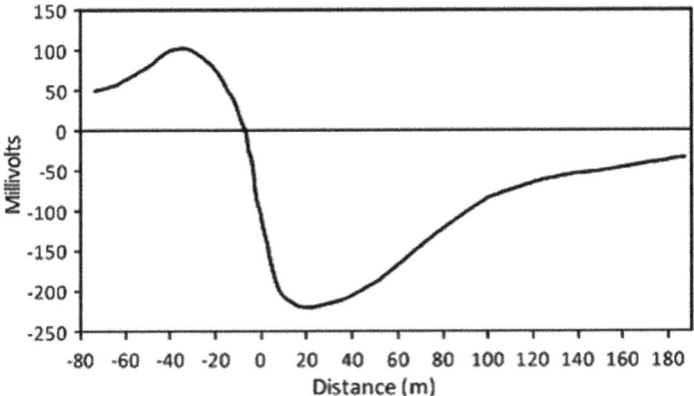

Fig. 1.15 The SP field of Sulleymonkey anomaly in Ergani copper district, Turkey

fitted function is represented by the network and do not have to be explicitly defined. Further, it has the ability to model highly nonlinear as well as linear input/output mapping with good generalization, i.e., it responds correctly to new data. The interpretation of SP anomalies based on ANN approach consists of two phases namely phase-I and phase-II, in phase I, a trial-and-error method is implemented for the analysis. The trial and error method starts from assuming (i.e., trial) an initial set of different ranges for required parameters (may be far from actual values), computes the predicted data values by using the self-potential effect and compares them with the observed data. Then, corrections are applied based on error to the range of parameters so that it minimizes the misfits between calculated and observed data. The procedure is repeated until a satisfying result is obtained. This process is carried out with a mathematical algorithm and implemented in Matlab which enables an efficient way of changing the model parameters. The main purpose of phase I is to obtain a suitable

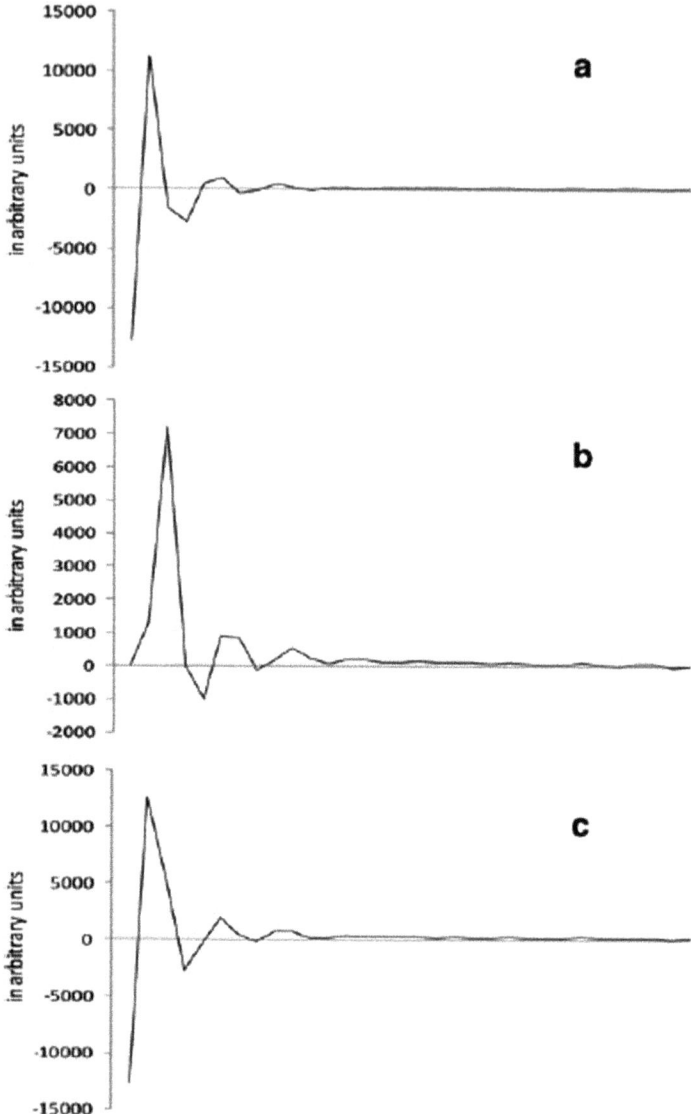

Fig. 1.16 Hartley spectral analysis of SP field of Sulleymonkey anomaly, Ergani copper district, Turkey. **a** even component, **b** odd component, **c** Hartley transform and **d** the amplitude spectrum

and close range of parameters which in turn ensures a very few training examples that are sufficient enough to train in order to extract the parameters of the model.

In phase-II, a committee machine is a type of ANN using the divide-and-conquer strategy in which the responses of multiple experts (MLPs) are combined into a single response. A committee machine is a method in which different experts sharing

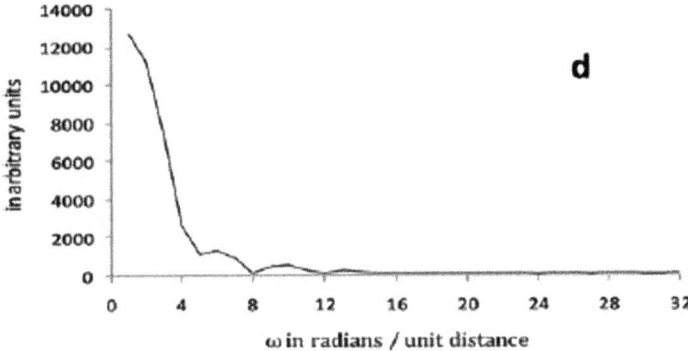

Fig. 1.16 (continued)

Table 1.4 Hartley spectral analysis of theoretical model and field self potential Sulleymonkey anomaly, Ergani copper district, Turkey

Parameters	Depth (z)	Polarization angle (α)	A = (I ρ)
Theoretical Example			
Assumed values	4.00 units	60°	1.00
Interpreted values	4.13 units	56.769°	0.92
	Field Example		
Interpreted values by the present method	35.80 m	47.7°	–
By Youngal (1950)	38.00 m	64°	–
Sundararajan and Srinivas (1996)	36.45 m	45.00°	–
Tlas and Asfahani (2008)	35.41 m	72.24°	…

a common input and whose individuals are combined to produce an overall output; such a technique is referred to as an ensemble averaging method. In phase II, initially training examples are created based on the close range of parameters obtained in phase I, and then an ANN-based committee machine is constructed by replacing each expert by MLP of the same topology (i.e., same number of layers, number of neurons in each layer). Each MLP is trained to extract exactly one parameter with examples, using the Levenberg–Marquart algorithm in batch mode. An extensive further mathematical details are given in Bhagwan Das and Sundararajan (2016).

Theoretical and Field Examples

The ANN based interpretation of SP anomalies are illustrated with a theoretical model in the case of 2-D horizontal circular cylinder and further studied on a field data of the Sulleymonkey anomaly in the Ergani copper district, Turkey. The interpreted results of both theoretical and field anomaly are presented in Table 1.5 and compared with the methods that are in vogue. All the computations are presented as illustrations in

Table 1.5 ANN based interpretation of theoretical model of 2-D horizontal circular cylinder and field self potential Sulleymonkey anomaly, Ergani copper district, Turkey

Parameters	Depth (z)	Polarization angle (α)	A = (I ρ)
Theoretical Example			
Assumed values	4.00 units	60.00°	150
Interpreted values	4.17 units	60.11°	17,519
	Field Example		
Interpreted values by the present method	38.13 m	51.39°	–
By Youngal (1950)	38.00 m	64°	–
Sundararajan and Srinivas (1996)	36.00 m	46.00°	
Hartley spectral analysis (Al-Garni and Sundararajan 2011)	35.00 m	48°	–

Fig. 1.17 (theoretical model) and Fig. 1.18 (Field anomaly). In addition, comparison of artificial neural network generated self potential response with that of generated by other techniques for self-potential field of the Sulleymonkey anomaly in Ergani Copper district, Turkey are shown in Fig. 1.19.

1.9 Noise Analysis

The effect of random noise is investigated on the interpretive process by adding various levels say 5%, 10% and 20% of white Gaussian noise (WGN) to the self-potential of the 2-D horizontal circular cylinder. In Hartely spectral analysis, the noisy anomalies were not subjected to smoothing using a statistical method such as moving average which is optional (Sundararajan and Srinivas 1996). But in Hilbert transform analysis, they show that this process should be carried out prior to the computation of Hilbert transform because their interpretation is based on the abscissa of the points of the intersection of the self-potential anomaly and the modified Hilbert transform. In the case of Hartley spectral analysis, the even and odd components, the Hartley, and the amplitude spectrum are computed from the noisy anomaly. The interpretation should be carried out earlier as in the case of prior to addition of noise. The interpreted results with and without noise did not differ much and presented (Table 1.6) for a specific case of 2-D horizontal circular cylinder model with identical assumed parameters based modified Hilbert transform and Hartley spectral analysis (Fig. 1.20). Therefore, the effect of the noise of 10% of WGN or even more on the interpretive process detailed in all the methods seem to be negligible and confirming that these methods of interpretation are not much prone to the presence of noise in the acquired data. It may be noted here that similar is the noise effect on the on interpretation of SP anomalies over spherical and 2-D cylindrical models based on

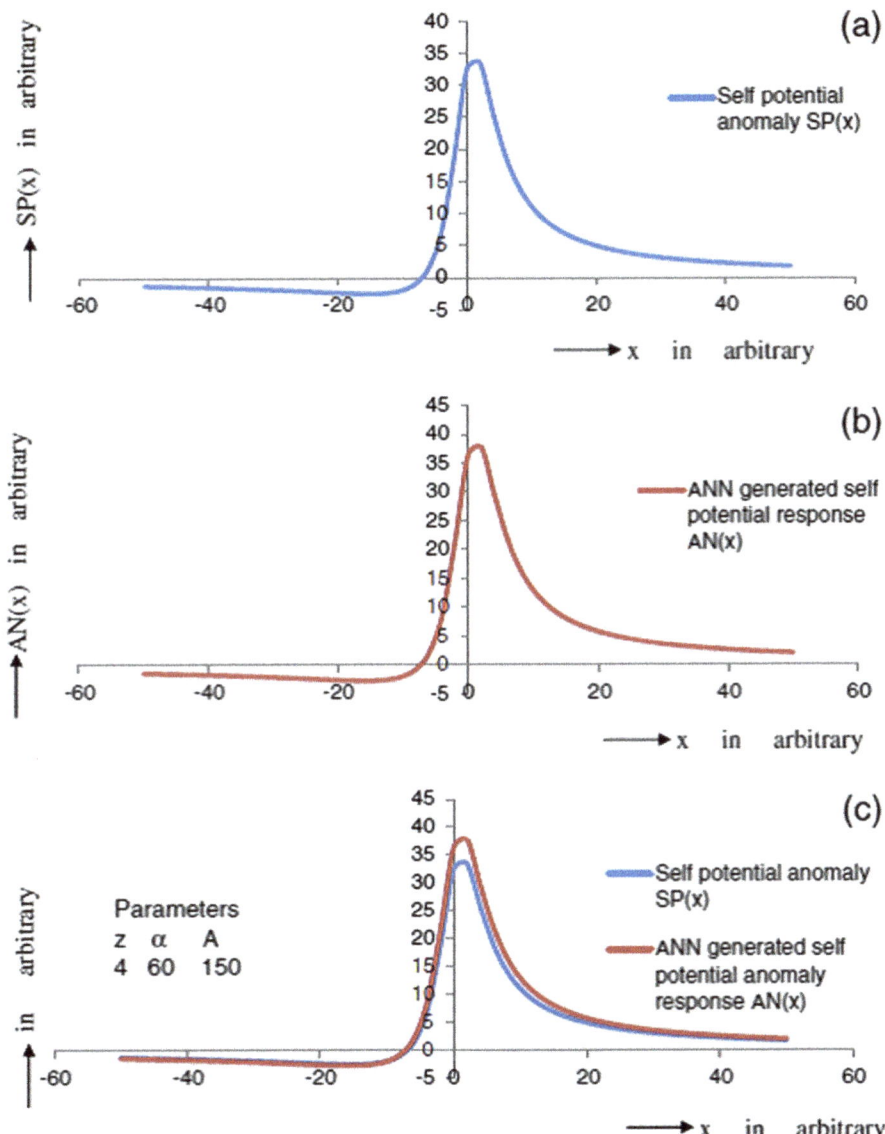

Fig. 1.17 ANN based interpretation of SP anomalies due to 2-D horizontal circular cylindrical model. **a** Self potential anomaly of the model, **b** ANN-generated self-potential response and **c** Self potential anomaly as in (**a**) and ANN-generated self potential response

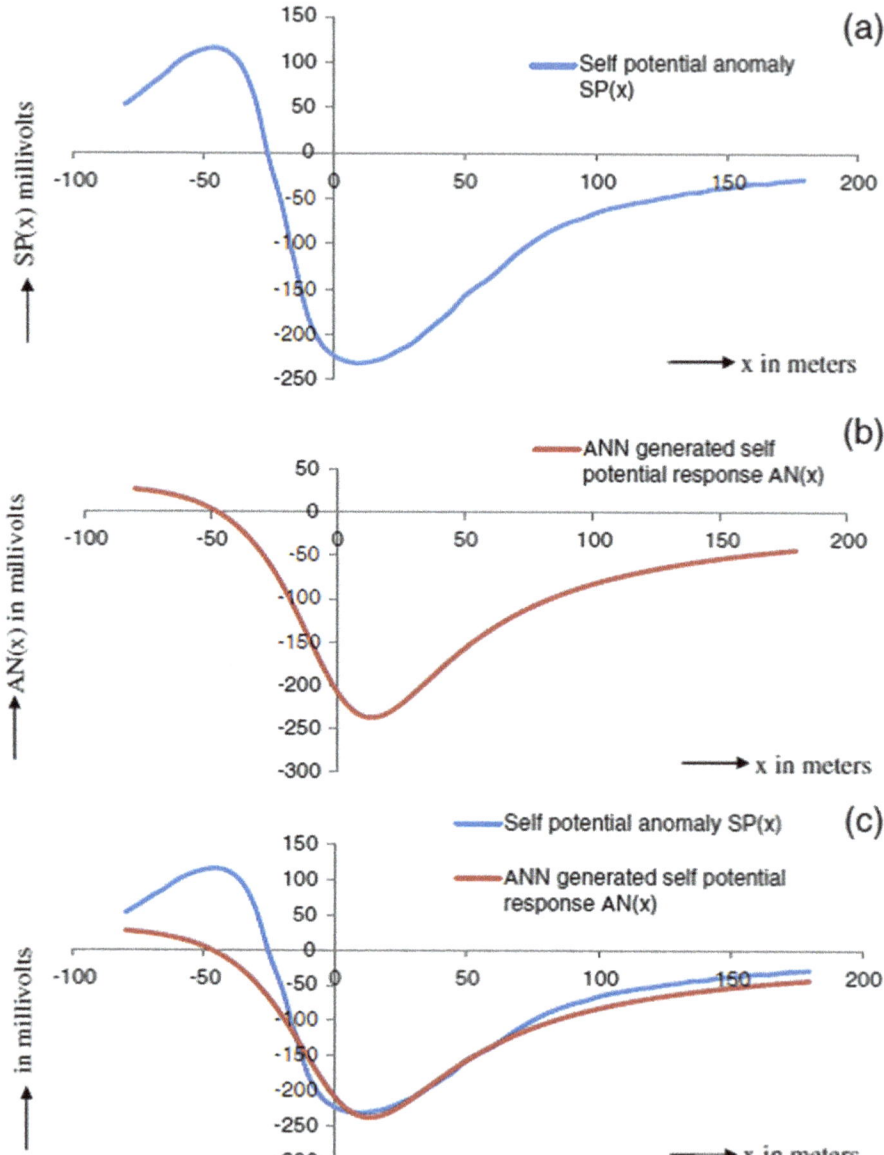

Fig. 1.18 ANN based interpretation of field data. **a** The self-potential field of the Sulleymonkey anomaly in Ergani Copper district, Turkey. **b** ANN-generated self-potential response and **c** The self potential anomaly as in (**a**) and ANN-generated self-potential response

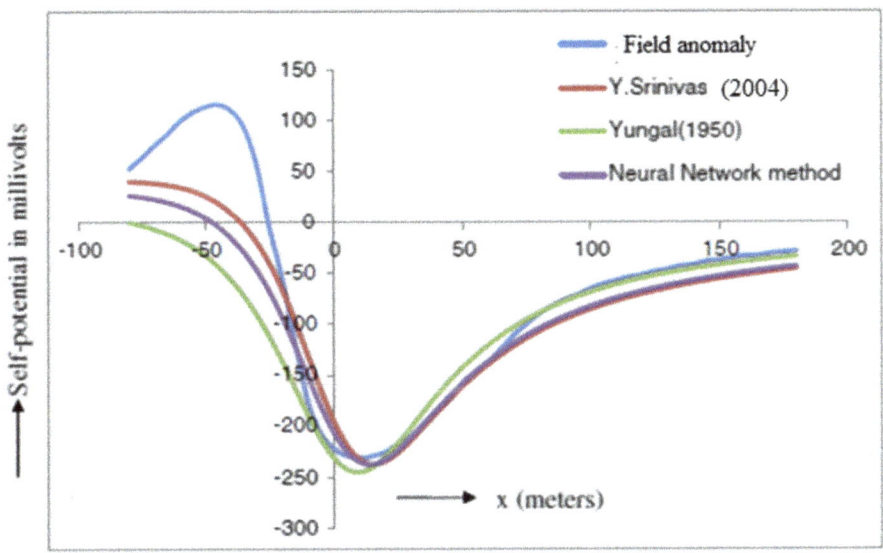

Fig. 1.19 Comparison of artificial neural network generated self potential response with other techniques of the self-potential field of the Sulleymonkey anomaly in Ergani Copper district, Turkey

Table 1.6 Effect of random noise on the interpretation of theoretical SP anomalies due to 2-D horizontal circular cylinder

Parameters	Depth (z)	Polarization angle (α)	A = (I ρ)
Modified Hilbert transform			
Assumed values	4.00 units	60.00°	1.00
Interpreted values	4.00 units	59.00°	0.98
Interpreted values with 10% random noise	4.30 units	49.00°	1.40
Hartley spectral analysis			
Interpreted values	4.13	56.80°	0.93
Interpreted values with 10% random noise	4.28	55.23°	0.91

the techniques of horizontal and vertical derivatives and artificial neural networks (ANN).

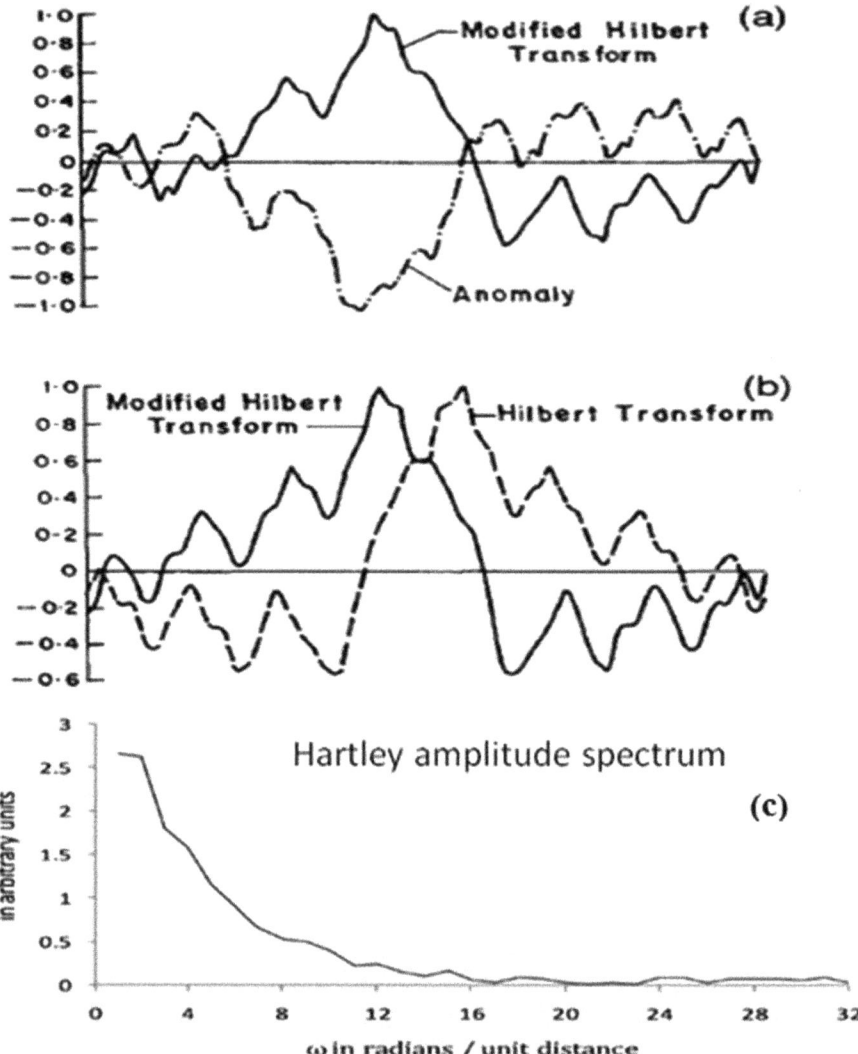

Fig. 1.20 Effect of random noise on the interpretation of SP anomalies due to 2-d horizontal circular cylinder **a** SP anomaly and its modified Hilbert transform. **b** Hilbert transform of SP anomaly and the modified Hilbert transform and **c** Hartley amplitude spectrum

1.10 Discussion

In general, all geophysical data invariably contaminated with various noise factors, including inappropriate interval measurements etc. SP data interpretation also is prone to error because of the choice of computation algorithm, assumptions etc. In this regards, the Hilbert transform/modified Hilbert transform based interpretation ensures maximum accuracy in extraction of body parameters as they are directly dependent on the real roots of the equations of SP potential of geometrical structures considered. However, the accuracy of this method depends on the accurate estimation of the abscissa of the points of intersections of the anomaly and its Hilbert transform/modified Hilbert transform or the horizontal and vertical derivatives of the SP anomaly.

The modified Hilbert transform is equal in magnitude to the Hilbert transform but differs in phase by 270°. The salient feature of the modified version of the Hilbert transform is that it facilitates in precise spatial location of subsurface targets in a couple of ways. As discussed and demonstrated, the point of intersection of the Hilbert transform of SP anomalies and modified Hilbert transform aid in determining the origin. Similar is the case with the, the amplitude of analytic signal $A1(x)$ and $A2(x)$ based on Hilbert transform and the modified Hilbert transform that the point of intersection of $A1(x)$ and $A2(x)$ confirm the exact location of in the targets.

This procedure is an analytical one without any assumptions; however, the inherent weakness of the method while obtaining the horizontal derivative of the SP anomaly by numerical differentiation which incorporates a bit of noise in the computation, and this can be minimized by any simple statistical filtering prior to computation of the vertical derivative by means of Hilbert transform. The method of interpretation is not influenced significantly by the presence of random noise in the data as evidenced by the noise analysis.

Hartley spectral analysis is not only similar to the traditional Fourier spectral analysis of geophysical data, but also numerically identical with all merits and demerits. However, Hartley transform being real function, unlike the complex Fourier transform, ensure efficiency in computation, particularly while using a large amount of data.

Applications of artificial neural network (ANN) is seen almost in every field of science and engineering including processing and interpretation of various geophysical data. It is elegant in its mathematical frame work, however, in the extraction of parameters from potential field data including SP data, the choice of training parameters of the targets may cost large computational time if the training set differs widely from the actual ones, else ANN techniques are simple to implement and does not require any prior knowledge about the input/output mapping that is required for model development.

References

Al-Garni MA, Sundararajan N (2011) Hartley spectral analysis of self potential anomalies caused by a 2-D circular cylinder. Arab J Geosci 3:27–32

Bescoby DJ, Cawley GC, Chroston PN (2006) Enhanced interpretation of magnetic survey data from archaeological sites using artificial neural networks. Geophysics 71(5):45–53

Bhagwan Das M, Sundararajan N (2016) Analysis of self-potential anomalies due to 2D horizontal cylindrical structures—an artificial neural network approach. Arab J Geosci 9:490. https://doi.org/10.1007/s12517-016-2492-9

Bhattacharya BB, Roy N (1981) A note on the use of a nomogram for self-potential anomalies. Geophy Prosp 29(1):102–107

Bracewell RN (1983) The discrete Hartley transform. J Opt Soc Am 73:1832–1835

Mohan NL, Sundararajan N, Seshagiri Rao SV (1982) Interpretation of some two dimensional magnetic bodies using Hilbert transform. Geophysics 47:376–387

Murthy BVS, Haricharan P (1984) Self-potential anomaly over double line of poles-interpretation through log curves. Proc Indian Acad Sci Earth Planet Sci 93:437–445

Murthy BVS, Haricharan P (1985) Nomograms for the complete interpretation of spontaneous potential profiles over sheet like and cylindrical 2-D structures. Geophysics 50:1127–1135

Nabhigian MN (1972) The analytical signal of two-dimensional magnetic bodies with polygonal cross section, its properties and use for automated anomaly interpretation. Geophysics 37:507–512

Paul MK (1965) Direct interpretations of self-potential extension anomalies caused by inclined sheets of infinite horizontal extension. Geophysics 30:418–423

Rao BSR, Murthy IVR, Reddy SJ (1970) Interpretation of self-potential anomalies of some simple geometric bodies. PAGEOPH 78:66–67

Saatcilar R, Ergintov S (1991) Solving elastic wave equations with the Hartley method. Geophysics 56:274–278

Sundararajan N, Arunkumar I, Mohan NL (1990) Use of the Hilbert transform to interpret self potential anomalies due to 2-D inclined sheets. Pure Appl Geophys 133:117–126

Sundararajan N, Narasimha Chary M (1993) Direct interpretation of self-potential anomalies due to spherical structures–a Hilbert transform technique. Geophys Trans 38:151–165

Sundararajan N (1995) 2-D Hartley transforms. Geophysics 60:262–267

Sundararajan N, Srinivas Y (1996) A modified Hilbert transform and its application to SP Interpretation. J Appl Geophys 36:137–143

Sundararajan N (1997) Fourier and Hartley transforms—a mathematical twin. Indian J Pure Appl Math 28:1361–1365

Sundararajan N, Srinivasa Rao P, Sunitha V (1998) An analytical method to interpret SP anomalies due to 2-D inclined sheets. Geophysics 63:1551–1555

Sundararajan N, Srinivas Y, Laxminarayana Rao T (2000) Sundararajan transform-a tool to interpret potentiaql field anomalies. Exploration Geophys 31:622–638

Sundararajan N, Al-Garni MA, Ramabrahmam G, Srinivas Y (2007) A real spectral analysis of the deformation of a homogenous electric field over a thin bed–a Hartley transform approach. Geophys Prospect 55(6):901–910

Sundararajan N, Srinivas Y (2010) Fourier–hilbert versus hartley–hilbert transforms with some geophysical applications. J Appl Geophys 71(4):157–161

Tlas M, Asfahani J (2008) Using of the Adaptive Simulated Annealing (ASA) for quantitative interpretation of self-potential anomalies due to simple geometrical structures. J King Abdulaziz Univ Earth Sci 19:99–118

Yungul S (1950) Interpretation of spontaneous polarization anomalies caused spherical ore bodies. Geophysics 15(2):237–246

Chapter 2
Metaheuristics Inversion of Self-Potential Anomalies

Mohamed Gobashy and Maha Abdelazeem

Abstract Artificial intelligence and metaheuristic approaches had gained a remarkable position in last-millennium geophysical inversion. The past two decades have witnessed the development of numerous metaheuristics in various communities that sit at the intersection of several fields including geophysics. Many inverse problems in geophysics are considered as constrained optimization, as the aim of the process is to find the best parameter estimates so as to minimize the differences between the predicted results and the observations while satisfying all known constraints or thresholds. Such optimization problems can thus be solved by efficient traditional optimization techniques (e.g.: Least-squares). However, as the number of degrees of freedom is usually very large, metaheuristic algorithms such as, Whale, Grey Wolf, particle swarm, genetic, Bat, and Cuckoo Search algorithms are particularly suitable for inverse problems of that kind, because metaheuristics are very efficient for solving non-linear global optimization problems. The inversion of spontaneous potential (SP) anomalies in particular, attracted many authors. This chapter provides a complete view of metaheuristics as effective tool for parameter estimation from the SP signal. We show the main design questions and search components for selected families of metaheuristics. Not only the design aspect of metaheuristics but also their implementation including the formulation of the Objective/target function. After covering the synthetic examples with the noise tests, many field examples will be presented to show its effectiveness and suitability for various geologic conditions and a diverse range of application domains.

Keywords Self-potential · SI · PSO · GWO · WOA · CSA · SSO · ABC

M. Gobashy (✉)
Faculty of Science, Department of Geophysics, Cairo University, Giza, Egypt

M. Abdelazeem
Lab of Geomagnetism, National Research Institute for Astronomy and Geophysics, NRIAG, Helwan, Egypt

© The Author(s), under exclusive license to Springer Nature Switzerland AG 2021   35
A. Biswas (ed.), *Self-Potential Method: Theoretical Modeling and Applications in Geosciences*, Springer Geophysics, https://doi.org/10.1007/978-3-030-79333-3_2

2.1 Introduction

The main critical problem of geophysical potential field inversion is the developing of a stable and plausible inverse problem solution which solves simple geophysical models and at the same time can resolve complicated geological structures. The potential field data analysis and inversion is of major interest because it is relatively fast, inexpensive and allow in most cases a full coverage for inaccessible areas. Several problems arose when dealing with the inversion. First, the inverse problem has no unique solution. For a given set of model parameters, the forward problem has a unique solution, but for a given set of observed field data, the inverse problem can have an infinite number of possible solutions. This property is one of the important consequences of the Green's third identity of the potential theory. Changing the number of parameters or observations or type of measurement does not solve this problem. Ill-posedness of the inverse problem is a consequent results since the forward model has to simplify the real world using a finite set of parameters. This means that for the inverse problem, minor changes in the observed field data can result in randomly large changes in the calculated model parameters. The second problem usually arose from the simplifying assumptions that most users put forward to constrain their inverse solutions. For example, the effects of transversely anisotropic background in self-potential inversion are not generally included as parameters in the inversion process, so any data containing effects when inverted using methods which does not incorporate them cannot possibly generate optimal correct plausible solution.

The third problem has an extreme importance such that whatever the formulation used for solving an inverse problem, a feasible model is generated. This is a simple consequence of the traditional iterative inversion process wrapped around a forward problem in which the forward model is successively modified until it fits the observed field data to the desired threshold. However, just because the model is feasible, it does not mean that it is either realistic or plausible (Paine 2007; Boschetti et al. 1999).

The above concise survey on the main controlling parameters of geophysical inversions and the non-linearity of most of the geophysical problems, fueled a greater desire to include much more sophisticated or complex algorithms rather than simple direct modelling techniques. Moreover, the improvements in hardware and modeling software have permitted a major increase in the problems dimensions which can be handled and initiated the interest to model even larger problems. The major advances in the last decade have been focused on providing methods for doing this (Paine 2007). Among this, is the introduction of voxel based inversion programs which use regularization techniques to generate geologically reasonable models (Li and Oldenburg 1996, 1998, 2000). The implementation of compression techniques to improve the computational performance of the algorithms (Li and Oldenburg 2003) and in the field of self-potential inversion, many progress can be observed where tomographic methods are presented (Di Maio and Patella 1994; Patella 1997; Di Maio et al. 2013, 2016a), least squares inversion (Abdelrahman et al. 2008; Agarwal and Srivastava 2009; Li and Yin 2012), spectral analysis (Rani et al. 2015), and global optimization approaches (Tlas and Asfahani 2008; Srivastava et al. 2014; Di Maio et al. 2016b).

2.2   Forward SP Model

Self-potential (SP) method is commonly used to explore metallic sulfides and graphite ore bodies. It is considered one of the oldest geoelectrical methods, that is still in use in many fields of applied geophysics, such as mining (Sato and Mooney 1960; Logn and Bolviken1974; Corry 1985), archaeological surveying (Wynn and Sherwood 1984; Cammarano et al. 1998), engineering geophysics (Bogoslovsky and Ogilvey 1977; Bogoslovsky et al. 1977; Corwin 1990) in dam and embankment (Bogoslovsky et al. 1977; Corwin 1990) in dam and embankment seepage control (Bogoslovsky and Ogilvy 1970a, b) and in many other geophysical fields. Many source mechanisms have been proposed to explain the genesis, time and space patterns of the SP field, either in the applied geophysics or in the tectonophysics. In general, the common aspect of many source models is that, an electric charge polarization is setup, which is responsible for the current flow in conductive rocks.

The current approaches to SP quantitative interpretation schemes are all based on the comparison between the SP field anomalies or signals and the computed responses due to simple sources like monopole, dipole, multipole or line sources located either in a homogeneous and isotropic half space or in layered or faulted geometries (Stern 1945; Rao et al. 1970; Telford et al. 1976; Demoully and Corwin 1980; Fitterman 1979, 1983; Patella 1997).

For few simple charge geometries where individual SP anomalies (Fig. 2.1) are found in a clear and resolvable form (i.e. can simply be separated from the regional and topographic effects), numerical estimates of the source parameters: depth to the source (z), electric charge polarization angle (ψ), electric dipole moment (k), and shape factor (q) can be directly derived from the analysis of the SP anomaly (Petrowsky 1928; Bhattacharya and Roy 1981; Fitterman and Corwin 1982; Rao and Babu 1984; Murty and Haricharan 1985).

Although, in many cases, the modeling of the phenomena associated with large and/or small-scale geophysical source, which may fully justify the estimated polarization state, cannot as a rule deduced from the specified source geometries

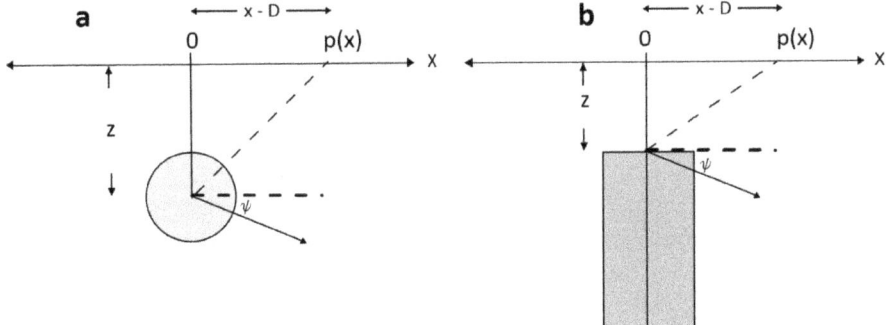

Fig. 2.1 2D model of **a** a sphere or horizontal cylinder and **b** a vertical cylinder model

Fig. 2.2 2D inclined sheet model

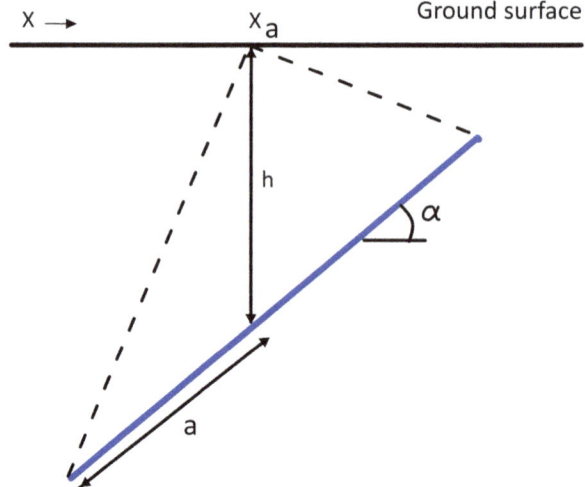

(Patella 1997). However, such simple geometries are still in use and their interpretation techniques are reported in many case studies. As an example, spherical and cylindrical-like models are used by Yüngül (1950), Murty and Haricharan (1985), Abdelrahman and Sharafeldin (1997), Abdelrahman et al. (2003), Abdelazeem and Gobashy (2006), Abdelrahman et al. (2019), and Abdelrahman and Gobashy (2021) as shown in Eq. (2.1). Sheet-like models (Fig. 2.2) studied by Paul (1965), Murty and Haricharan (1985), as shown in Eq. (2.2).

$$V(x, D, \psi, z, q) = K \frac{(x - D)sin\psi - zcos\psi}{\left((x + D)^2 + z^2\right)^q} \qquad (2.1)$$

where, K is defined as the electric dipole moment, D and h are the horizontal location of the center axis of the buried body and its depth, respectively, ψ defines the polarization angle, x denotes the surface measurement point and q refers to the shape factor, q = 0.5 for the vertical cylinder, q = 1.0 for the horizontal cylinder and q = 1.5 for sphere.

$$V(x, D, a, h, \alpha) = kln \frac{[(x - D) - acos\alpha]^2 + (h - asin\alpha)^2}{[(x - D) + acos\alpha] + (h + asin\alpha)^2} \qquad (2.2)$$

The formula expressing the self-potential anomaly at any surface point V(x) along a line normal to the strike of a 2D inclined sheet model where k is the polarization amplitude, D is the horizontal location of the sheet center, h refers to the depth to the sheet center, a denotes the half-width of the sheet and a defines the inclination angle. If the sought model consists of fewer parameter than the number of field data points, then the inverse problem is said to be OVERDETERMINED and can be formally solved using methods based on achieving a best fit to the data.

2.3 Optimization Methods

Based on the complexity of the given problem, it may be solved by an exact method or an approximate method (Fig. 2.3). Exact methods obtain optimal solutions and guarantee their optimality, while approximate (or heuristic) methods give high quality solutions in an acceptable time for practical use, but there is no guarantee of finding a global optimal solution.

In more detail, the Exact Methods include, dynamic programming that divides the problem into simpler Sub-problems, the branch and X family of algorithms (branch and bound, branch and cut, branch and price) that support an implicit enumeration of solutions of the considered optimization problem and developed in the field of operations research, constraint programming which is a language built around concepts of tree search and logical implications and A* family of search algorithms (A*, IDA*-iterative deepening algorithms) (Korf 1985) developed in the field of artificial intelligence (Russell and Norvig 1995).

Under the approximate strategies, two subclasses of algorithms could also be distinguished: approximation algorithms and heuristic algorithms. The approximation algorithms give verifiable solution quality and provable run-time bounds and there is a guarantee on the bound of the obtained solution from the global optimum (Hochbaum 1996). The heuristics typically find "good" solutions in a very affordable time on large-size problem. They allow to get acceptable answer performance at acceptable costs in a very wide range of problems. In general, heuristics don't have an approximation guarantee on the obtained solutions. They may be classified into two families: specific heuristics and metaheuristics.

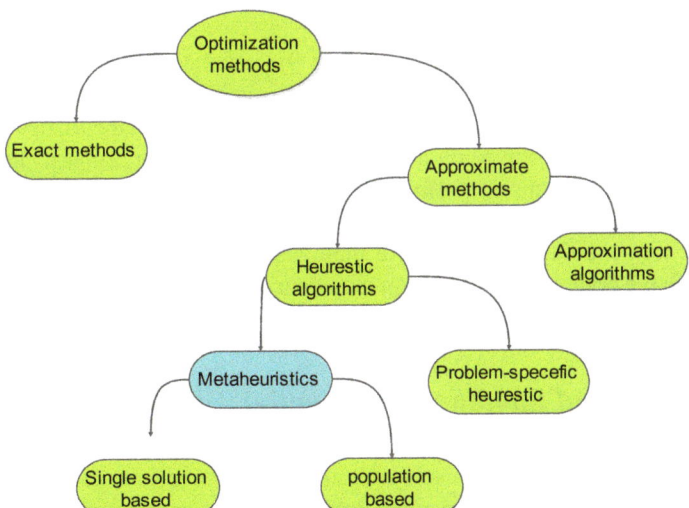

Fig. 2.3 General classification of optimization methods (modified after El-Ghazali 1965)

The first who coined the term "metaheuristic" was (Glover 1986) so as to differentiate and generalize heuristic method without problem-specific characteristic. The reason behind the robust searching mechanism of metaheuristic is the synchronization between two search schemas: exploration (diversification) and exploitation (intensification) (Blum and Roli 2003). However, the heuristic concept in solving optimization problems was firstly introduced by Polya (1945). The simplex algorithm, introduced by Dantzig (1947), may be recalled as a local search algorithm for linear programming problems. J. Edmonds was the first to discuss the greedy heuristic in the combinatorial optimization literature in 1971 (Edmonds 1971). The original references of the following metaheuristics are based on their application to optimization and/or machine learning problems.

Since that time, many metaheuristics applications are introduced, among them: AIS (artificial immune systems) (Bersini and Varela 1990; Farmer et al. 1986), CA (cultural algorithms), CMA-ES (covariance matrix adaptation evolution strategy) (Hansen and Ostermeier 1996), GLS (guided local search) (Voudouris and Tsang 1995; Voudouris 1998), GP (genetic programming) (Koza, 1992), evolutionary programming (EP) (Fogel 1962), DE (differential evolution) (Pric 1994; Storn and Price 1995), GA (genetic algorithms) (Holland 1962, 1975), SM (smoothing method) (Glover and McMillan 1986), SS (scatter search) (Glover 1977), ILS (iterated local search) (Martin et al. 1991), EDA (estimation of distribution algorithms) (Baluja 1994), NM (noisy method) (Charon and Hudry 1993), PSO (particle swarm optimization) (Kennedy and Eberhart 1995), ACO (ant colonies optimization) (Dorigo 1992), CEA (coevolutionary algorithms) (Hillis 1990; Ho et al. 2005), BC (bee colony) (Shaw 1998), GDA (great deluge) (Dueck 1993), ES (evolution strategies) (Rechenberg 1965), CSO (Cuckoo search optimization) is a recent optimization algorithm suggested by (Xin-She Yang 2010), WOA (Whale optimization) (Mirjalili and Lewis 2016), and many others.

Many classification criteria may be used for metaheuristics. Following Abdel-Basset et al. (2018), a classification of metaheuristics that divides metaheuristics into metaphor primarily based and non-metaphor based metaheuristics controlled by the nature of the algorithm could be accepted. The metaphor primarily based metaheuristics are algorithms that simulate natural phenomena, human behavior or maybe mathematic, etc. On the other hand, non-metaphor based metaheuristics didn't use any simulation for decisive their search strategy. Figure 2.4 provides a comprehensive review of the common metaheuristic taxonomies.

2.4 Inversion of Spontaneous Potential (SP) Anomalies

The inversion of spontaneous or self-potential (SP) anomalies in particular, attracted many authors as it can be used in many fields of applied geophysics, such as mining, archaeology, dam seepage and many other problems. Also the simplification of the causative body to sheet, sphere or cylinder guarantee that the problem is over determined and consequently, with constraints, there is a plausible solution.

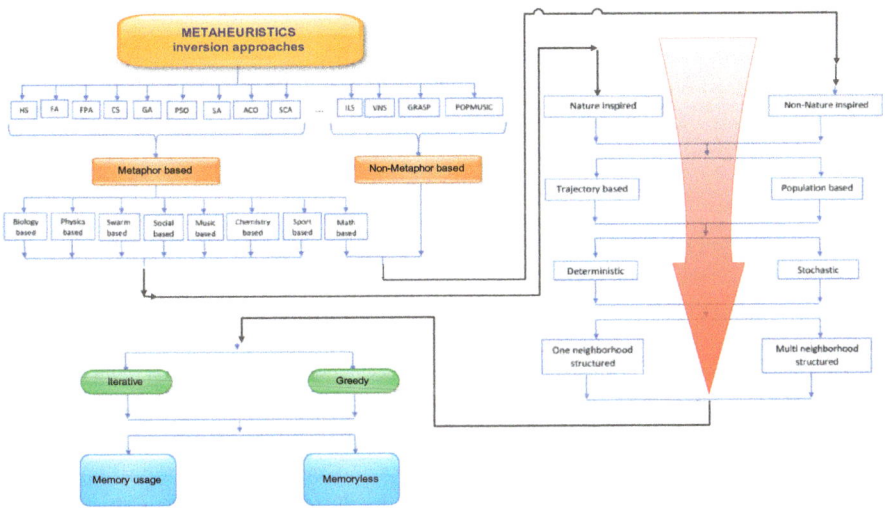

Fig. 2.4 Classification of metaheuristics (modified after Abdel-Basset et al. 2018)

2.4.1 Ambiguity and Non-uniqueness of SP Inverse Solutions

It is understood that the potential within any sub-region of a region R can be related to an infinite variety of surface distributions (Green's third identity). Hence, no unique boundary conditions exist for a given harmonic function (Blakely 1995), and an infinite number of possible solutions can be found that satisfy the data equally well. This property of non-uniqueness constitutes an important limitation that faces any interpretation of a measured potential field in terms of its causative sources and consequently affects all numerical inversion techniques used in interpretation of potential fields. However, many authors attempted to overcome this critical problem by different strategies. Among these is forcing a non-unique problem to behave as a unique one. This can be done by looking for solutions with unique features. However, this leaves the actual extent of the ambiguity unknown (Boschetti et al. 1999), and a critical question arose dealing with the effectiveness of parameterization in the final results? Many authors dealt with such problem that apparently affects the reliability of the overall inversion process. Among them Parker (1974, 1975), Ander and Huestis (1987), Huestis and Parker (1977), Al-Chalabi (1971), Vasco et al. (1993) and Boschetti et al. (1997). These techniques suffer from the inherent limitation that the extent of the ambiguity domain is so large, and its shape so complicated, that reasonable results may be obtained only for very simple problems (Boschetti et al. 1999). Fedi and Abbas (2013) and earlier Boschetti et al. (1999), modified the above techniques dealing with the self-potential fields. In Boschetti et al. (1999), a tomographic map of the distribution of electrical source strength from which the possible location of discrete sources can be estimated in the ambiguity domain using the source imaging methods. This technique is thought to reduce ambiguity, but actually they are only useful in problems where we have limited knowledge of the source mechanism, or where multiple source mechanisms (electrochemical, electrokinetic or thermoelectric) are at work. However, the above mentioned approaches do not give direct interpretation of the unknown associated with a certain source mechanism (Biswas 2017a). This specific problem requires inversion of SP data considering simple geometrical sources according to Biswas and Sharma (2014).

2.4.1.1 Transverse Anisotropy as a Source of Ambiguity

Anisotropy is a major source of uncertainty or ambiguity in inversion of SP fields. Where models of dipole current distribution in a homogeneous and isotropic medium is usually considered in the formulation of most of the interpretation techniques of SP field (Skianis and Hernandez 1999). Ground electrical inhomogeneity or ground electrical anisotropy is rarely taken into consideration. Anisotropy, either micro or macro, however, distorts the electric current flow and, consequently, significant errors in the interpretation of the SP anomaly may be introduced (Skianis et al. 2000). The former occurs in rocks, which contain mineral grains, preferentially oriented with respect to their internal crystal structure. While, the later, occur in sedimentary formations, where thin layers of different resistivity may alternate. In geoelectrical exploration, the term anisotropy is generally employed, including all cases of micro- and macro-anisotropy. When the resistivity in the plane of schistosity is constant and different from the resistivity in the perpendicular transverse plan, a rock or a formation is said to be transversely anisotropic. This is the most dominant type of anisotropy encountered in geoelectrical prospecting (Bhatacharya and Patra 1968; Skianis and Hernandez 1999). We summarize some of the important techniques that are used to control the potential field ambiguity problem, particularly, of the SP fields in the following section.

2.4.2 Ambiguity Control

As mentioned above, potential-field interpretation is characterized by an inherent ambiguity in the determination of the source from field data, which may lead to a loss of depth information (Fedi et al. 2005). Blakely (1995) proved mathematically using Green's third identity that any potential field in a sub-region can be reproduced by an infinite variety of surface (shallow) distributions. Also, the annihilator or the source distribution which produces a null field, couldn't be determined (Parker 1977).

Mathematically, if the system is underdetermined, i.e. with more unknowns than data, this leads to algebraic ambiguity. Many authors solved such ambiguity problems by using additional information about the problem (Gobashy and Abdelazeem 2005). The direct and most easy way to assume simple geometric shapes with homogeneous source distributions to represent the solution using the parametric discretization (Menke 1989; Cordell and Grauch 1985). Consequently, the algebraic ambiguity can be solved assuming a 2- or 3-Diemnsional array of cubes or prisms or simply vertical prisms variable depth to top, and the position of the prism source's boundaries is contained in the field derivatives (Blakely and Simpson 1986).

Other additional a priori information is implementing lower and upper bounds, as example in gravity field inversion, for a density monotonically increasing with depth (Fisher and Howard 1980), constraining the source to have minimum momentum of inertia (Guillen and Menichetti 1984), assuming a condition of compact volume (Last and Kubik 1983) to the causative body requiring compactness along several axes using a priori information about the axes' length using approximate equality (linear) constraints and many other ideas to reduce the ambiguity.

Stable inversion is not by default a well-posed one. All kinds of data smoothing, introduction of different kinds of constraints, application of data and model variance–covariance matrices for bringing stability in an inversion algorithm are the members of the regularization family. In the next sections we present the formulation of few commonly used types of ambiguity controllers: the linear and nonlinear constraints and the probability density function, active set strategy, Depth weighting matrix, and Covariance matrix.

2.4.2.1 Linear and Nonlinear Constraints

We can incorporate previously obtained information about the sought model parameters in our problem formulation. This external information could be results from previous experiments or quantified expectations dictated by the physics of the problem. Generally, these external data help to single out a possible unique solution from all equivalent ones and the process is said to be constraint.

2.4.2.2 Probability Density Function

The usefulness of this technique depends, in part, on the complexity of p(m), the complete probability density function for model parameters. If the probability density functions p(m$_i$) for an individual model parameter $\mathbf{m}_i$ has only one peak (Fig. 2.5A), then it provides little more information than an estimate based on the position of the peak's center with error bounds based on the peak's shape. On the other hand, if the probability density function is very complicated (Fig. 2.5C), it is basically uninterpretable (except in the sense that it implies that the model parameter cannot be well estimated). Only in those exceptional instances in which it has some intermediate complexity (Fig. 2.5B) does it really provide information toward the solution of an inverse problem.

2.4.2.3 Active Set Strategy

In source imaging methods and tomographic inversion of SP signals, the concept of active set strategy (Gill and Murray 1974) can be used to control ambiguity

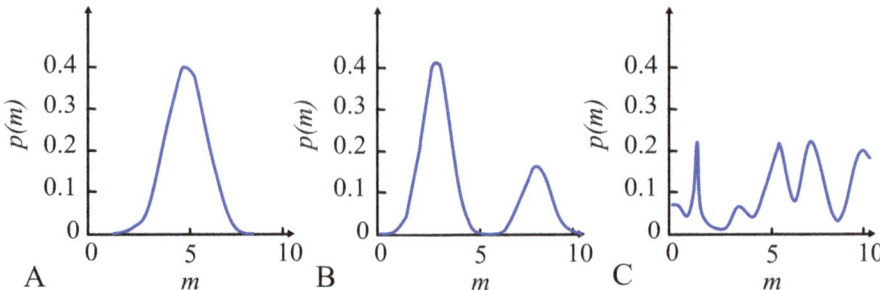

Fig. 2.5 Synthetic probability density functions (PDF) for a specific model parameter, m. **A** In the first properties can be given by its position, at m = 5, and the width of its peak. **B** The second suggests that the model parameter has two probable ranges of values, one near m = 3 and the other near m = 8. **C** The third is a composite one. It provides complex interpretable information about the model parameter. Narrow ranges are usually used for accurate data, while wide ranges are used for noisy data

and constraining solutions during inversion and for the minimization of a target (objective) function of the form:

$$\min \, f(x), \quad l < x < u, \quad x \subset R_n \tag{2.3}$$

where, an active set IA, that contains the indices of the variables at their bounds, is built for a given starting vector x_c. A variable is called a *"free variable"* if it is not in the active set. The free variables search directions is calculated according to the formula:

$$S = -A^{-1}\nabla f \tag{2.4}$$

where, (A) is the Hessian and (∇f) is the gradient evaluated at x_c; both are computed relative to the free variables.

Figure 2.6 Shows a complete flowchart to the FD-Newton method with active set strategy. At Each iteration, a line search method is used to find a new point (x_n), such that:

$$x_n = x_c + \lambda S, \quad \lambda \in (0, 1] \tag{2.5}$$

Finally, the optimality conditions are:

$$
\begin{aligned}
\|\nabla f(x_i)\| \leq \varepsilon, \quad & l_i < x_i < u_i \\
\nabla f(x_i) < 0, \quad & x_i = u_i \\
\nabla f(x_i) > 0, \quad & x_i = l_i
\end{aligned}
\tag{2.6}
$$

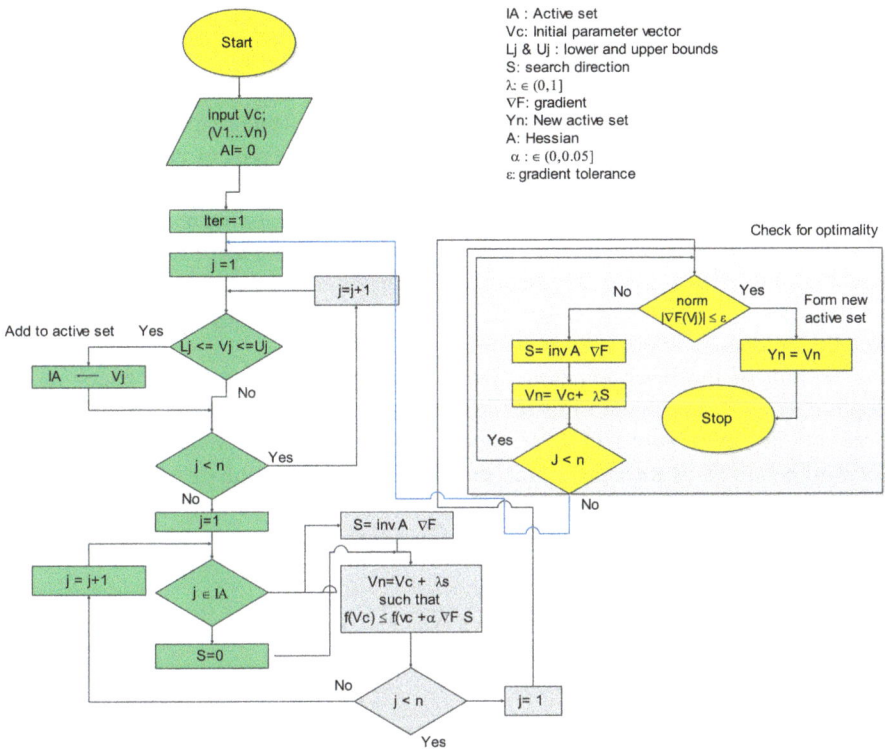

Fig. 2.6 A proposed flowchart for Active set Strategy as a constraint tool of SP signal (modified after Abdelazeem et al. 2003)

Such conditions are checked: where, ε is a gradient tolerance. When optimality is not achieved, another search direction is computed to begin the next iteration. This process is repeated until the optimality criterion is met.

During the course of iteration, if a free variable hits the bounds, or the stopping criteria is met for the free variables, the active set is changed. Moreover, the variable that violates the optimality conditions will be excluded from the IA set (Dennis and Schnabel 1996; Gill and Murray 1974).

2.4.3 Formulation of the Objective/target Function for SP Problem

Methodologies based on simple geometric source can interpret SP data when the number of the unknown is very limited (1–6 unknown parameters), forward modeling, imaging methods, and inversion. In the inversion class, the number of unknown parameters vary based on the proposed anomaly source. Consequently, the quantitative interpretations can be grouped into two major categories (Biswas

2017a): The first category is based on the multi-dimensional SP inversion of some arbitrary structures, which can be classified as 2D or 3D subsurface structures. The second category is based on fixed simple geometrical methods of which 2-D and 3-D continuous modeling (Guptasarma 1983; Furness 1992; Minsley et al. 2007a, b). Based on the dimensionality of the energy surface and type of unknown physical parameters, the choice of the form of the error formula is becoming critical. Where, as the number of the parameters increase, the possibility of trapping in a local minimum also increases (Abdelrahman et al. 1998). This is due to the ill-posedness and non-linearity of the SP problem. These problems can significantly be less sensitive when a nonlinear algorithm is used to estimate the unknowns using a suitable objective function (Gobashy 2000).

The aim of the inversion is to find a parameter vector p (share factor q, and depth z, or more parameters) of length M which minimize the disagreement/error between the observations (f_i) and theoretical prediction (V_i) calculated from the forward equation. A measure of the misfit (objective function or target function or misfit function) can be defined in various ways. Following Silva and Hohmann 1983; Gobashy 2000). We can express the following forms:

$$\varphi(p) = \frac{1}{N} \sum_{i=1}^{N} |f_i - V_i|, \tag{2.7}$$

where, (f_i) and (V_i) the observed and computed responses, respectively for L_1 norm or:

$$\varphi(p) = \frac{1}{N} \left[\sum_{i=1}^{N} (f_i - V_i)^2 \right]^{1/2}, \tag{2.8}$$

which, is equivalent to the form in Eq. 2.5 except in using L_2 norm. The objective function in both definitions is a dimensional quantity with a function space of dimension depends on the length of p and as a result, depends upon the absolute values of both synthetic and field data. This dependency may cause difficulties in defining a general criterion for a good fit between both fields. Moreover, the contribution of each observation to the objective function depends upon the magnitude of the observations. This may cause an exaggerated influence of one part of the curve with respect to all others. A more dynamic form of the objective function may be expressed as the sum of squares of the deviations between the observed and theoretical responses normalized by the observed data (Eq. 2.9 or by the synthetic field, Eq. 2.10).

$$\varphi(p) = \sum_{1}^{N} \{(f_i - V_i(p))/f_i\}^2 \tag{2.9}$$

$$\varphi(p) = \sum_{1}^{N} \{(f_i - V_i(p))/V_i(p)\}^2 \tag{2.10}$$

The above definitions (Eqs. 2.9 and 2.10) are usually called variable weighting (after Bevington 1969). They provide an equivalent contribution of each observation to the objective function. Another form of φ may be given as,

$$\varphi(p) = \left(\frac{1}{N} \sum_{i=1}^{N} (f_i - V_i(p))^2 / f_i^2 \right)^{1/2} \tag{2.11}$$

The above equation defines φ as the root mean squares (rms) relative error (Jupp and Vozoff 1975).

Monteiro Santos (2010) and Abdelazeem et al. (2019) used a misfit between observed and calculated SP data as in Eq. (2.12) and the average relative error percentage evaluated by Eq. (2.13).

$$\varphi(p) = 2||V_i^0 - V_i^c|| / \left[||V_i^0 - V_i^c|| + ||V_i^0 + V_i^c|| \right] \tag{2.12}$$

$$MisfitErr(\%) = \left(\frac{100}{N} \right) \sqrt{\sum_{i=1}^{N} \left[(V_i^0 - V_i^c) / V_i^0 \right]^2} \tag{2.13}$$

where, N are the number of measured or observed Self potential readings, V_i^0 and V_i^c are the measured and calculated fields respectively. The above expression is to be highly stable and convergent in inverting the ill-posed SP problem.

Di Maio et al. (2016a, b) expressed an objective function for minimizing the variance of the fitting:

$$\sigma^2_{p/sheet} = \frac{1}{(N_{obs} - m)} \left(\sum_{i=1}^{N_{obs}} V_i^*(\lambda)_{p/sheet} - V_{obs} \right)^2 \tag{2.14}$$

where, N_{obs} is the number of measured samples, m is the number of parameters, $V_i^*(\lambda)_{p/sheet}$ is the output of the model corresponding to the forward function, for a given λ, and V_{obsi} are the measured SP values. The above expressions are some of many various forms that can be used (Menke 1989).

Traditional algorithms have been successfully used and tested with SP data by many authors. Fourier transform (Roy and Mohan 1984), Least-square fitting (Fitterman and Corwin 1982; El-Araby 2004; Abdelrahman et al. 2008). Derivative analysis and gradient method (Abdelrahman et al. 2003), moving average (Abdelrahman et al. 2009), and spectral method developed by Sundararajan et al. (1990).

However, the effectiveness and applicability of these techniques have many drawbacks (Biswas 2017a). It is well known that by increasing the number of model parameters the objective function shows several local minima in the ambiguity domain (Ramillien and Mazzega 1999) contrarily to a continuous inverse solution where the number is infinite. Although, the entrapment in a local minimum has been partially treated in Mehanee (2015) who solved non-linear SP and gravity inverse problem using gradient type techniques, but the problem of ill-posedness due to error propagation in is solutions when using derivatives and gradient is still present. Moreover, the problem of transverse anisotropy as a major source of ambiguity which entails -if present- increase the dimensionality of the problem is not solved.

The above uncertainty and drawbacks with the applications of traditional algorithms together with the advances in the computation, hardware efficiency, and the increase in the number of unknown parameters led to the development of another class of optimization strategies named global optimizers. These are in many cases naturally inspired metaheuristic and heuristic artificial intelligence based algorithms that overcome the common limitations of the conventional techniques. They played a major role in enhancing the solution quality and better treating the ambiguity. In the next section we focus on the use of metaheuristics and naturally inspired algorithm to solve SP problem. We start with briefly demonstrate the mechanisms of most commonly used metaheuristic algorithms in SP inversion.

2.5 Metaheuristics Inversion of SP Anomalies

As we previously mentioned, the majority of metaheuristics are based on biological evolution principles, i.e., they are concerned with simulating various biological metaphors that differ in the nature of the representation schemes (structure, components, etc.) (Abdel-Basset et al. 2018). There are three main models: evolutionary, swarm, and immune systems which are at the same time metaphor based metaheuristics.

2.5.1 Evolutionary Algorithms (EAs)

Mimic the biological progression of evolution at the cellular level employing selection, crossover, mutation, and reproduction processes to generate increasingly better chromosomes. There are four historic paradigms for evolutionary computation: programming, strategies, genetic algorithms, and genetic programming.

2.5.1.1 Swarm Intelligence (SI)

Simulate the communal behavior of agents in a community, such as birds, fish, and insects. SI mainly depends on updating the candidate solutions through the local interaction with their environment and with each other. The most popular SI algorithms are Particle Swarm Optimizer (PSO) and Ant Colony Optimizer (ACO).

2.5.1.2 Artificial Immune Systems (AIS)

Originated from observed immune functions and theoretical immunology principles and models. The antibodies represent possible solutions in optimization process and iteratively develop through iterating the operators: cloning, mutation and selection. The objective function is represented by the antigen and accepted solutions are stored in memory cells.

Application of metaheuristics in SP inversion is mainly belongs to the first and second categories, EA's (Gobashy et al. 2019; Abdelazeem and Gobashy 2006; Li and Yin 2012; Göktürkler and Balkaya 2012; Di Maio et al. 2016b; Balkaya et al. 2017) and SI's (Sweilam et al. 2007; Monteiro Santos 2010; Srivastava et al. 2014; Singh and Biswas 2016). The application of AIS in SP or geophysical inversion is not yet notices in any literature.The physics based metaheuristics are presented in SP inversion by the simulated annealing (SA), as in (Sharma and Biswas 2013; Biswas and Sharma 2014, 2015).

All of the above algorithms use similar mechanisms this is shown in Fig. 2.7. This mechanism is summarized in two main steps: exploration and exploitation. The first is responsible for searching in the best solution surrounded areas while the latter tends to invade new searching areas. These two steps are the inherent core of robustness of metaheuristics.

Broadly speaking, artificial intelligence (AI) is the intelligence exhibited by machines presently common approaches of AI include traditional statistical methods (Agarwal 2006), computational intelligence (CI) (Voges and Pope 2006) and traditional symbolic AI. CI is a relatively new research area. These are computational methodologies, naturally inspired, used to solve complex problems, where traditional or conventional approaches are infeasible. CI includes fuzzy logic, evolutionary computation/algorithms (EC), and artificial neural network (ANN). Swarm intelligence (SI) is a part of EC.

Fig. 2.7 Generalized algorithmic framework for metaheuristics (modified after Abdel-Basset et al. 2018)

```
Create initial solution/s
    While (stopping criterial is not met) do
        If exploit then
         Create new solution by exploitation;
        Else
         Create updated solution by exploration;
        End
     Update best found solution
    End
Return
```

2.5.2   When Using Metaheuristics?

An important question may arise during practices. When using metaheuristics? Generally, the complexity of the problem and the dimensionality is the key answer to this question. It is not recommended to use metaheuristics to solve problems where efficient exact algorithms are available. Hence for easy and small (low number of unknown parameters) optimization problems which can be solved by non-iterative simple techniques, metaheuristics are seldom used. So the first guideline in solving a problem is to analyze first its complexity and dimensionality. If the problem can be transformed to a classical, then a better choice to one of the working optimization algorithms will help in solving the problem. Otherwise, if there are related problems, the same methodology must be applied (El-Ghazali 1965; Gobashy et al. 2020).

2.5.3   Examples of the Metaphor Based Algorithms

In this section, a brief review on the mechanism of some selected algorithms are given together with the main design and search components for selected families of metaheuristics.

2.5.3.1   Genetic Algorithm (GA)

Genetic algorithms are biological based metaheuristic algorithms that simulate the process of biological evolution and natural selection to reach their power, and their operational characteristics are typically equivalent to the evolution theory. GAs initially work with a group of individuals, each representing a likely problem solution. Each candidate solution, or individual, is generally represented as a string of bits (a set of binary character strings) equivalent to chromosomes and genes in evolution theory. GAs assign to each individual a fitness value based on its solution quality it shows, and highly fit and promising individuals are reproduced. To reach the fitness of the objective function, the GA go through three stages: selection, crossover, and mutation (Fig. 2.8). These are called reproduction operators.

In the selection phase, a stochastic sampling mechanism of the parents that is mainly based on the Roulette wheel algorithm (Goldberg 1989) is used. This procedure is such that individual's fitness is proportional to the probability of selection.

During crossover phase, some parts of two selected chromosomes are swapped and a portion of two fit parent individuals combine to produce two child individuals. There are three types of crossover operations typically used: single-point, two-point and uniform crossover. Figure 2.9 is an example of two point cross-over mechanism. Mutation is the third operator Fig. 2.8 that offers a theoretical guarantee that no bit value is ever permanently fixed in all strings. Mutation introduces random adjustments, thereby inducing a random walk through the search space. During mutation, with a low probability, a portion of the new individuals will be flipped to generate a new bit (Abdelazeem and Gobashy 2006).

Fig. 2.8 A flowchart for the GA (modified after Abdelazeem and Gobashy 2006)

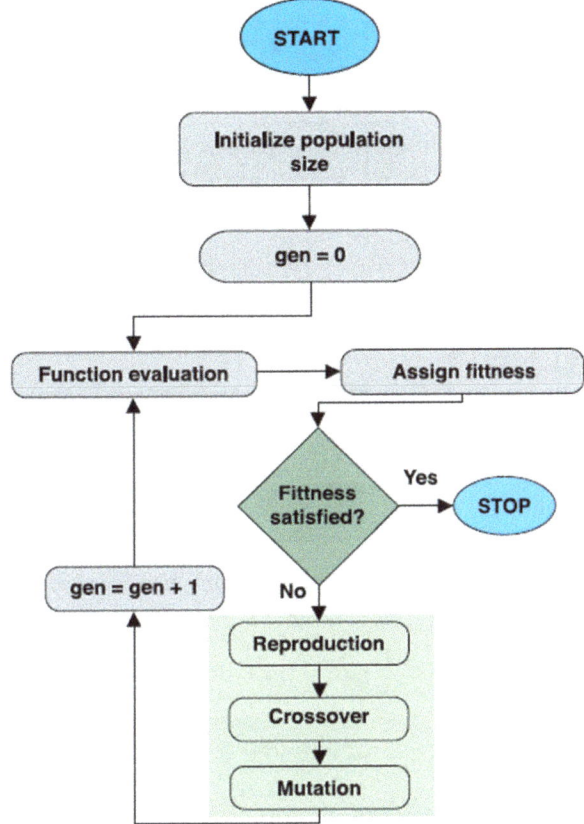

2.5.3.2   The Particle Swarm Optimization (PSO)

A key feature of SI systems is self-organization, where some regional or global order arises out of the local relations between the components of an initially random system. Bonabeau et al. (1999) implemented these feature into swarm systems in three steps: (1) Strong dynamical nonlinearity, (2) Balance between exploration and exploitation, and (3) Multiple interactions of agents. Another important principals or features proposed by Millonas (1994) that SI must satisfy are: quality principle, proximity principle, diverse response principle, adaptability principle, and stability principle. Particle swarm (PSO) search via a swarm or population of particles that change during iterations. To seek the optimal solution, each particle moves in the direction to its previously best (p_{best}) position and the global best (g_{best}) position in the swarm (Sweilam et al. 2007, 2008; Zhang et al. 2014) (Fig. 2.10). In more detail: by assuming a D-dimensional search space $S \subset R^D$ and a swarm consisting of N particle, the algorithm begins with a population of potential solutions (particles or agents) to the problem under consideration and uses them to probe the search space. By definition, a point of position coordinate $X_i = (x_{i1}, x_{i2}, \ldots, x_{iD})^T$ moving in a

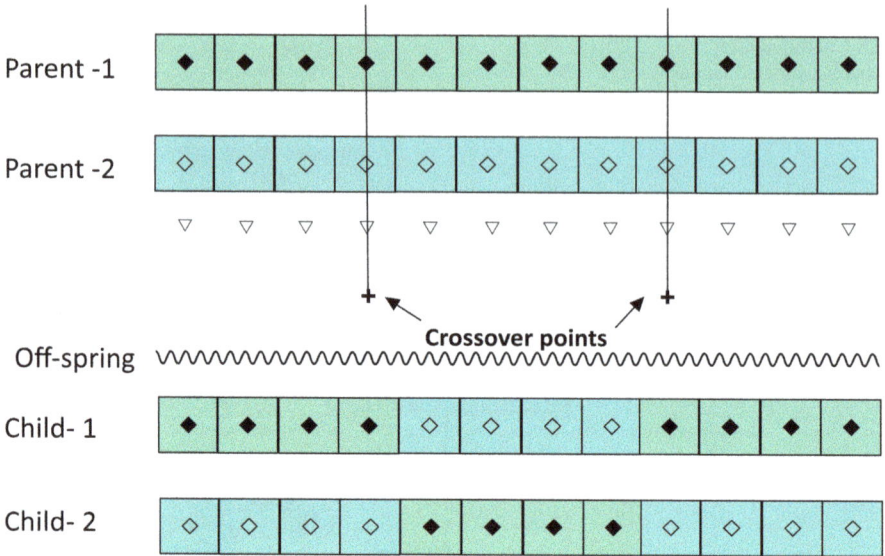

Fig. 2.9 Two point cross-over mechanism (modified after Abdelazeem and Gobashy 2006)

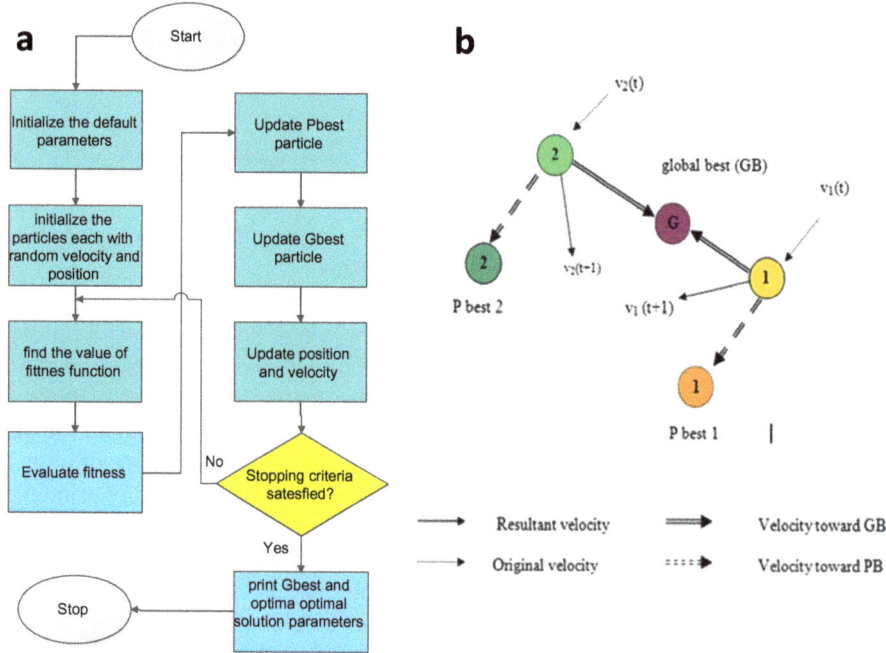

Fig. 2.10 **a** A flowchart for PSO and **b** Mechanism of P_{best} and G_{best} particles in standard Particle swarm optimization PSO

Initialize the population, the velocities randomly, and the best value
While stopping criterion is not met do
    If the best value is not changed for long time
        compute the population new positions, new velocities, and new best
        value using the objective function V(k, D, ψ, z, q)
    Else
        Compute the population new positions, new velocities, and new best
    value using the objective function V(k, D, ψ, z, q)
        End if
End while
Print the result

Fig. 2.11 Pseudo code for solving SP inverse problem with standard PSO

D-dimension hyperspace is a particle 'i'. Each individual or particle of the swarm has a variable velocity (position change) $V_i = (v_{i1}, v_{i2}, \ldots, v_{iD})^T$, accordingly it travels in the function search space. Each individual particle performance is assessed by the objective or target function. The best particle position of all previous iterations, the global best, P_{gb}, and the optimum particle position of the current swarm, the local best, P_{lb}, are both stored and used for adaptation of the new particle speed and position (Eberhart and Kennedy 1995; Eberhart and Shi 1998a, b). The adjustment of the particles speeds and position is given by the following equations:

$$v_{id}^{n+1} = w \cdot v_{id}^n + \alpha(P_{lb}^n - x_{id}^n) + \beta(P_{gb}^n - x_{id}^n) \tag{2.15}$$

$$x_{id}^{n+1} = x_{id}^n + v_{id}^{n+1}, \tag{2.16}$$

where, i is current particle, n is the iteration number, d is the current dimension index, α, β are bounded positive random numbers, uniformly distributed, that control particle global minimum approaching mechanism. The speed of the particle is constraint to $\pm V_{max}$ to avoid search explosion. A critical source of the swarm's capability to search is the interrelations among particles as they react to one another's findings. A pseudo code for the algorithm shown in Fig. 2.11.

Inversion of Self potential anomalies using PSO started with Sweilam et al. (2007, 2008) in a classical two papers where the l_1–norm is chosen instead of the quadratic norm where large noise are usually expected in the field data. Discretization of the forward formula of the response due to simple geometrical forms (Spheres, cylinders) can be obtained at $x = x_i$, $i = 1, 2, \ldots, n$ as follows (Sweilam et al. 2007):

$$J(z_i, \theta_i, q_i, K) = \sum_{i=1}^{n} | \ Vobs(x_i) - \frac{k \ (x_i \cos(\theta_i) + z_i \sin(\theta_i))}{(x_i^2 + z_i^2)^{q_i}} |, \tag{2.17}$$

where J represents the chosen form of objective or target functions to be minimized by Particle swarm technique. Several authors implemented the PSO to invert

SP anomalies (Monteiro Santos 2010; Göktürkler and Balkaya 2012). Detailed application on synthetic and real cases will be shown in Sect. 6.

2.5.3.3 Grey Wolf Optimization

Grey Wolves are typical social animals having clear separation of social work. This Grey Wolves algorithm simulates and describes the actions and rules of grey wolf behavior in nature and how to implement in optimization problems. To understand the algorithm, social dominant hierarchy of Grey wolfs should be understood. Grey wolf (Canis lupus) belongs to Canidae family and are considered as apex predators, meaning that they are at the top of the food chain. They tend to live in a small group. The group size ranges from 5–12 on average. Their social dominant hierarchy is very strict as given in Fig. 2.12. The leaders are a male or a female, called alphas. The alpha member is responsible for making decisions about hunting, … etc. The beta member is in the second level in the grey wolf's hierarchy. The betas are wolves that help the alpha in making decisions, i.e. subordinate. They can be either male or female. The beta wolf strictly respect the alpha, but can command the other lower-level wolves as well. The lowest ranking grey wolf is Omega member.

They always have to submit to all the other dominant wolves. Although Omega is not an important individual in the pack, most pack face internal fighting and problems in case of losing the omega. Delta wolves (he/she) are those who are not omega, beta, or alpha. They are subordinate. Delta wolves are in lower level than alphas and betas, have to obey alphas and betas, but they lead the omega group. Hunting is another

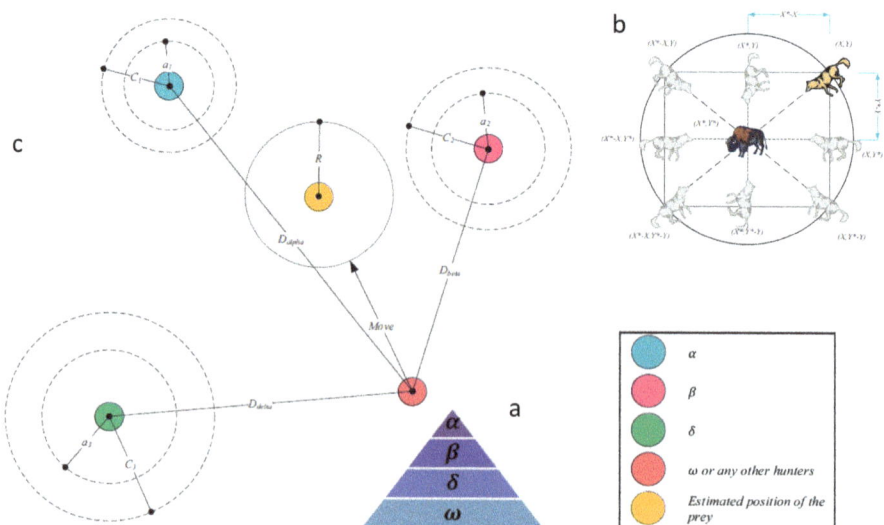

Fig. 2.12 Hierarchy of grey wolves **a** dominance decreases from top down, **b** 2D position vectors and their possible next locations, and **c** Position updating in GWO (compiled from Mirjalili et al. 2014)

interesting social action of grey wolves, main phases of grey wolf hunting are as follows (Mirjalili et al. 2014; Kuliev et al. 2019):

- Tracking, chasing, and approaching the prey.
- harassing and encircling and the prey until it stops.
- Attack towards the prey.

The design of GWO and its optimization properties are inspired from the social hierarchy and hunting of grey wolves. Briefly, this is mathematically modelled in the following steps:

1   Social hierarchy

Here, the fittest solution as the alpha (α). Beta (β) and delta (δ) are the next best solutions, respectively. The other solutions are omega (ω). The hunting (optimization) in the present algorithm is led by α, β and δ. The ω wolves follow these wolves.

2   Encircling prey

This is mathematically may be expressed as (Mirjalili et al. 2014):

$$\vec{D} = |\vec{C} \cdot \vec{X}_p(t) - \vec{X}(t)| \tag{2.18}$$

$$\vec{X}(t+1) = \vec{X}_p(t) - \vec{A} \cdot \vec{D} \tag{2.19}$$

t is the iteration number, $\vec{A}$ and $\vec{C}$ are coefficient, $\vec{X}_p(t)$ is the prey position, and $\vec{X}(t)$ indicates the position of a grey wolf. All quantities are vectors.

The vectors $\vec{A}$ and $\vec{C}$ are calculated as follows:

$$\vec{A} = 2\vec{a} \cdot \vec{r}_1 - \vec{a} \tag{2.20}$$

$$\vec{C} = 2 \cdot \vec{r}_2 \tag{2.21}$$

where, components of $\vec{a}$ are linearly decreased from 2 to 0 over the course of iterations and $\vec{r}_1$ and $\vec{r}_2$ are random vectors in [0, 1] (Mirjalili et al. 2014; Kuliev et al. 2019).

3   Hunting

The hunt is guided by the alpha. Moreover, to model mathematically the hunting mechanism of grey wolves, we assume that alpha, beta, and delta have the knowledge about the best location of prey. Therefore, we save the first three best solutions and these solutions lead all other search agents. The new updated positions of the agents follow the following equations:

$$\vec{D}_\alpha = \vec{C}_1 \cdot \vec{X}_\alpha - \vec{X}|, \ \vec{D}_\beta = \vec{C}_2 \cdot \vec{X}_\beta - \vec{X}|\vec{D}_\gamma = \vec{C}_3 \cdot \vec{X}_\gamma - \vec{X}|, \tag{2.22}$$

$$\vec{X}_1 = \vec{X}_\alpha - \vec{A}_1 \cdot (\vec{D}_\alpha), \ \vec{X}_2 = \vec{X}_\beta - \vec{A}_2 \cdot (\vec{D}_\beta), \ \vec{X}_3 = \vec{X}_\gamma - \vec{A}_3 \cdot (\vec{D}_\gamma), \tag{2.23}$$

$$\vec{X}(t+1) = \frac{\vec{X}_1 + \vec{X}_2 + \vec{X}_3}{3}, \tag{2.24}$$

4   Attacking prey (exploitation)

The grey wolves finish the hunt by attacking the prey when it stops moving. This is expressed mathematically by decreasing the value of $\vec{\alpha}$ and consequently, the fluctuation range of $\vec{A}$ is also decreased by $\vec{\alpha}$. In other words $\vec{A}$ is a random value in the interval $[-2a, 2a]$ where 'a' is decreased from 2 to 0 over the course of iterations. When random values of $\vec{A}$ are in $[-1, 1]$, the next position of a search agent can be in any position (Mirjalili et al. 2014). The search agents update their position based on the current location of the delta, beta, and alpha, and attack towards the prey.

5   Search for prey (exploration)

The Grey wolf's algorithm system operates according to agents' location. They diverge from each other to search for prey and converge to attack prey. Mathematically, divergences is simulated by utilizing $\vec{A}$ with random values greater than 1 or less than − 1 to oblige the search agent to diverge from the prey. This support exploration and allow the global behavior of the algorithm.

2.5.3.4   Whale Optimization WOA

This algorithm is a naturally inspired meta-heuristic optimization algorithm that imitates the hunting behavior of humpback whales. It is firstly presented by Mirjalili and Lewis (2016). Its performance benefits from modeling the hunting behavior using random or the best quest agent to pursue the prey and use a spiral to mimic the humpback whale bubble-net attacking process. Whales are smart, they have twice the number of spindle cells that a human adult have, that makes them different from other creatures. This is the main cause of their smartness. It makes them think, learn, judge, communicate, become even emotional as a human does, and some of them can develop their own dialect. Some of them can live in family over their entire life period. Like the humpback whales (Megaptera novaeangliae). These type of whales have the most interesting way of hunting called bubble-net feeding method. During hunting, they create two maneuvers associated with bubble: 'upward-spirals' and 'double- loops'. Goldbogen et al. (2013) discovered such tricks. They are using sensors to investigate their

3-Dimensional maneuver during hunting. In the upward spiraling maneuver, hump-back whales plunge around 12 m below and then continue to build a spiral-shaped bubble around the prey and swim up to the surface. The double loop technique consists of three separate stages: the coral loop, the lobtail and the capture loop. This spiral bubble net feeder maneuver is mathematically modeled for optimization.

Encircling prey

The position of the optimal design in the search space usually is not known a priori, the WOA algorithm assumes that the current best candidate solution is the target or is close to the best. After the best search agent is defined, the other search agents will then try to update their positions towards the best search agent. This behavior is represented by the following equations (Mirjalili and Lewis 2016):

$$\vec{D} = |\vec{C} \cdot \vec{X}^*(t) - \vec{X}(t)| \tag{2.25}$$

$$\vec{X}(t+1) = \vec{X}*(t) - \vec{A} \cdot \vec{D} \tag{2.26}$$

where t is the current iteration, $\vec{X}^*(t)$ the position vector of the best solution obtained so far, $\vec{X}(t)$ is the position vector, $\vec{A}$ and $\vec{C}$ are coefficient vectors, $||$ is the absolute value, and—is an element-by-element multiplication. Notice that $\vec{X}^*(t)$ should be updated in each iteration if there is a better solution. The vectors $\vec{A}$ and $\vec{C}$ are calculated as follows:

$$\vec{A} = 2\vec{\alpha} \cdot \vec{r}_1 - \vec{\alpha} \tag{2.27}$$

$$\vec{C} = 2 \cdot \vec{r}_2 \tag{2.28}$$

where, $\vec{\alpha}$ is linearly decreased from 2 to 0 over the course of iterations in both phases and $\vec{r}$ is a random vector in [0,1]. It is worth mentioning here that this procedure is similar to the same *Encircling prey* used in Grey Wolf Optimization (Eqs. 2.20–2.21)
    Bubble-**net attacking method** (exploitation phase).
    Here two approaches are designed as follows (Mirjalili and Lewis 2016):

1   **Shrinking encircling mechanism**:

$\vec{A}$ is a random value in the interval $[-a, a]$, where a is decreased from 2 to 0 over the course of iterations. Setting random values for $\vec{A}$ in $[-1, 1]$, the new position of a search agent can be defined anywhere in between the original position of the agent and the position of the current best agent (Fig. 13a).

2   **Spiral updating position***:*

This approach first calculates the distance between the whale located at (X, Y) and prey located at (X^*, Y^*). A spiral equation is then created between the position of whale and prey to mimic the helix-shaped movement of humpback whales as follows (Fig. 2.13b):

$$\overrightarrow{X}(t+1) = \overrightarrow{D}' \cdot e^{bi} \cdot \cos(2\pi l) + \overrightarrow{X}^*(t) \qquad (2.29)$$

$\overrightarrow{D}' = |\overrightarrow{X}^*(t) - \overrightarrow{X}(t)|$ is the distance of the ith whale to the best solution obtained so far or the prey, l is a random number in $[-1, 1]$, b is a constant for defining the shape of the logarithmic spiral, and. is an element-by-element multiplication.

$$\overrightarrow{X}(t+1) = \begin{cases} \overrightarrow{X}^*(t) - \overrightarrow{A} \cdot \overrightarrow{D} \, if \, p < 0.5 \\ \overrightarrow{D}' \cdot e^{bl} \cdot \cos(2\pi l) + \overrightarrow{X}^*(t) \, if \, p \geq 0.5 \end{cases} \qquad (2.30)$$

where, p is a random number in the rang [0,1].

Search for prey (exploration phase):

Actually, humpback whales search randomly according to the position of each other. Therefore, we use $\overrightarrow{A}$ with the random values as $-1 < \overrightarrow{A} < 1$ to force search agent to move far away from a reference whale. Figure 2.14 is a flowchart describing the main steps in the algorithm.

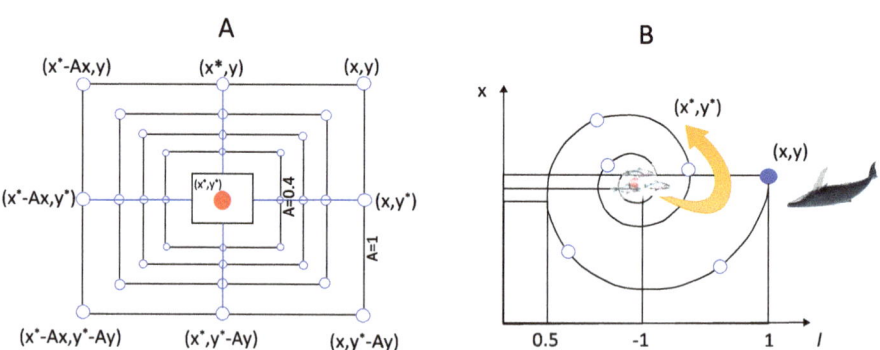

Fig. 2.13 a A flowchart for the WOA heuristic algorithm, **b** Bubble-net search mechanism. (x* is the best solution obtained so far): (**A**) Shrinking encircling mechanism and (**B**) spiral updating position (modified after Abdel-Basset et al. 2018)

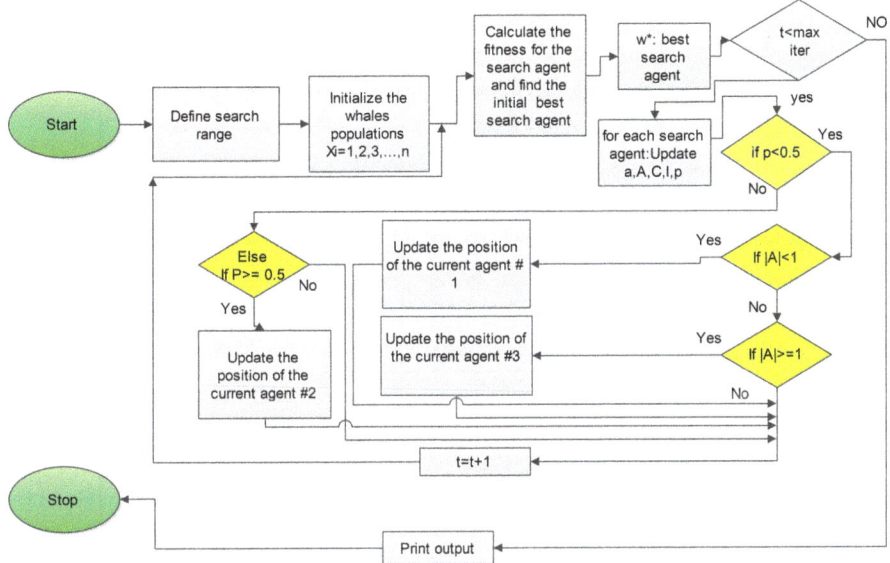

Fig. 2.14 A flowchart for the WOA (modified after Gobashy et al. 2019)

2.5.3.5   Cuckoo Search Algorithms

The reproduction strategy of Cuckoo birds have an interesting aggressive behavior. The the ani and Guira cuckoos species lay their eggs in an obligate brood parasitism (Joshi et al. 2016; Yang and Deb 2009), i.e., communal nests. Though they may remove others' eggs and by this way, they increase the probability of hatching of their own eggs. Generally, there are three basic types of brood parasitism: cooperative breeding, intraspecific brood parasitism, and nest takeover. If a host bird discovers the eggs are not their own, they will either throw these alien eggs away or simply abandon its nest and build a new nest elsewhere. In some species, the female parasitic cuckoos are often very specialized in the mimicry in shape of the eggs of a few chosen host species. This reduces the probability of their eggs being abandoned and thus increases their reproductively. Another two interesting actions are observed, the first is that some parasitic cuckoos often choose a nest where the eggs are just laid. In general, the cuckoo eggs hatch earlier than their host eggs. Once the first cuckoo chick is hatched, the first instinct action it will take is to evict the host eggs by blindly propelling the eggs out of the nest. The second is the Lévy-flight-style intermittent scale free search pattern that is usually observed during the flight of Cuckoo birds. Both actions are used to design promising optimization algorithm (CS). A flowchart for the standard Cuckoo is shown in Fig. 2.15. The basic steps of the Cuckoo Search (CS) can be summarized as: When generating new solutions $x(t + 1)$ for, say, a cuckoo i, a Lévy flight is performed (Joshi et al. 2016),

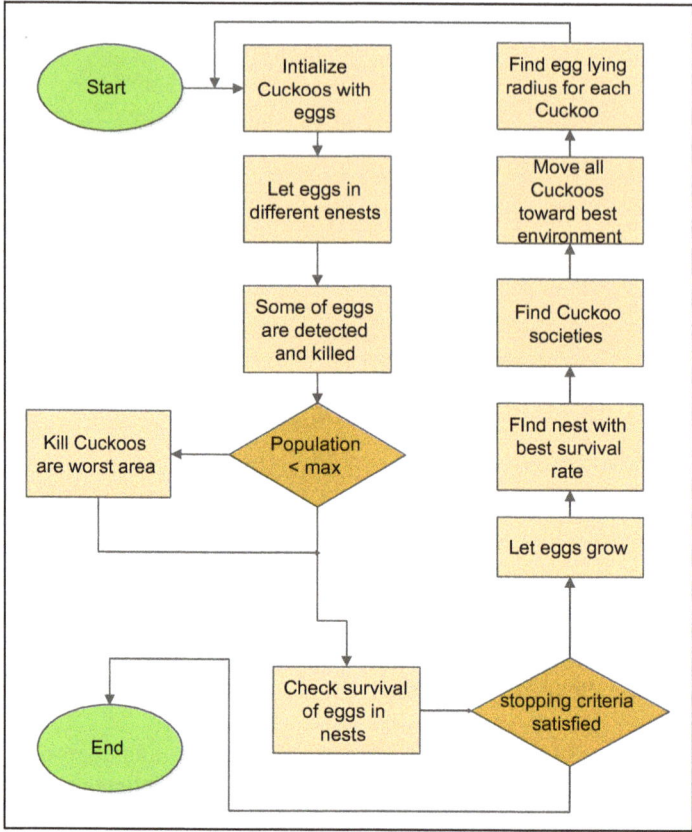

Fig. 2.15 A flowchart for the standard Cuckoo (after Reddy 2017)

$$X_i^{(t+1)} = X_i^t + \alpha \bigoplus \text{Lévy}\,(\lambda), \qquad (2.31)$$

where, $\alpha > 0$ is the step size which should be related to the scales of the problem of interests. In general, $\alpha = 1$ is used. The above equation is the stochastic form for random walk. A random walk is a Markov chain whose next status/location only depends on the present location (X_i^t) and the transition probability ($\alpha \bigoplus \text{Lévy}\,(\lambda)$). The product $\bigoplus$ means entry wise multiplications; the random walk via Lévy flight is more efficient in exploring the search space as its step length is much longer in the long run. This entry wise product is similar but more efficient than those used in PSO. The Lévy flight provides a random walk while the random step length is drawn from Lévy distribution (Yang and Deb 2009):

$$\text{Lévy}\,u = t^{-\lambda}, (1 < \lambda \leq 3), \qquad (2.32)$$

2.5.3.6 Salp Swarm Optimization SSO

Salp swarm optimization (Mirjalili et al. 2017) is different from other metaheuristics in terms of inspiration, mathematical formulation, and real-world application. Briefly, Salps are from the family of Salpidae, have transparent and body barrel-shaped. They are similar to the jelly fish in their movement by pumping water through the body to move forwards as sort of propulsion (Madin 1990). The shape of a salp is shown in Fig. 2.16a. An interesting trait is their swarming behavior. Salps, in deep oceans, often form a swarm called salp chain Fig. 2.16b. Researchers believe that this is done for achieving better locomotion using rapid coordinated changes and foraging. Mathematically, the population is divided to two parts: leader and followers. The salp at the front of the chain is the leader, while all other salps are called followers. A two dimensional matrix x is used to store the positions of each salp. The food source F (the swarm target) is located in the n-dimensional search space.

$$x_j^1 = \begin{cases} F_j + c_1\big((ub_j - lb_j)c_2 + lb_j c_3 \geq 0 \\ F_j - c_1\big((ub_j - lb_j)c_2 + lb_j c_3 < 0 \end{cases} \tag{2.33}$$

where x_j^1 represent the first salp position (leader), ub_j indicates the upper bound of jth dimension, lb_j indicates the lower bound of jth dimension, $c_1, c_2,$ and c_3 are random numbers, and F_j is the food source position in the jth dimension, c_1 it balances between the exploration and exploitation phases of the algorithms. Parameter c_3 and

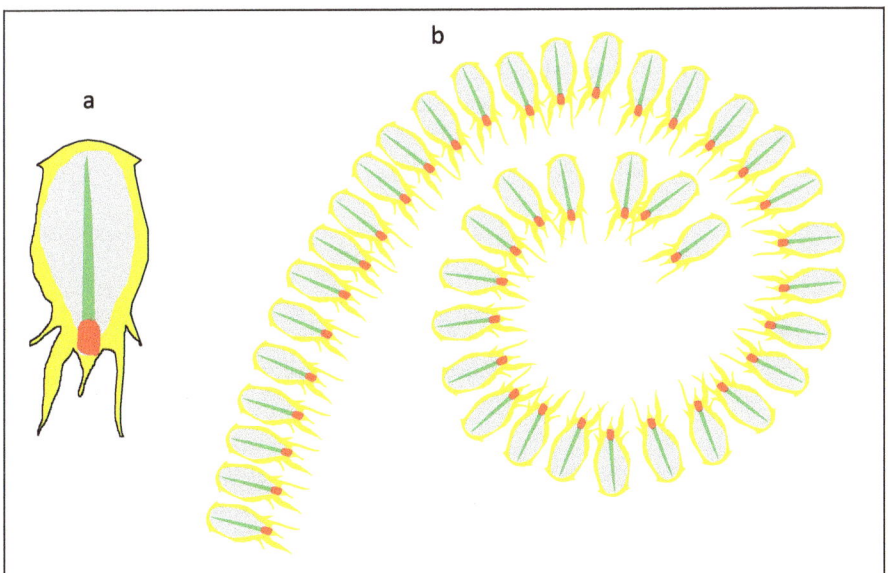

Fig. 2.16 **a** Individual salp, **b** swarm of salps (salps chain)

c_2 are uniformly generated random numbers in the interval of [0, 1]. Newton's law is used to change the position of the followers in the form (Mirjalili et al. 2017):

$$x_j^i = \frac{1}{2}at^2 + v_0t \tag{2.34}$$

Considering $v_0 = 0$, the above equation may be expressed as:

$$x_j^i = \frac{1}{2}(x_j^i + x_j^{i-1}) \tag{2.35}$$

where, jth is the dimension, $i \geq 2$, x_j^i shows the ith salp position, v_0 is the starting velocity, and t is time,

2.5.3.7 Artificial Bee Colony Optimizer (ABC)

The specific intelligent behaviors of the honey bee swarms have been suited by many authors (Tereshko 2000; Tereshko and Lee 2002; Teodorovi´c 2003; Tereshko and Loengarov 2005; Lucic and Teodorovi´c 2002; Wedde et al. 2004; Drias et al. 2005; Benatchba et al. 2005) and applied to solve several combinatorial problems. In general, the colony of artificial bees contains three groups of bees: *employed bees,* onlookers and scouts. A bee waiting on the dance area for making decision to choose a food source, is called an *onlooker,* a bee going to the food source visited by itself previously is named an employed bee, and a bee carrying out random search is called *a scout.*

The ABC algorithm is designed such that the first half of the colony consists of employed artificial bees and the second half constitutes the onlookers. Hence, for every food source, there is only one employed bee. The main steps of the algorithm are (Karaboga and Basturk 2007):

- Initialize.
- REPEAT (until requirements are met).

  (a) Place the employed bees on the food sources in the memory;
  (b) Place the onlooker bees on the food sources in the memory;
  (c) Send the scouts to the search area for discovering new food sources.

- UNTIL.

These are shown in detail in the flowchart given in Fig. 2.17. Following Karaboga and Basturk (2007), in the ABC algorithm, three steps constitute each cycle, these are: sending the employed bees onto the food sources and then measuring their nectar amounts of foods; the nectar amounts is determined by selecting of the food sources by the onlookers after sharing the information of employed bees; determining the scout bees and then sending them onto possible optimum food sources. At the

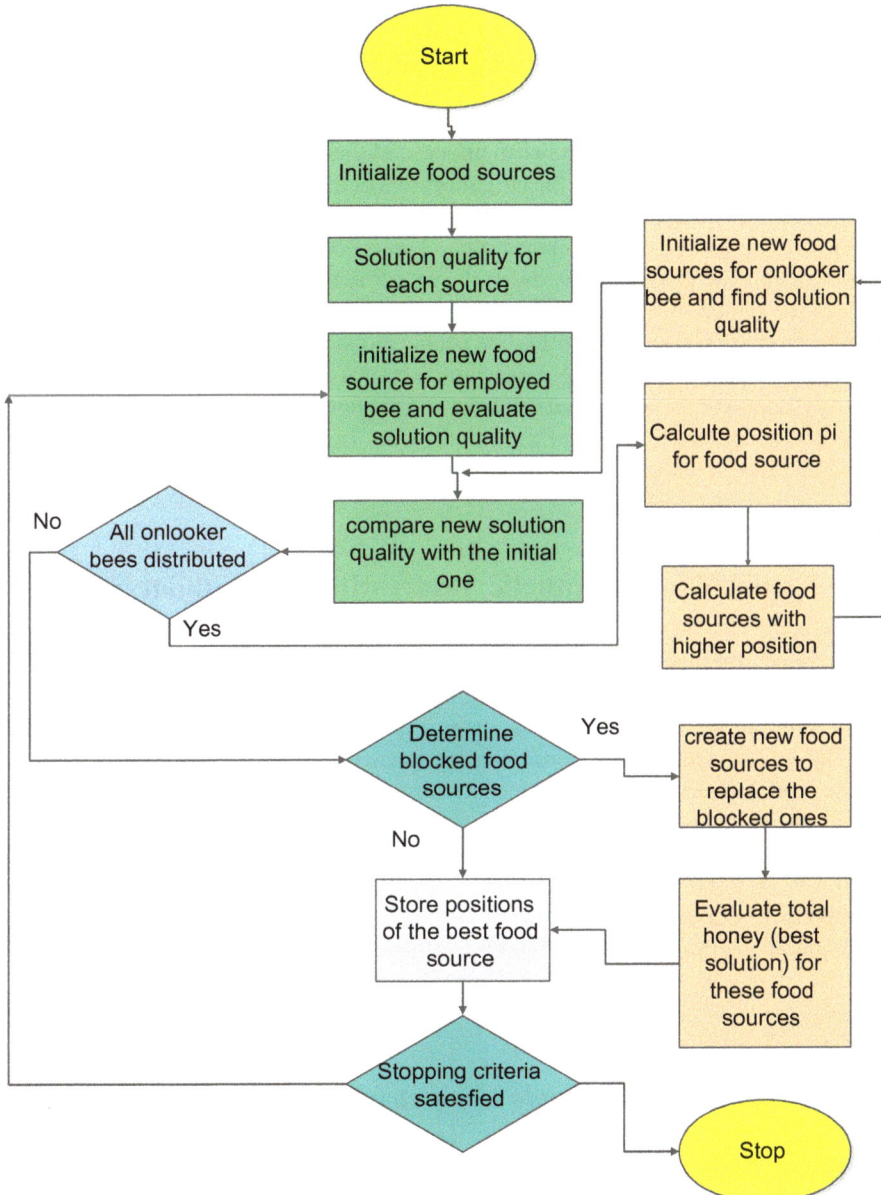

Fig. 2.17   Artificial bee colony ABC algorithm flowchart (modified after Shah et al. 2017)

initialization stage, food sources and the nectar amounts are determined randomly
by the bees. These information are transferred to the colony when they come back
on the dance area. After transferring the information and in the second stage, every
employed bee goes to the food source area visited by herself at the previous cycle
since that food source exists in her memory, and then chooses a new food source by
means of visual information in the neighborhood of the present one. At the third stage,
an onlooker favors a food source area depending on the nectar information distributed
by the employed bees on the dance area. The probability with which an onlooker
chooses that food source increases, as the nectar amount of a food source increases.
Hence, the dance of employed bees carrying higher nectar recruits the onlookers
for the food source areas with higher nectar amount. In the Artificial Bee colony
algorithm, mathematically, the nectar amount of a food source corresponds to the
quality (fitness) of the associated solution, and the position of a food source represents
a possible solution of the optimization problem. The number of the employed bees
or the onlooker bees is equal to the number of solutions in the population.

2.6   Application to Synthetic Model with and Without Noise

To examine stability and effectiveness of the above selected meta-heuristic algorithms, they have all been applied to the same synthetic examples with and without
noise added to the calculated fields, using Eq. (2.1) for simple geometrical models
and Eq. (2.2) for thin sheet model. Search space for each parameter changes based
on the model used but is fixed with each run for comparison. For a vertical cylinder
model, the range for the different parameters are: K is $-10,000$ to $10,000$ mV, D:
$0–80$, ψ: $0–90$, z: $1–30$, and q: $0.5–1.5$. For a horizontal cylinder, K is -2000 to
2000 mV, D: $0–80$, ψ: $1–180$, z: $1–30$, and q: $0.5–1.5$. For a spherical model, K is
$-5,000$ to $15,000$ mV, D: $0–80$, ψ: $0–180$, z: $0–30$, and q: $0.5–1.5$. For the 2D thin
sheet model the range for the different parameters are: K is -200 to 200, D: $0–100$,
a: $1–30$, alpha: $30–180$ and h: $0–20$.

2.6.1   Application to Simple Geometrical Models

The selected metaphor have been used to invert a group of simple models using identical lower and upper bounds and the same number of iterations. The first synthetic
example is a vertical cylinder with parameters (K $= -2000$ mV, $X_0 = 2$ m, $\psi° =$
$30°$, z $= 8$ m and q $= 0.5$) and profile length is 101 units with (1 unit interval)
without adding noise. Table 2.1 shows the results from all methods, the upper and
lower limits used, misfit between inverted and real fields, number of iterations and
the elapsed time for each method. The calculated and inverted fields are shown in
Fig. 2.18 for each method (Fig. 2.19).

Table 2.1 Comparison between the WOA and ABC, SSO, CSO, GWA, and PSO algorithms, without adding noise to invert SP field due to vertical cylinder model

Noise free model

Algorithm	K (mVm$^{(2q-1)}$)	D (m)	ψ°	Z (m)	q	Misfit error (%)	Fval	Iter.	Elapsed time (s)
WOA	−2014.8	2.2226	29.618	7.9257	0.5	0.95716	0.0055292	1000	2.3125
ABC	−2257.2	1.2512	34.911	9.134	0.53176	0.46436	0.017877	1000	6.0313
SSO	−2001.6	1.9976	30.027	8.0077	0.50022	0.0039592	8.6786e-05	1000	3.625
CSO	−2000.2	1.9999	30.002	8.0007	0.50002	**0.0004997**	8.5731e-06	1000	6.1719
GWO	−2393.7	0.4801	35.767	9.4887	0.54246	6.6778	0.026538	1000	3.125
PSO	−2006.6	1.9872	30.102	8.0258	0.50085	0.0069469	0.00033495	1000	28.016
Parameter range	−10000 :10000	0:80	0:90	1:30	0.5: 1.5				

WOA Whale optimization algorithm, ABC Artificial Bee Colony, SSO Salp Swarm Optimization, CSO Cuckoo Search Optimization, GWA Grey wolf Algorithm and PSO Particle swarm optimization

F_{val} Minimum objective function value at optimum solution

Iter Allowed number of iterations

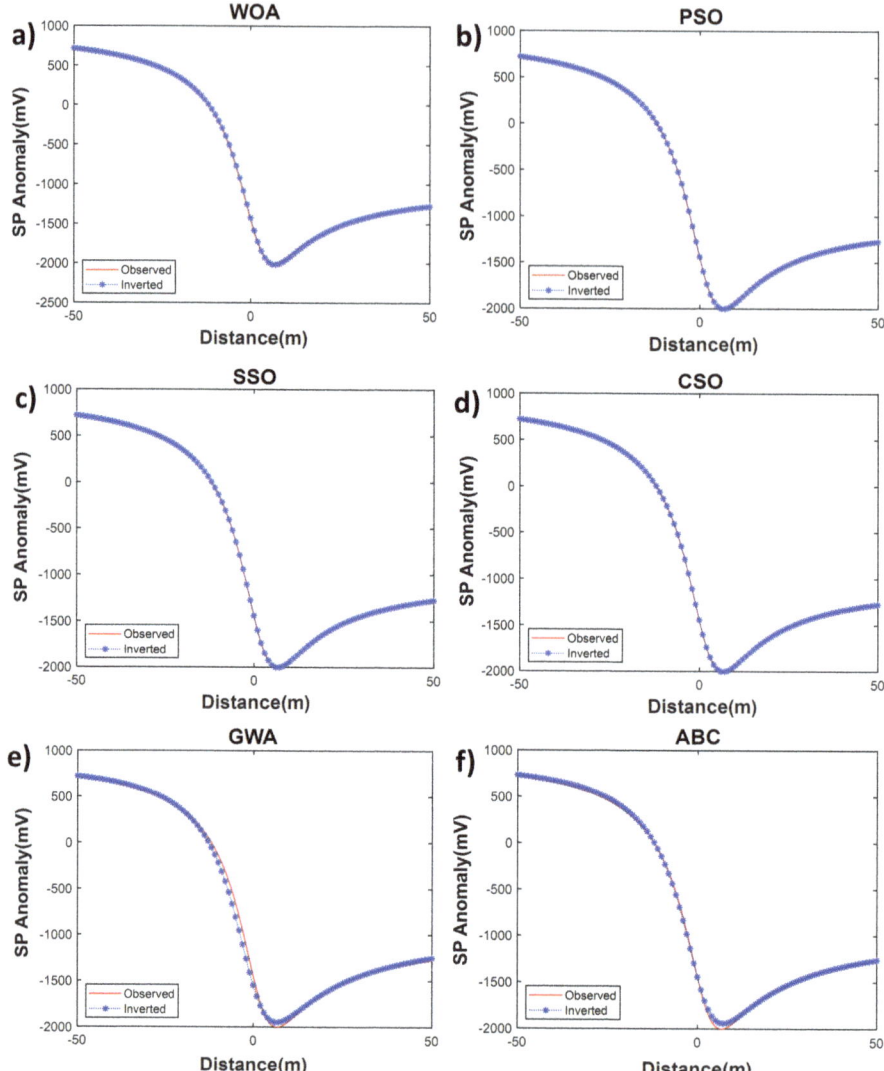

Fig. 2.18 Inversion of Vertical cylinder model (noise free) with **a** WOA, **b** ABC, **c** SSO, **d** CSO, **e** GWA, and **f** PSO methods

The second synthetic example is a horizontal cylinder with parameters ($K = -500$ mV, $X_0 = 5$ m, $\psi^\circ = 90^\circ$, z $= 3$ m and q $= 1$), profile length $= 101$ units with (1 unit interval) and with adding noise of 10%. Table 2.2 shows the complete inverted parameters, misfit, iterations and elapsed time taken by each method. Inverted fields are drawn together with real one for each method (Fig. 2.21).

The third and fourth synthetic examples are SP fields due to a sphere with parameters ($K = -0,000$ mV, $X_0 = 40$ m, $\psi^\circ = 60^\circ$, z $= 10$ m and q $= 1.5$), profile length $=$

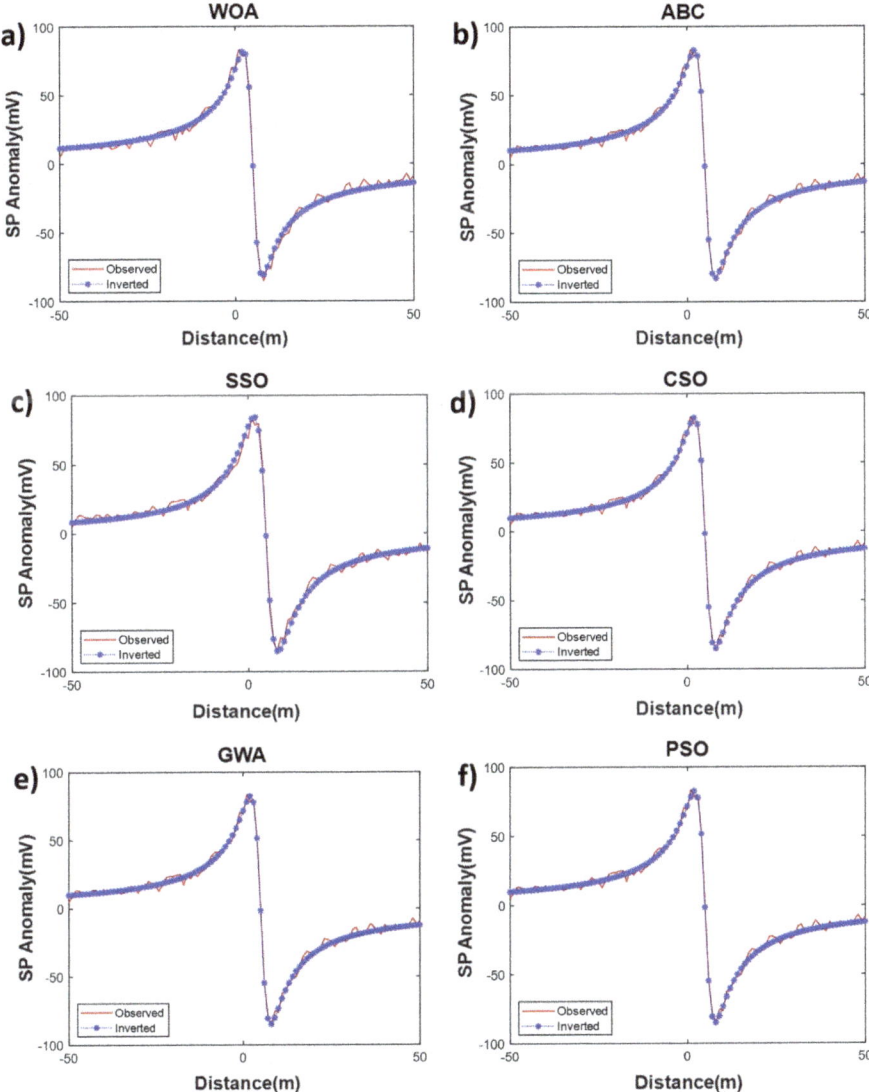

Fig. 2.19 Inversion of horizontal cylinder model (noise 10%) with **a** WOA, **b** ABC, **c** SSO, **d** CSO, **e** GWA, **f** PSO methods

101 units with 1-unit interval) and noise added is 20 and 30%. Results are presented in Tables 2.3 and 2.4 and Figs. 2.20 and 2.21.

Table 2.2 Comparison between the WOA and ABC, SSO, CSO, GWA, and PSO algorithms, to invert SP field due to a horizontal cylinder model with (10%) noise added

Noisy model (10%)

Algorithm	$K\ (mVm^{(2q\_1)})$	D (m)	ψ^{o}	Z (m)	q	Misfit error (%)	F_{val}	Iter.	Elapsed time (s)
WOA	−302.05	4.9617	90.434	2.3078	0.90635	2.3885	0.08626	2500	13.875
ABC	−378.4	4.9738	90.081	2.6156	0.94723	1.9916	0.07668	2500	14.406
SSO	−668.48	4.9811	89.861	3.4244	1.0435	**1.7424**	0.09599	2500	7.7188
CSO	−408.2	5.0164	89.163	2.7113	0.9602	1.9193	0.07579	2500	14.656
GWO	−408.5	5.0167	89.147	2.7141	0.96033	1.9183	0.07580	2500	7.4844
PSO	−408.2	5.0164	89.163	2.7113	0.9602	1.9193	0.07579	2500	63.594
Parameter range	−2000 :2000	0:80	1:180	1:30	0.5:1.5				

WOA Whale optimization algorithm, ABC Artificial Bee Colony, SSO Salp Swarm Optimization, CSO Cuckoo Search Optimization, GWA Grey wolf Algorithm and PSO Particle swarm optimization

F_{val} Minimum objective function value at optimum solution

Iter Allowed number of iterations

Table 2.3 Comparison between the WOA and ABC, SSO, CSO, GWA, and PSO algorithms, to invert SP field due to spherical model with (20%) noise added

Noisy model (20%)

Algorithm	$K(mVm^{(2q-1)})$	D (m)	ψ^o	z (m)	q	Misfit error (%)	Fval	Iter	Elapsed time (s)
WOA	−4340	39.386	64.375	9.2916	1.3332	79.996	0.21086	4000	20.813
ABC	−13,937	39.068	67.328	11.388	1.5	**56.2**	0.20752	4000	23.156
SSO	−9237.8	39.359	66.593	10.52	1.446	73.1	0.20676	4000	11.891
CSO	−12,822	39.215	67.598	11.041	1.4918	68.94	0.20653	4000	24.531
GWO	−13,381	39.22	67.633	11.062	1.4993	68.975	0.20659	4000	11.219
PSO	−12,113	39.253	67.381	10.927	1.4848	69.962	0.20655	4000	91.859
Parameter range	−15,000:15,000	0:80	0:180	0:30	0.5:1.5				

WOA Whale optimization algorithm, ABC Artificial Bee Colony, SSO Salp Swarm Optimization, CSO Cuckoo Search Optimization, GWA Grey wolf Algorithm and PSO Particle swarm optimization

F_{val} Minimum objective function value at optimum solution

Iter Allowed number of iterations

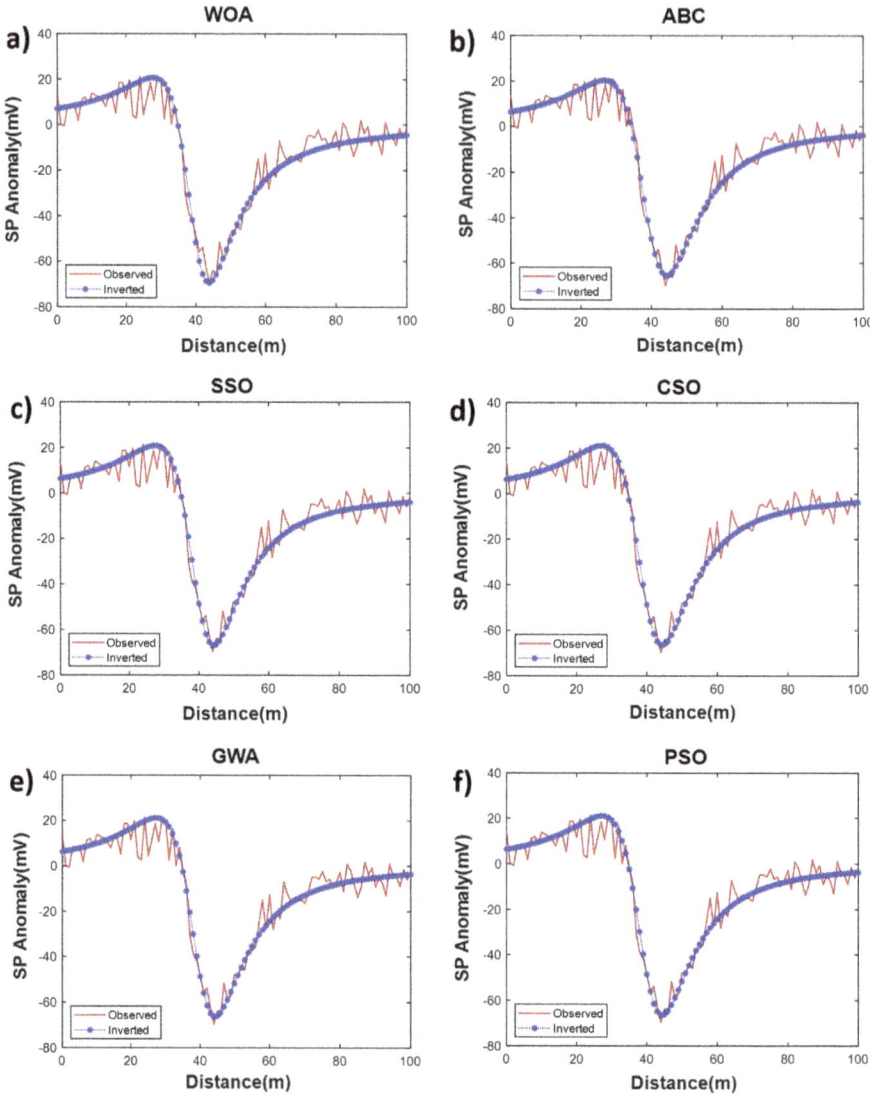

Fig. 2.20 Inversion of spherical model (noise 20%) with **a** WOA, **b** ABC, **c** SSO, **d** CSO, **e** GWA, **f** PSO methods

2.6.2 Application to Thin Sheet Model

The same six chosen methods are also applied to 2D inclined thin sheet model (Eq. 2.2) with and without noise. The parameters of the calculated field are (K = 50 mV, D = 55 m, a = 12 m, α = 150° and h = 10 m) and profile length = 101 units

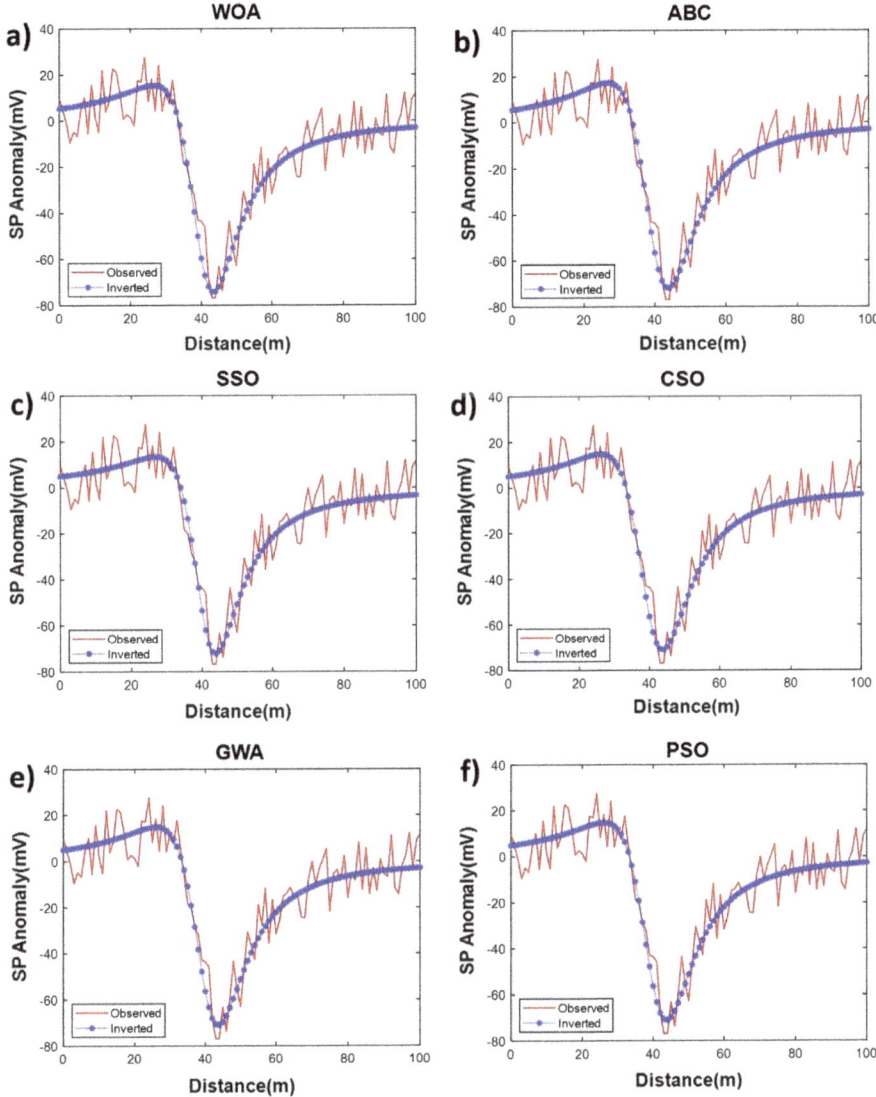

Fig. 2.21 Inversion of Vertical cylinder model (with 30% noise) with **a** WOA, **b** ABC, **c** SSO, **d** CSO, **e** GWA, **f** PSO methods

with 1-unit interval). Tables 2.5, 2.6, 2.7 and 2.8 show results of such model without noise, with 10, 20 and 30% noise added respectively. Figures 2.22, 2.23, 2.24 and 2.25 presents the results for the same levels of noise tested by all methods.

Table 2.4 Comparison between the WOA and ABC, SSO, CSO, GWA, and PSO algorithms, to invert SP field due to spherical model with (30%) noise added to sphere model

Algorithm	Noisy model (30%)								
	$K(mVm^{(2q-1)})$	D (m)	ψ^o	z (m)	q	Misfit error (%)	Fval	Iter	Elapsed time (s)
WOA	−7271	39.539	58.726	9.6388	1.4311	58.349	0.34471	5500	26.328
ABC	−12,156	39.351	62.41	10.511	1.5	60.487	0.34421	5500	28.219
SSO	−4603.7	40.253	55.717	9.1493	1.3669	58.377	0.34612	5500	15.375
CSO	−11,824	39.529	59.863	10.625	1.5	58.121	0.34196	5500	28.984
GWO	−11,821	39.528	59.864	10.623	1.5	**58.108**	0.34196	5500	14
PSO	−13,440	39.53	60.104	10.73	1.5214	56.953	0.34167	5500	124.25
Parameter range	−15,000:15,000	0:80	0:180	1:30	0.5: 1.5				

WOA Whale optimization algorithm, ABC Artificial Bee Colony, SSO Salp Swarm Optimization, CSO Cuckoo Search Optimization, GWA Grey wolf Algorithm and PSO Particle swarm optimization

F_{val} Minimum objective function value at optimum solution

Iter Allowed number of iterations

Table 2.5 Comparison between the WOA and GWO, WOA, CSA, SSO, and BCO algorithms, to invert SP field due to a thin sheet model (noise free)

Noise free model

Algorithm	K(mV)	D (m)	a (m)	a (m)	h (m)	Misfit error (%)	F_{val}	iterations	Elapsed time (s)
WOA	48.812	55.187	12.308	149.9	10.045	1.1148	0.053836	500	7.3438
ABC	59.502	54.202	10.424	148.91	10.803	0.50014	0.025075	500	2.25
SSO	55.515	54.636	10.846	150.59	10.807	0.37034	0.014952	500	2.1406
CSO	49.782	55.162	12.059	150.53	9.9269	**0.11606**	0.0064641	500	2.1563
GWO	54.514	54.578	11.073	150.2	10.51	0.89377	0.013093	500	1.7656
PSO	81.316	53.779	14.011	118.07	3.6613	1.0504	0.040114	500	2.0313
Parameter range	−200 :200	0:100	0:30	0:180	0: 20				

WOA Whale optimization algorithm, ABC Artificial Bee Colony, SSO Salp Swarm Optimization, CSO Cuckoo Search Optimization, GWA Grey wolf Algorithm and PSO Particle swarm optimization

F_{val} Minimum objective function value at optimum solution

Iter Allowed number of iterations

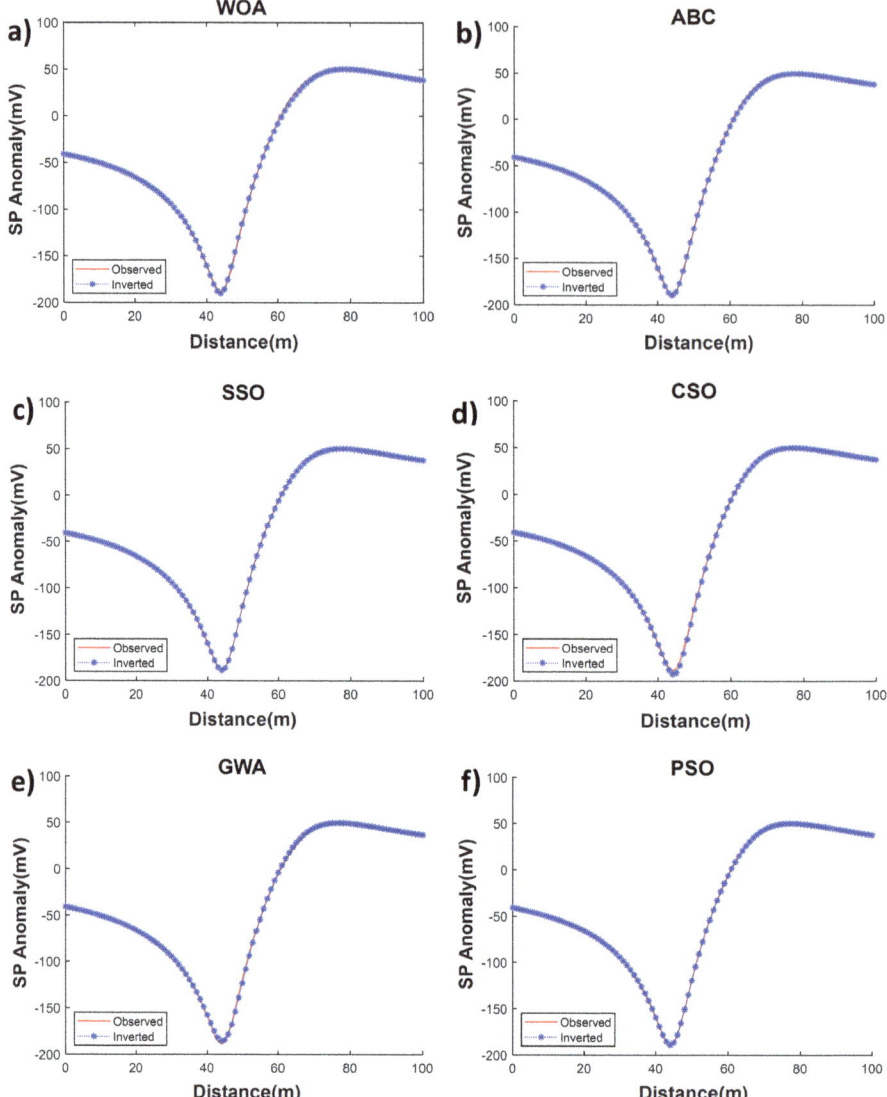

Fig. 2.22 Inversion of 2D inclined sheet model (without noise) with **a** WOA, **b** ABC, **c** SSO, **d** CSO, **e** GWA, **f** PSO methods

Even though each algorithm yielded very similar results with the noise-free data (Figs. 2.18 and 2.22), the minimum misfit error value (0.0004997 mV) was obtained by CSO in case of vertical cylinder model, while the other ranges between 0.0039592 (SSO) and 6.6778% for (GWO). In this later case the minimum function value reached is 8.5731e-06, and the elapsed time is 6.1719s during 1000 iterations using Inter® core ™ i7-8705G CPU @ 3.10 GHz with 16.0 GB RAM and 64-bit operating system.

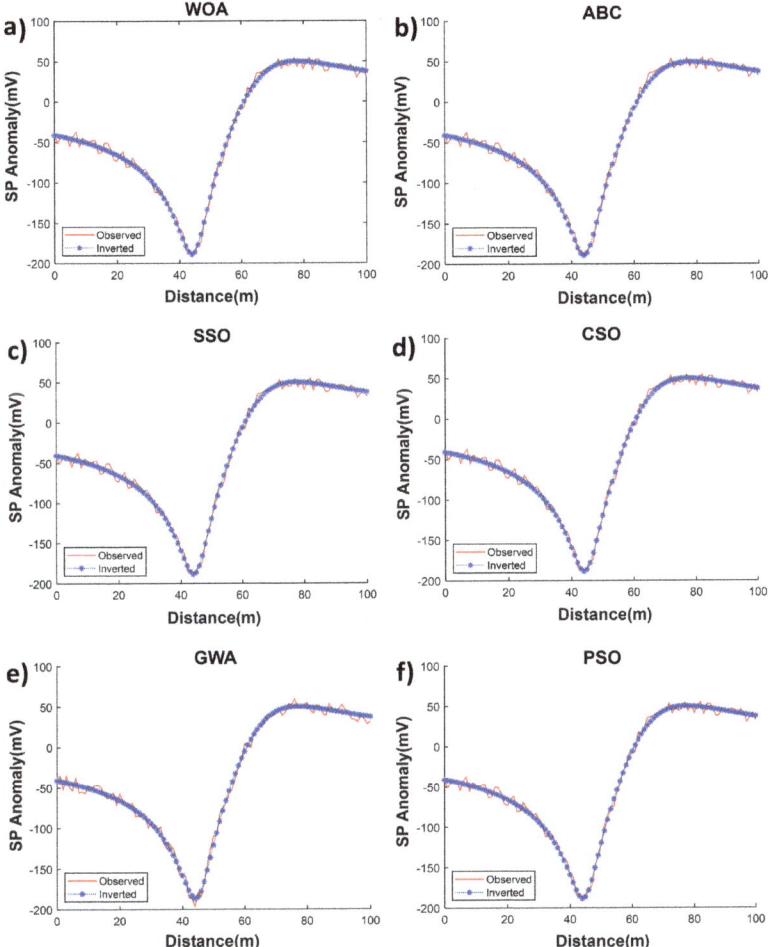

Fig. 2.23 Inversion of 2D inclined sheet model (with 10% noise) with **a** WOA, **b** ABC, **c** SSO, **d** CSO, **e** GWA, **f** PSO methods

Results with noise added varies from method to another and with the noise percentage added, e.g.: for the spherical model with 30% noise added, in this particular case, the allowed number of iterations is 5500, and the optimum parameters are obtained with misfit error ranges between 56.9 and 60.34%. The elapsed time ranges between 124 s (PSO) and 14 s (GWO).

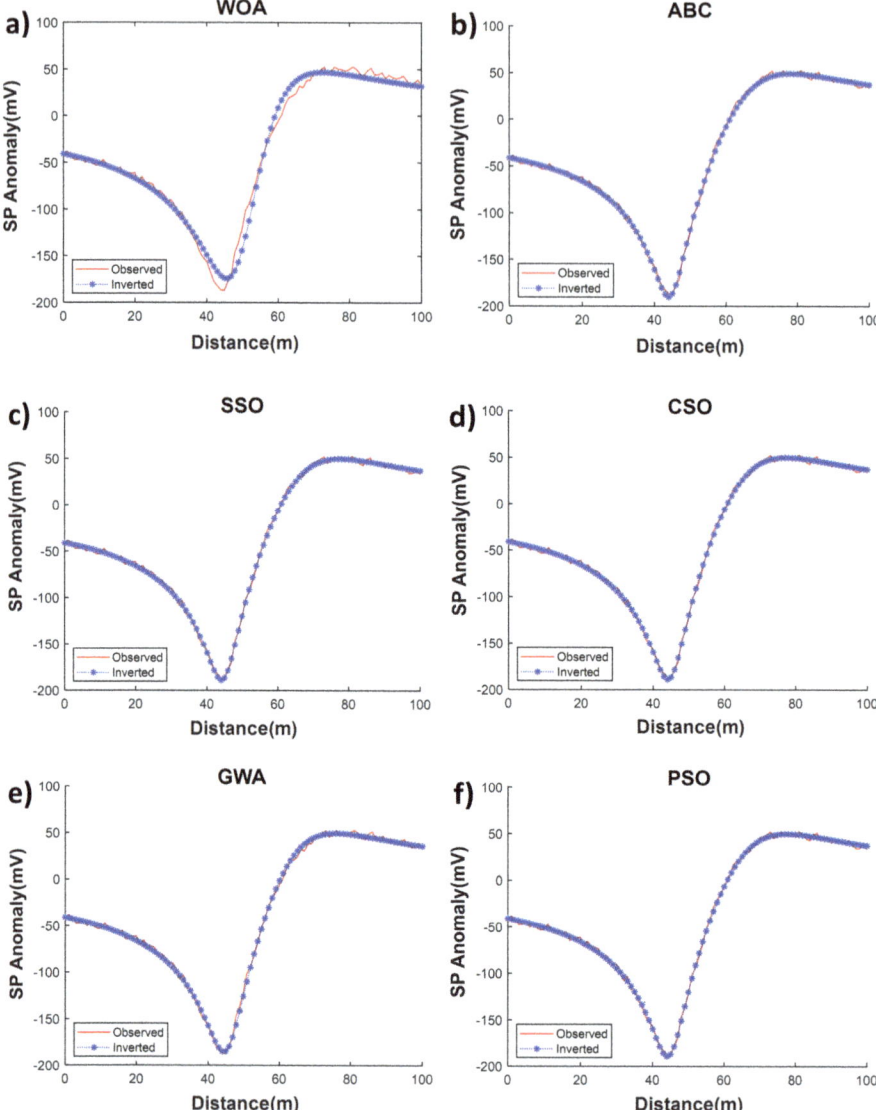

Fig. 2.24 Inversion of 2D inclined sheet model (with 20% noise) with **a** WOA, **b** ABC, **c** SSO, **d** CSO, **e** GWA, **f** PSO methods

Although, the overall misfit errors between the observed and calculated anomalies are relatively high, because of the outliers, but the errors in the inverted parameter are very acceptable (Table 2.9).

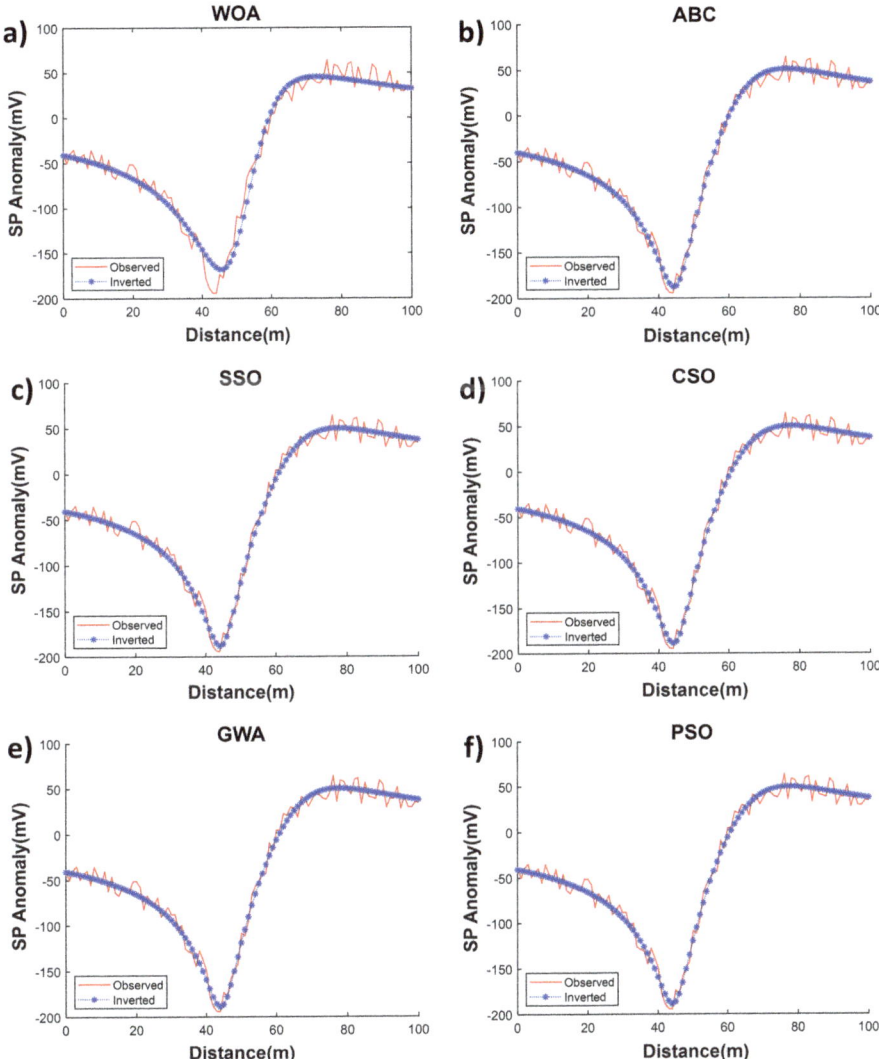

Fig. 2.25 Inversion of 2D inclined sheet model (with 30% noise) with **a** WOA, **b** ABC, **c** SSO, **d** CSO, **e** GWA, **f** PSO methods

For a thin sheet model (Tables 2.5, 2.6, 2.7 and 2.8), the iterations allowed varies from 500 to 1200. The misfit error in all cases in very low and the obtained results are close to true one even with higher percentage of noise.

Table 2.6 Comparison between the WOA and GWO, WOA, CSA, SSO, and BCO algorithms, to invert SP field due to thin sheet model with (10%) noise added

Noisy model (10%)

Algorithm	K (mV)	D (m)	*a* (m)	α (m)	h (m)	Misfit error (%)	F_{val}	iterations	Elapsed time (s)
WOA	84.108	53.711	7.3091	150.92	12.449	0.56643	0.041213	800	11.938
ABC	52.626	54.64	11.377	150.36	10.012	0.4947	0.010024	800	2.125
SSO	108.82	53.478	5.7151	151.03	12.979	0.62907	0.047772	800	1.9219
CSO	49.98	55.011	12.014	150.02	9.9999	0.45006	0.00072483	800	1.7813
GWO	50.733	54.943	14.36	135.54	5.8871	**0.42256**	0.0028156	800	1.4844
PSO	49.991	55.001	14.423	136.11	6.001	0.44691	3.5022e-05	800	1.7188
Parameter Range	−200:200	0:100	0:30	0:180	0:20				

WOA Whale optimization algorithm, ABC Artificial Bee Colony, SSO Salp Swarm Optimization, CSO Cuckoo Search Optimization, GWA Grey wolf Algorithm and PSO Particle swarm optimization

F_{val} Minimum objective function value at optimum solution

Iter Allowed number of iterations

Table 2.7 Comparison between the WOA and GWO, WOA, CSA, SSO, and BCO algorithms to invert SP field due to a thin sheet model with (20%) noise added

Noisy model (20%)

Algorithm	K (mV)	D (m)	a (m)	a (m)	h (m)	Misfit error (%)	F_{val}	Iter	Elapsed time (s)
WOA	83.53	53.749	7.36	150.84	12.398	0.894	0.041101	1000	15
ABC	50.69	54.856	11.851	150.05	10.229	**0.74476**	0.0066596	1000	2.3906
SSO	50	55	14.422	136.1	6	0.75553	1.3522e-07	1000	2.3281
CSO	49.99	55.003	12.003	150.01	9.9995	0.75597	7.5788e-05	1000	1.9375
GWO	49.65	55.059	12.096	150.02	10.005	0.77394	0.0016968	1000	1.875
PSO	85.21	53.675	13.949	116.86	3.4987	0.87806	0.041873	1000	2.2813
Parameter range	−200:200	0:100	0:30	0:180	0: 20				

WOA Whale optimization algorithm, ABC Artificial Bee Colony, SSO Salp Swarm Optimization, CSO Cuckoo Search Optimization, GWA Grey wolf Algorithm and PSO Particle swarm optimization

F_{val} Minimum objective function value at optimum solution

Iter Allowed number of iterations

Table 2.8 Comparison between the WOA and GWO, WOA, CSA, SSO, and BCO algorithms to invert SP field due to a thin sheet model with (30%) noise added

Noisy model (30%)

Algorithm	K(mV)	D (m)	a (m)	α(m)	h (m)	Misfit error (%)	F_{val}	Iter	Elapsed time (s)
WOA	43.141	56.312	13.668	150.08	9.3206	3.6172	0.025095	1200	18.078
ABC	50.697	54.856	11.851	150.05	10.229	**1.9219**	0.0066596	1200	2.8281
SSO	50	55	12	150	10	1.9738	2.4064e-07	1200	2.2656
CSO	50	55	12	150	10	1.974	7.9419e-06	1200	2.1094
GWO	−21.303	61.512	22.142	0.16194	0.76125	3.9891	0.24658	1200	1.625
PSO	50.002	55	12	150	10	1.9736	5.5806e-06	1200	2.2031
Parameter range	−200:200	0:100	0:30	0:180	0: 20				

WOA Whale optimization algorithm, ABC Artificial Bee Colony, SSO Salp Swarm Optimization, CSO Cuckoo Search Optimization, GWA Grey wolf Algorithm and PSO Particle swarm optimization

F_{val} Minimum objective function value at optimum solution

Iter Allowed number of iterations

Table 2.9 Relative error percentage in the calculated parameters for the spherical model example with 30% noise added

Parameter	K(mV)	X_0 (m)	ψ°	Z (m)	q
True	− 10,000	40	60	10	1.5
Inverted	− 10,185.9	39.62	59.44	10.21	1.46
Percentage error %	1.85	0.95	0.95	2.1	2.6

The above synthetic cases shows the stability of the metaheuristics based algorithms in dealing with SP fields. To test efficiency in real case, the AI techniques are applied to different field examples with different complexities.

2.7 Application to Field Examples

2.7.1 Application to Field Data Approximated by Simple Geometrical Models

2.7.1.1 Bender Anomaly

The Bender SP anomaly (Orissa, India) has been studied by many authors with different methods (Table 2.9). The main lithological unit in this area belongs to the Archaen cyrstalline complex. The exposed rocks are mainly garnetiferous and mostly rocks are covered by laterite and alluvium (Agarwal and Srivastava 2009). Comparison between the inverted solutions, obtained by different metaheuristic methods, are presented (Table 2.10 and Fig. 2.26). The average values of the inverted parameters are (K $(\text{mVm}^{(2q-1)})$ = −8271, D(m) = 87.3, ψ° = 55.9, z(m) = 22.8, q = 0.98). The minimum misfit error is 22.36% between the observed and inverted anomalies using the WOA algorithm.

2.7.1.2 Ahirwala Deposit Anomaly

The SP anomaly was measured in the Ahirwala deposit of the Neem-Ka-Thana copper belt, Rajasthan, India (Reddi et al. 1982). The anomaly was observed along a profile of 300 m; it has amplitudes ranging from −20 to −85 mV. Six different metaheuristic approaches are run to determine the model parameters, using the same upper and lower limits as shown in Table 2.11. The low misfit error percentage proves the efficiency of such methods in determining the optimum parameters within the desired range. Figure 2.27 presents the inverted fields and the measured one.

The average values of the inverted parameters are (K $(\text{mVm}^{(2q-1)})$ = −111.67, D(m) = 74.87, ψ° = 4.2, z(m) = 21.1, q = 0.48). The minimum misfit error is 0.4784% between the observed and inverted anomalies using the GWO algorithm.

Table 2.10 Interpretation of Bender anomaly using different heuristic methods, with SP field approximated by a simple geometrical model

Bender anomaly

Algorithm	$K(mVm^{(2q-1)})$	D (m)	ψ°	Z (m)	q	Misfit error (%)	*Fval*	Iter	Elapsed time (s)
WOA	−8625.7	85.146	58.615	24.695	0.99212	**22.366**	0.3452	2000	16.484
ABC	−10,000	87.616	56.623	23.205	1.0247	23.533	0.34399	2000	11.891
SSO	−3448	88.154	51.902	20.371	0.88421	24.225	0.34578	2000	6.0938
CSO	−10,000	87.751	56.571	23.257	1.0241	23.723	0.34383	2000	6.5781
GWO	−9695	87.841	56.298	23.207	1.02	23.664	0.34386	2000	5.5781
PSO	−7857.5	87.838	55.476	22.514	0.99415	23.599	0.34414	2000	24.25
Biswas (2017b)	−9850.3 ± 29.5	99.6 ± 0.2	72.5 ± 0.2	15.7 ± 0.1	1.0	–	–	–	–
Di Maio et al. (2016a)	−5950	105.5	81.62	18.1	1.0	–	–	–	–
Di Maio et al. (2016b)	−10,000	101.37	72.66	21.8	1.32	–	–	–	–
Parameter range	−10,000: 10,000	0:100	0:100	0:40	0.5: 1.5				

Fig. 2.26 Inversion of Bender anomaly (Orissa, India)

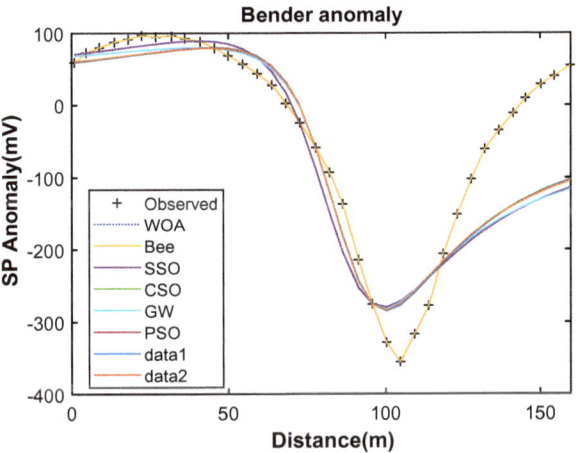

2.7.1.3 Malachite Anomaly

Malachite anomaly is associated with the Malachite mine in Jefferson County, Colorado, USA. This anomaly was also interpreted by several authors as a vertical cylindrical structure (e.g.: Tlas and Asfahani 2007, 2013; Biswas and Sharma 2015). Biswas and Sharma (2015) also interpreted the same anomaly using multiple structures. However, the main anomaly was caused by such vertical cylinder. Table 2.12 illustrates the inverted model parameters from various published results together with the heuristic algorithms results. Figure 2.28 Displays the fittings between the inverted fields from different heuristic algorithms and the observed one. The misfit percentage is about 2% or less.

The average values of the inverted parameters are (K (mVm$^{(2q-1)}$)) $= -217.22$, $D(m) = 89.56$, $\psi^\circ = 9.04$, $z(m) = 15.51$, $q = 0.55$). The minimum misfit error is 1.3708% between the observed and inverted anomalies using the SSO algorithm.

2.7.1.4 Süleymanköy Anomaly

This Anomaly is from Eastern Turkey and lies in the Ergani Copper District, 65 km. SE of Elazig (Bhattacharya and Roy 1981). It represents a polarized copper ore body and is approximated as a simple geometrical model to be solved by the heuristic techniques. Results are shown in Table 2.13 and compared to previous results processed by many authors. Figure 2.29 shows the inverted fields using different methods together with the measured data. For comparison, same number of iterations (3000) and same upper and lower bounds for parameters are used. The misfit error percentage is less than 2%, proving reliable solutions.

Table 2.11 Interpretation of Ahirwala deposit anomaly using different heuristic methods, with SP field approximated by a simple geometrical model

Ahirwala deposit anomaly

Algorithm	$K(mVm^{(2q-1)})$	D (m)	ψ°	Z (m)	q	Misfit error (%)	Fval	Iter	Elapsed time (s)
WOA	−373.13	70.33	10.623	30	0.7165	1.283	0.07783	2000	19.328
ABC	−57.768	74.947	3.0922	18.843	0.4328	0.5087	0.04001	2000	13.625
SSO	−61.589	75.761	2.9948	19.954	0.4469	0.4916	0.03794	2000	6.9688
CSO	−60.171	75.835	2.9279	19.67	0.4427	0.4868	0.03791	2000	7.3281
GWO	−55.787	76.682	2.5883	18.719	0.4283	**0.4784**	0.03807	2000	6.4063
PSO	−61.594	75.711	3.0106	19.937	0.4469	0.4917	0.03794	2000	29.656
Göktürkler and Balkaya (2012) (SA)	−83.49	76.26	86.48	23.4	0.5	2.78	–	–	–
Göktürkler and Balkaya (2012) (PSO)	−49.53	76.77	88	17.56	0.4	2.4	–	–	–
Parameter range	−1000:1000	0:100	0:100	0:30	0:1.5				

Fig. 2.27 Inversion of
Ahirwala deposit anomaly
(Rajasthan, India)

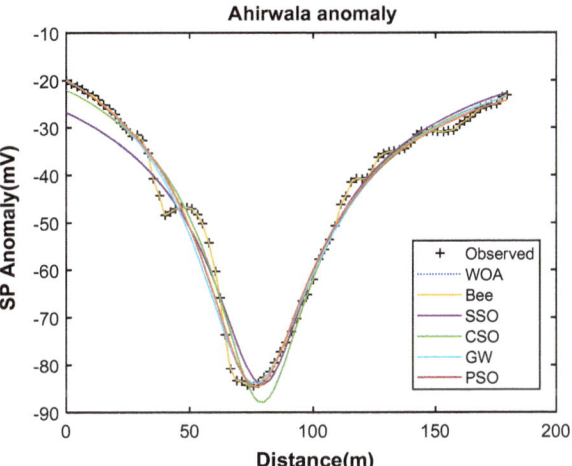

The average values of the inverted parameters are: (K ($mVm^{(2q-1)}$)) $= -2771.9$, $D(m) = 77.7$, $\psi° = 53.98$, $z(m) = 24.01$, $q = 0.842$). The minimum misfit error is 1.4928% between the observed and inverted anomalies using the PSO algorithm.

The genetic algorithm GA is also used as an EA technique to invert the same anomaly (Abdelazeem and Gobashy 2006; Göktürkler and Balkaya 2012). The inverted parameters are shown in Table 2.14 and compared with previous conventional results. The measured and inverted fields are shown in Fig. 30a, b.

2.7.2 Anomalies Approximated by Thin Sheet Model

2.7.2.1 Surda Anomaly

The Surda SP anomaly of Rakha mines, Singhbhum cooper belt, Jharkhand, India are inverted by some metaheuristic methods and compared to previous inversions (Murthy et al. 2005; Di Maio et al. 2016a, b). The inverted parameters are shown in Table 2.15 and the inverted field are drawn in Fig. 2.31. The average values of the inverted parameters are ($K(mV = 88.89$, $D(m) = 146.92$, $a = 28.19$, $\alpha° = 47.12$, $h(m) = 31.29$). All parameters are found to be in agreement with published results. The minimum misfit error is 5.79% between the observed and inverted anomalies using the GWO method.

2.7.2.2 Bavarian Anomaly

The graphite ore deposits in Bavarian Woods, Germany (Meiser 1962; Abdelazeem and Gobashy 2006). These deposits are situated in a hercynic gneissic complex.

Table 2.12 Interpretation of Malachite deposit anomaly using different heuristic methods, with SP field approximated by a simple geometrical model

Malachite deposit anomaly

Algorithm	$K(mVm^{(2q-1)})$	D (m)	ψ°	Z (m)	q	Misfit error (%)	Fval	Iter	Elapsed time (s)
WOA	−298.97	88.748	11.04	18.009	0.5533	1.4236	0.05971	2000	8.5625
ABC	−192.79	89.679	9.0187	15.633	0.4782	1.8542	0.05601	2000	6.5781
SSO	−257.26	88.955	10.312	16.571	0.5252	**1.3708**	0.05794	2000	3.5938
CSO	−181.21	90.195	7.7289	14.079	0.4573	2.3435	0.05493	2000	3.5
GWO	−182.05	90.143	7.8029	14.14	0.4584	2.2922	0.05495	2000	3
PSO	−191.09	89.643	8.3875	14.68	0.4695	1.9739	0.05512	2000	13.719
Biswas and Sharma (2015)	−224.4	89.04	80.7	15.2	0.5	4.62	–	–	–
Tlas and Asfahani (2013)	−220.6	–	80.2	15.6	0.501	11.36	–	–	–
Tlas and Asfahani (2007)	−229.28	–	79.98	12.79	–	18.25			
Parameter range	−300: 100	0:100	0:100	0:20	0:1.5				

Fig. 2.28 Inversion of
Malachite deposit anomaly
(Colorado, USA)

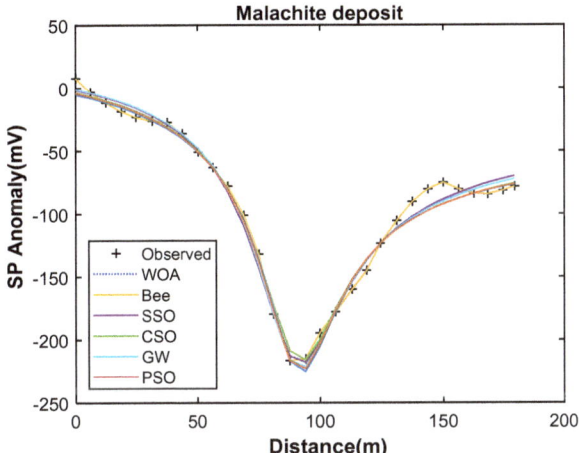

Conformably intercalated between paragneiss and crystalline limestone of the same
age, they form seams that are to be interpreted as bituminous sediments of presumably
Precambrian age. The expected ore form a parallel sequence of lenses where it
is expanded between limestone and gneisses, variable in thickness, and can be in
some localities approximated by simple geometrical models. The Bavarian woods SP
anomaly has been studied by various authors using different interpretation methods
(Table 2.16).

The inverted parameters, misfit error, elapsed time and the minimum value of the
function after 2000 iterations are shown in the Table 2.16. A good congruence among
the measured and inverted fields are clarified in (Fig. 2.32). The average values of
the inverted parameters are $(K(mV) = -5030.48, D(m) = 297.35, a = -22.74, \alpha^o = 38.50, h(m) = 0.75)$. The high percentage of misfit error is due to the difference
between the inverted fields and the observed at the right end of the anomaly.

Bavarian woods anomaly was also interpreted by many authors as simple
geometric model using Genetic algorithm, GA, the resulted inverted parameters
obtained are: depth $h = 35.50$ m, polarization angle $\psi = -62.99°$ The, and shape
factor $q = 0.792$ (Fig. 33a, b). The (rms) error calculated between the observed and
calculated SP anomalies from the obtained parameters is $= 1.7043$ mV. The resulted
shape factor also suggests that a 2D horizontal cylinder model buried at a depth of
35.5 m can represent the shape of the source body (Table 2.17).

2.7.2.3 Pinggirsari Anomaly

The SP field data were measured in Pinggirsari village, West Java, Indonesia16. The
acquired profile was laid in S–N direction to cross the fault based on the cross section
from the geological map. The profile length was about 1040 m with a separation of
25 m between the measuring electrodes. In all cases, 200 search agents and 3000

Table 2.13 Interpretation of Süleymanköy anomaly using different heuristic methods, with SP field approximated by a simple geometrical model

Süleymanköy anomaly

Algorithm	K(mVm$^{(2q-1)}$)	D (m)	ψ°	Z (m)	q	Misfit error (%)	Fval	Iter	Elapsed time (s)
WOA	−1974.2	80	48.684	22.036	0.8027	1.8592	0.13754	3000	15.313
ABC	−3000	80	51.874	24.778	0.8548	1.7808	0.13028	3000	18.297
SSO	−3000	80	51.897	24.798	0.8546	1.7833	0.13027	3000	9.2344
CSO	−2690.7	73.273	59.428	23.441	0.8388	1.5275	0.1057	3000	24.609
GWO	−3000	80	51.87	24.776	0.8548	1.7803	0.13028	3000	8.2188
PSO	−2966.5	73.224	60.143	24.243	0.8503	**1.4928**	0.10364	3000	25.406
Abdelrahman and Sharafeldin (1997)	−229.28	–	79.98	12.79	–	–	–	–	–
El-Araby (2004)	−2661.2	–	14.74	47	1.468	–	–	–	–
Srivastava and Agarwal (2009)	–	64.1	–	28.9	1.0	–	–	–	–
Agarwal and Srivastava (2009)	1560	68	165	27	1.0	–	–	–	–
Parameter range	−3000: 100,000	80:120	0:180	0:80	0.5:1.5	–			

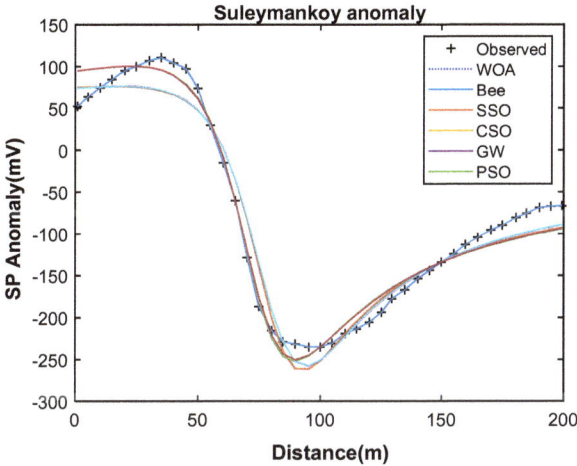

Fig. 2.29 Inversion of Süleymanköy deposit anomaly (eastern Turkey)

Table 2.14 Results of inversion of Süleymanköy SP anomaly results of inversion of graphite ore body, southern Bavarian woods, Germany using GA compared with other conventional techniques

Parameter	Yungul (1950)	Bhattacharya and Roy (1981)	Asfahani et al. (2001)	Gobashy (2000)	Abdelazeem and Gobashy (2006), GA	Göktürkler and Balkaya (2012), GA
Depth (h) m	38.8	40	27	33.6	29.999	32.68
Polarization angle (α) deg	21	15	17.25	15	11.549	22.85
Shape factor (q)	H. cylinder	H. cylinder	H. cylinder	0.99	0.961	1.16
K mv. m $^{2q-1}$	–	–	–	–	–	– 36,724.18
X_0 m	–	–	–	–	–	79.24

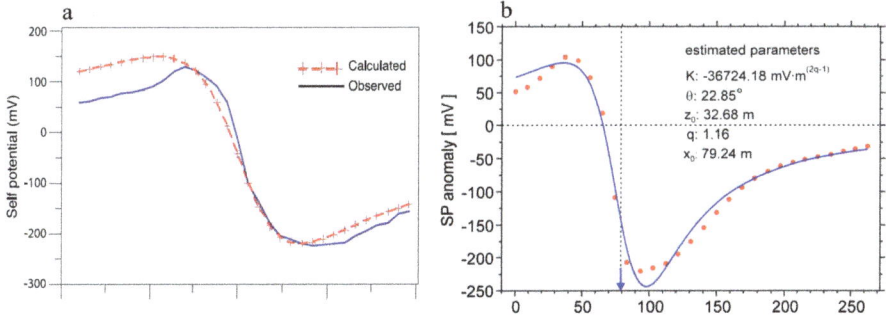

Fig. 2.30 Inversion of Süleymanköy SP anomaly using genetic algorithm **a** after Abdelazeem and Gobashy (2006) and **b** after Göktürkler and Balkaya (2012)

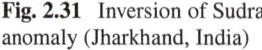

Fig. 2.31 Inversion of Sudra anomaly (Jharkhand, India)

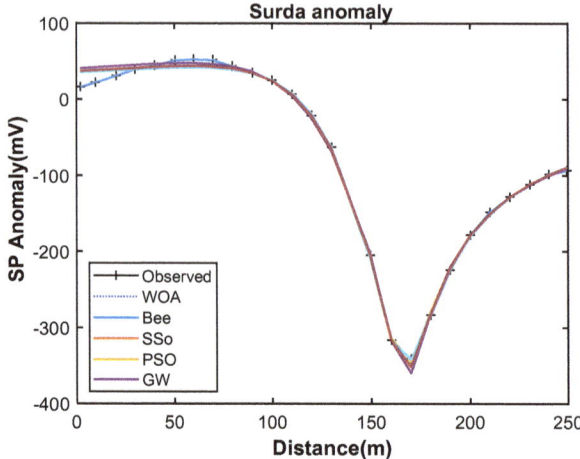

iterations were used. Inversion results of this anomaly is shown in Table 2.18 and Fig. 2.34. The average values of the inverted parameters are (K(mV) = 44.26, D(m) = 494.45, a = 37.60, α° = 235.17, h(m) = 13.43). Although the minimum misfit error is 50.5% (relatively high) between the observed and inverted anomalies using the GWO method due to the outliers in the observed data and the complex nature of the surrounding structures, but the range of the inverted parameters is in agreement with published results.

2.8   Conclusions

The problem of determining the controlling parameters of a buried structures or mineralization zones from self-potential data can be solved with a high degree of stability and efficiency using a bundle of metaheuristic algorithms (e.g.: WOA, ABC, SSO, CSO, GWO, PSO, and GA). Results show a great stability of the AI metaphor based techniques against high levels of noise (geologic or artificial). The complexity of the geologic conditions does not affect the convergence behavior of the all global optimizers. The advantages of the AI based techniques over conventional methods are: (1) no computation of analytical or numerical derivatives with respect to the model parameters, (2) less sensitive to errors in data and (3) the techniques are independent of the base line determination. It is also emphasized that the AI inversion can be used to gain geological insight concerning the subsurface, as illustrated in the field examples.

Table 2.15 Interpretation of Sudra anomaly using different heuristic methods, with SP field approximated by a 2D inclined thin sheet model

Sudra anomaly

Algorithm	K(mV)	D (m)	a (m)	α°	h (m)	Misfit error (%)	Fval	Iter	Elapsed time (s)
WOA	88.941	144.75	31.318	45.487	31.281	6.5961	0.04097	2000	4.9531
ABC	96.795	146.9	28.448	47.011	31.247	5.9499	0.03813	2000	3.75
SSO	90.423	144.55	30.801	45.032	31.322	6.7885	0.03990	2000	2.125
GWO	100.51	147.68	27.275	47.568	31.09	**5.7941**	0.03821	2000	0.84375
PSO	95.357	146.55	28.909	46.826	31.208	6.0217	0.03811	2000	3.0781
Sharma and Biswas (2013)	121.3	151.1	22.4	50.8	30.4	10.7	–	–	–
Di Maio et al. (2016a)	111	11.08	19.8	52.47	28.5	35.2	–	–	–
Di Maio et al. (2016b)	128.67	151.72	21.92	66.93	31.94	20	–	–	–
Parameter range	70:150	80:170	20:45	30:50	20:40				

Table 2.16 Interpretation of Bavarian anomaly using different heuristic methods, with SP field approximated by a 2D inclined thin sheet model

Bavarian anomaly

Algorithm	$K(mVm^{(2q-1)})$	D (m)	ψ^o	Z (m)	q	Misfit error (%)	Fval	Iter	Elapsed time (s)
WOA	−7299.6	298.34	−24.74	41.306	0.80679	**104.37**	0.09082	3500	20.344
ABC	−3873.7	296.96	−21.64	36.772	0.72973	115.07	0.07492	3500	25.859
SSO	−7875.5	298.59	−26.19	43.814	0.81554	107.71	0.08918	3500	11.922
CSO	−3573.6	296.55	−20.88	36.09	0.71945	114.1	0.07440	3500	33.922
GWO	−3816.9	296.87	−21.59	36.647	0.72785	115.61	0.07464	3500	12.031
PSO	−3743.6	296.83	−21.41	36.377	0.72539	115.64	0.07452	3500	34.031
Abdelazeem and Gobashy (2006)	–	–	−62.9	35.0	0.792	–	–	–	–
Göktürkler and Balkaya (2012)	21,272.9	–	−51.29	45.03	0.97	–	–	–	–
Parameter range	−10,000:10,000	250:300	−90:90	0:55	0:2	–	–	–	–

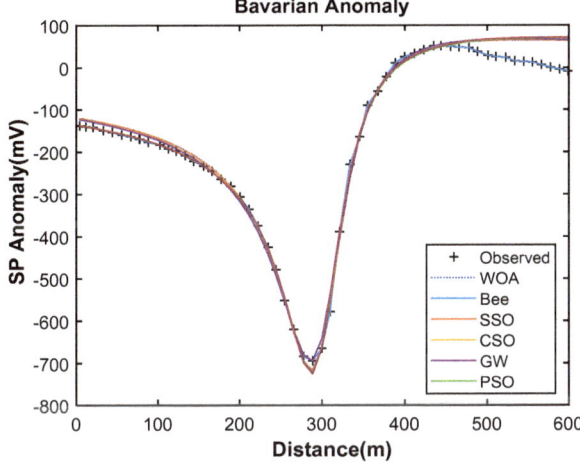

Fig. 2.32 Inversion of Bavarian deposit anomaly (Bavarian woods, Germany)

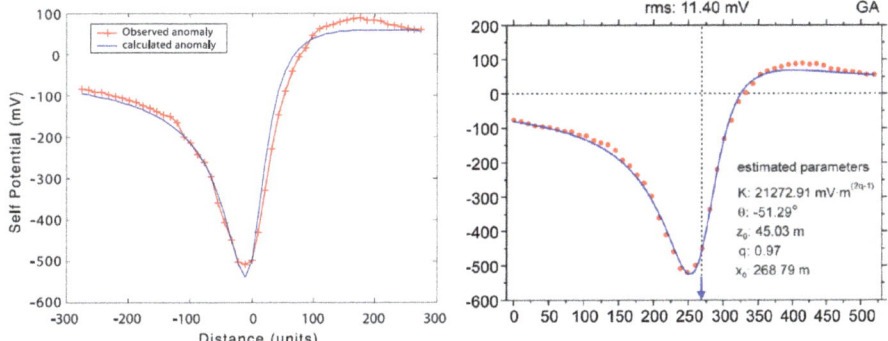

Fig. 2.33 Measured and calculated SP anomaly over a graphite ore body, southern Bavarian woods, Germany (Meiser 1962). **a** Modified after Abdelazeem and Gobashy (2006), and **b** modified after Göktürkler and Balkaya (2012)

Table 2.17 Results of inversion of graphite ore body, southern Bavarian woods, Germany using GA compared with other conventional techniques

Parameter	Meiser (1962)	Abdelrahman et al. (2003) Higher derivatives	Abdelrahman et al. (2003) Lest-squares	Abdelazeem and Gobashy (2006) GA	Göktürkler and Balkaya (2012) GA
Depth (h) m	53	53	49.3	35	45.03
Polarization angle (ψ) deg	–	–	−55.7	−62.9	−51.29
Shape factor (q)	–	0.9	0.91	0.792	0.97
K mv. m $^{2q-1}$		–	–	–	21,272.91
Rms	–	–	25.3	**1.704**	–

Table 2.18 Interpretation of Pinggirsari anomaly using different heuristic methods, with SP field approximated by a 2D inclined thin sheet model

Pinggirsari anomaly

Algorithm	K(mV)	D (m)	a (m)	α°	h(m)	Misfit error (%)	Fval	Iter	Elapsed time (s)
WOA	56.053	483.22	33.416	220.68	12.318	54.686	0.29332	3000	2.4688
ABC	44.419	479	35.72	211.32	15.275	51.367	0.28646	3000	5.5781
SSO	47.06	479.5	35.951	213.99	14.316	53.286	0.28766	3000	3.2031
CSO	31.214	567.47	50.688	348.9	4.9807	87.659	1e + 10	3000	8.7656
GWO	43.248	478.66	35.354	208.99	16.021	**50.582**	0.28615	3000	2.6719
PSO	43.612	478.86	34.509	207.15	17.72	50.704	0.28616	3000	4.2969
Fajriani et al. (2017)	41.5	478.25	34	334.52	14.63		–	–	–
Abdelazeem et al. (2019)	47.38	479.62	35.85	−149.98	15.68		–	–	–
Parameter range	10:100	400: 600	20:60	30:50	20:40				

Fig. 2.34 Inversion of
Pinggirsari deposit anomaly
(South Bandung, Indonesia)

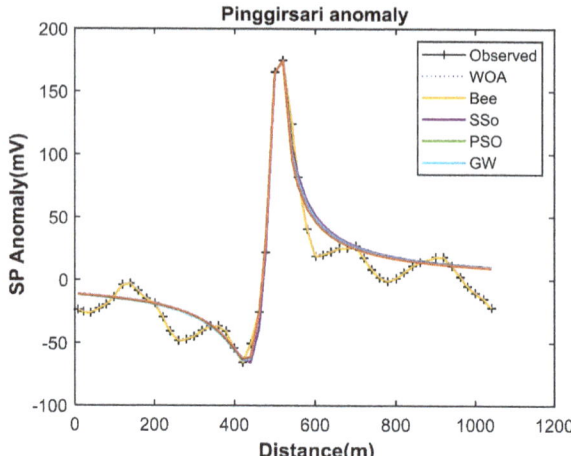

References

Abdelrahman EM, Sharafeldin SM (1997) A least squares approach to depth determination from residual self-potential anomalies caused by horizontal cylinders and spheres. Geophys 62:44–48

Abdelrahman EM, Ammar AA, Hassanein HI, Hafez MA (1998) Derivative Analysis of SP Anomalies. Geophys 63:890–897

Abdelazeem M, Sweilam N, Bayoumi AI (2003) A hybrid technique for solving gravity inverse problem. Proc Math Phys Soc Egypt 78:39–66

Abdelrahman EM, El-Araby HM, A ft G Hassaneen ARG, Hafez MA (2003) New methods for shape and depth determinations from SP data: Geophys 68:1202–1210

Abdelazeem M, Gobashy MM (2006) Self-potential inversion using genetic algorithm. JKAU: earth. Science 17:83–101

Abdelrahman EM, Gobashy MM (2021) A Fast Method for Interpretation of Self-Potential Anomalies Due to Buried Bodies of Simple Geometry. Pure Appl. Geophys. https://doi.org/10.1007/s00 024-021-02788-x

Abdelrahman EM, Essa KS, Abo-Ezz ER, Sultan M, Sauck WA, Gharieb AG (2008) New least-square algorithm for model parameters estimation using self-potential anomalies. Comput Geosci 34:1569–1576

Abdelrahman EM, Soliman K, Essa KS, Abo-Ezz ER, El-Araby TM (2009) A least-squares minimization approach to depth determination from numerical second horizontal self-potential anomalies Explor. Geophys 40:214–221

Abdel-Basset M, Abdel-Fatah L, Kumar A (2018) Metaheuristic algorithms: a comprehensive. Review. https://doi.org/10.1016/B978-0-12-813314-9.00010-4

Abdelrahman EM, Abdelazeem M, Gobashy MM (2019) Minimization approach to depth and shape determination of mineralized zones from potential field data using the Nelder-Mead simplex algorithm. Ore Geol Rev 114. https://doi.org/10.1016/j.oregeorev.2019.103123

Abdelazeem M, Gobashy MM, Khalil MH, Abdrabou M (2019) A complete model parameter optimization from self-potential data using Whale algorithm. J Appl Geophys 170.https://doi.org/10.1016/j.jappgeo.2019.103825

Agarwal BL (2006) Basic statistics, 4th edn, New age international publishers, New Delhi, 3 & 5. ISBN-81–1814–7

Agarwal BNP, Srivastava S (2009) Analyses of self-potential anomalies by conventional and extended Euler deconvolution techniques. Comput Gosci 35:2231–2238

Al-Chalabi M (1971) Some studies relating to non-uniqueness in gravity and magnetic inverse problems. Geophys 36(5):835–855

Ander ME, Huestis SP (1987) Gravity ideal bodies. Geophys 52:1265–1278

Asfahani J, Tlas M, Hammadi M (2001) Fourier analysis for quantitative interpretation of self-potential anomalies caused by horizontal cylinder and sphere. JKAU Earth Sci 13(1): 41–53. https://doi.org/10.4197/ear.13-1.3

Balkaya Ç, Ekinci YL, Göktürkler G, Turan S (2017) 3D non-linear inversion of magnetic anomalies caused by prismatic bodies using differential evolution algorithm. J Appl Geophys 136:372–386

Baluja S (1994) Population based incremental learning: a method for integrating genetic search based function optimization and competitive learning. technical report CMU-CS-94–163, Carnegie Mellon University, Pittsburgh, PA

Benatchba K, Admane L, Koudil M (2005) Using bees to solve a data-mining problem expressed as a max-sat one, artificial intelligence and knowledge engineering applications: a bioinspired approach. In: Proceedings of the first international work-conference on the interplay between natural and artificial computation, IWINAC 2005, Las Palmas, Canary Islands, Spain, pp 15–18

Bersini H, Varela FJ (1990) Hints for adaptive problem solving gleaned from immune networks. In: Parallel problem solving from nature, Dortmund, Germany, LNCS, vol 496. Springer, pp 343–354

Bevington PR (1969) Data reduction and error analysis for the physical sciences. McGraw-Hill Book Co

Bhattacharya BB, Roy N (1981) A note on the use of a nomogram for self-potential anomalies. Geophys Prosp 29:102–107

Bhattacharya PK, Patra HP (1968) Direct current geoelectric sounding: Principles and interpretation; Methods in Geochem Geophys, Series-9. Elsevier Publishing Company, 135 p

Biswas A (2017a) A review on modeling, inversion and interpretation of self-potential in mineral exploration and tracing paleo-shear zones. Ore Geol Rev 91(2017):21–56

Biswas A (2017b) Inversion of source parameters from magnetic anomalies for mineral/ore deposits exploration using global optimization technique and analysis of uncertainty. Nat Resour Res. https://doi.org/10.1007/s11053-017-9339-2

Biswas A, Sharma SP (2014) Optimization of self-potential interpretation of 2-D inclined sheet-type structures based on Very fast simulated annealing and analysis of ambiguity. J Appl Geophys 105:235–247

Biswas A, Sharma SP (2015) Interpretation of self-potential anomaly over idealized body and analysis of ambiguity using very fast simulated annealing global optimization. Near Surf Geophys 13(2):179–195

Blakely RJ (1995) Potential theory in gravity and magnetic applications. Cambridge University Press

Blakely RJ, Simpson RW (1986) Approximating edges of source bodies from magnetic or gravity anomalies. Geophys 51:1494–1498

Blum C, Roli A (2003) Metaheuristics in combinatorial optimization: overview and conceptual comparison. ACM Comput Surv (CSUR) 35(3):268–308

Bogoslovsky VA, Ogilvy AA (1977) Geophysical methods for the investigation of landslides. Geophys 42:562–571

Bogoslovsky VA, Ogilvlt AA, Strukhovu NA (1977) Magnetometric and electrometric methods for the investigation of the dynamics of landslide processes. Geophys Prospect 25:280–291

Bogoslovsky VA, Ogilvy AA (1970a) Application of geophysical methods for studying the technical status of earth darns: Geophys Prospect 18(1):758–773

Bogoslovsky VA, Ogilvy AA (1970b) Natural potential anomalies as a quantitative index of the rate of water seepage from reservoirs. Geophys Prospect 18:261–268

Boschetti F, Horowitz FG, Hornby P (1999) Ambiguity analysis and the constrained inversion of potential field data. Australian geodynamics cooperative research centre report

Boschetti F, Dentith M, List R (1997) Inversion of potential field data by Genetic Algorithm. Geophys Prospect 45:461–478

Bonabeau E, Dorigo M, Theraulaz G (1999) Swarm intelligence: from natural to artificial systems. Oxford University Press, New York, NY, USA

Cammarano F, Mauriello P, Pattella D, Piro S, Rosso F, Versino L (1998) Integration of high-resolution geophysical methods. detection of shallow depth bodies of archaeological interest. Ann Geofis 41:359–368

Charon I, Hudry O (1993) The noising method: a new method for combinatorial optimization. Oper Res Lett 14:133–137

Cordell L, Grauch VJS (1985) Mapping basement magnetization zones from aeromagnetic data in the San Juan basin, New Mexico. In: Hinze WJ (ed) the utility of regional gravity and magnetic anomaly maps. Soc Explor Geophys 181–197.https://doi.org/10.1190/1.0931830346.ch16

Corry CE (1985) Spontaneous polarization associated with porphyry sulphide mineralization. Geophysics 50:1020–1034

Corwin RF (1990) The self-potential method for environmental and engineering applications geotechnical and environmental geophysics. 1 Society exploration geophysics investigations in geophysics, vol 5, pp 127–145

Dantzig GB (1947) Maximization of a linear function of variables subject to linear inequalities, T.C. Koopmans (ed.): Activity Analysis of Production and Allocation, New York-London 1951 (Wiley & Chapman-Hall), pp 339–347

Demoully GT, Corwin RF (1980) Self-potential survey results from the Beowawe KGRA, Nevad transactions of the geothermal resources council, vol 4, pp 33–36

Dennis JE, Schnabel RB (1996) Numerical methods for unconstrained optimization and nonlinear equations. Classics in applied mathematics, SIAM 16

Di Maio R, Patella D (1994) Self-potential anomaly generation in volcanic areas. The Mt. Etna Case History, Acta Vulcanologica 4:119–124

Di Maio R, Piegari E, Rani P, Avella A (2016a) Self-Potential data inversion through the integration of spectral analysis and tomographic approaches. Geophys J Int 206:1204–1220

Di Maio R, Rani P, Piegari E, Milano L (2016b) Self-potential data inversion through a genetic-price algorithm. Comput Geosci 94:86–95

Di Maio R, Cecere G, De Martino P, Piegari E (2013) Electric effects induced by artificial seismic sources at Somma-Vesuvius volcano. Ann Geophys 56:S0445

Dorigo M (1992) Optimization, learning and natural algorithms. PhD thesis, Politecnico di Milano, Italy

Drias H, Sadeg S, Yahi S (2005) Cooperative bees swarm for solving the maximum weighted satisfiability problem, computational intelligence and bio-inspired systems. In: Proceedings of the 8th international workshop on artificial neural networks, IWANN 2005, Vilanovai la Geltr, Barcelona, Spain, pp 8–10

Dueck G (1993) New optimization heuristics: the great deluge algorithm and the record-torecord travel. J Comput Phys 104(1):86–92

Eberhart RC, Kennedy JA (1995) New optimizer using particle swarm theory. In: Proceedings sixth symposium on micro machine and human science, pp 39–43. IEEE Service Center, Piscataway, NJ

Eberhart RC, Shi YH (1998a) Parameter selection in particle swarm optimization. In: Proceedings of annual conference on evolutionary programming, San Diego

Eberhart RC, Shi YH (1998b) A modified particle swarm optimizer. In: Proceedings of IEEE international conference on evolutionary computation, Anchorage, Alaska

Edmonds J (1971) Matroids and the greedy algorithm. Math Program 1(1):127–136

El-Araby H (2004) A new method for complete quantitative interpretation of self-potential anomalies. J Appl Geophys 55(3–4):211–224

El-Ghazali T (1965) Metaheuristics: from design to implementation. Wiley

Farmer JD, Packard N, Perelson A (1986) The immune system, adaptation and machine learning. Physica D 2:187–204

Fedi M, Abbas MA (2013) A fast interpretation of self-potential data using the depth from extreme points method. Geophys 78:E107-116

Fedi M, Cella F, Florio G, Rapolla A (2005) Multiscale derivative analysis of the gravity and
    magnetic fields of southern Apennines Italy. In: Finetti I (ed) CROP project deep seismic explo-
    ration of the central Mediterranean and Italy Atlases in geoscience, vol 1. Elsevier Science, pp
    281–318
Fisher NJ, Howard LE (1980) Gravity interpretation with the aid of quadratic programming.
    Geophys 45(3):403–419
Fitterman DV (1983) Modelling of self-potential anomalies near vertical dykes. Geophys 48:l7l-180
Fitterman DV, Corwin RF (1982) Inversion of self-potential data from the Cerro Prieto geothermal
    field, Mexico. Geophys 47:938–945
Fitterman DV (1979) Calculations of self-potential anomalies near Vertical contacts. Geophys
    44:195–205
Fajriani V, Srigutomo W, Pratomo PM (2017) Interpretation of self-potential anomalies for inves-
    tigating fault using the Levenberg–Marquardt method: a study case in Pinggirsari West Java
    Indonesia. In: IOP conference series: earth and environmental science. https://doi.org/10.1088/
    1755-1315/62/1/012004
Fogel LJ (1962) Toward inductive inference automata. In Proceedings of the international federation
    for information processing congress, Munich, pp 395–399
Furness P (1992) Modelling spontaneous mineralization potentials with a new integral equation. J
    Appl Geophys 29(2):143–155
Gill PE, Murray W (1974) Quasi-Newton methods for linearly constrained optimization. In: Gill
    PE, Murray W (eds) Numerical methods for constrained optimization. Academic Press, New
    York, pp 67–92
Glover F (1977) Heuristics for integer programming using surrogate constraints. Decis Sci 8:156–
    166
Glover F (1986) Future paths for integer programming and links to artificial intelligence. Comput
    Oper Res 13(5):533–549
Glover F, McMillan C (1986) The general employee scheduling problem: an integration of MS and
    AI. Comput Oper Res 13(5):563–573
Gobashy MM, Abdelazeem M, Abdrabou M, Khalil MH (2019) Estimating model parameters from
    self-potential anomaly of 2D inclined sheet using whale optimization algorithm: applications to
    mineral exploration and tracing shear zones. Nat Resour Res (NRR) 1–21. DOI: https://doi.org/
    10.1007.2Fs11053-019-09526-0
Gobashy M, Abdelazeem M (2005) Delineation of basement surface relief from its magnetic
    anomaly using hybrid numerical approach. . J King Abdulaziz Univ Earth Sci (JKAU) 16:39–49
Gobashy MM (2000) Constraint inversion of residual self-potential anomalies. Delta J Sci 24. Tanta
    University, Egypt
Gobashy M, Abdelazeem M, Abdrabou M (2020) Minerals and ore deposits exploration using
    meta-heuristic based optimization on magnetic data. Contributions to Geophysics and Geodesy,
    50(2):161–199
Göktürkler and Balkaya (2012) Inversion of self-potential anomalies caused by simple geometry
    bodies using global optimization algorithms. J Geophys Eng 9:498–507
Goldbogen JA, Friedlaender AS, Calambokidis J, Mckenna MF, Simon M, Nowacek DP (2013)
    Integrative approaches to the study of baleen whale diving behavior, feeding performance, and
    foraging ecology. Biosci 63:90–100
Guillen A, Menichetti V (1984) Gravity and magnetic inversion with minimization of a specific
    functional. Geophys 49:1354–1360
Guptasarma D (1983) Effect of surface polarization on resistivity modeling. Geophys 48:98–106
Hansen N, Ostermeier A (1996) Adapting arbitrary normal mutation distributions in evolution
    strategies: the covariance matrix adaptation. In: IEEE conference on evolutionary computation
    (ICEC'96), pp 312–317
Hillis WD (1990) Co-evolving parasites improve simulated evolution as an optimization procedure.
    Physica D 42(1):228–234

Ho SL, Shiyou Y, Guangzheng N, Lo EWC, Wong HC (2005) A particle swarm optimization based method for multiobjective design optimizations. IEEE Trans Magn 41(5):1756–1759

Hochbaum DS (1996) Approximation algorithms for NP-hard problems. International Thomson Publishing

Holland JH (1975) Adaptation in natural and artificial systems. University of Michigan Press, Ann Arbor, MI

Holland JH (1962) Outline for a logical theory of adaptive systems. J ACM 3:297–314

Huestis SP, Parker RL (1977) Bounding the thickness of the oceanic magnetized layer. J Geophys Res 82:5293–5303

Joshi AS, Kulkarni O, Kakandikar GM, Nandedkar VM (2016) Cuckoo Search Optimization- A Review. International Conference on Advancements in Aeromechanical Materials for Manufacturing (ICAAMM-2016): Organized by MLR Institute of Technology, Hyderabad, Telangana, India, Editor: M. Satyanarayana Gupta

Jupp DLB, Vozoff K (1975) Stable iterative methods for the inversion of geophysical data. Geophys J R Astr Soc 42:957–976

Karaboga D, Basturk B (2007) A powerful and efficient algorithm for numerical function optimization: artificial bee colony (ABC) algorithm. J Glob Optim 39:459–471. https://doi.org/10.1007/s10898-007-9149-x

Kennedy J, Eberhart RC (1995) Particle swarm optimization. In IEEE international conference on neural networks, Perth, Australia, pp 1942–1948

Korf R (1985) Depth-first iterative-deepening: an optimal admissible tree search. Artif Intell 27(1):97–109

Koza JR (1992) Genetic programming. MIT Press, Cambridge, MA

Kuliev EV, Zaporozhets DYu, Kureichik VV, Kursitys IO (2019) Wolf pack algorithm for solving VLSI design tasks. J Phys Conf Ser 1333. IOP Publishing. https://doi.org/10.1088/1742-6596/1333/2/022009

Last BJ, Kubik K (1983) Compact gravity inversion. Geophysics 48:713–721

Logn O, Bolviken B (1974) Self-potentials at the joma pyrite deposit. Geoexplor 12:11–28

Li Y, Oldenburg DW (2003) Fast inversion of large-scale magnetic data using wavelet transforms and a logarithmic barrier method. Geophys J Int 152(2):251–265. https://doi.org/10.1046/j.1365-246X.2003.01766.x

Li Y, Oldenburg DW (2000) 3-D inversion of induced polarization data. Geophysics 65(6):1931–1945. https://doi.org/10.1190/1.1444877

Li Y, Oldenburg DW (1998) 3-D inversion of gravity data. Geophysics 63:109–119

Li Y, Oldenburg DW (1996) 3-D inversion of magnetic data. Geophysics 61(2):394–408

Li X, Yin M (2012) Application of differential evolution algorithm on self-potential data. PLoS ONE. https://doi.org/10.1371/journal.pone.0051199

Lucic P, Teodorović D (2002) Transportation modeling: an artificial life approach. ICTAI, Washington DC, pp 216–223

Madin L (1990) Aspects of jet propulsion in salps. Can J Zool 68:765–777

Martin O, Otto SW, Felten EW (1991) Large-step markov chains for the traveling salesman problem. Complex Syst 5(3):299–326

Mehanee S (2015) Tracing of paleo-shear zones by self-potential data inversion: Case studies from the KTB, Rittsteig, and Grossensees graphite-bearing fault planes. Earth Planets Space 67:14–47

Meiser P (1962) A method for quantitative interpretation of self potential measurements. Geophys Prospect 10(2):203–218. https://doi.org/10.1111/j.1365-2478.1962.tb02009.x

Menke W (1989) Geophysical data analysis: discrete inverse theory. Int Geophys Ser 45

Minsley BJ, Sogade J, Morgan FD (2007a) Three-dimensional source inversion of self-potential data. J Geophys Res 112:B02202. DOI: https://doi.org/1029/2006JB004262

Minsley BJ, Sogade J, Morgan FD (2007b) Three-dimensional selfpotential inversion for subsurface DNAPL contaminant detection at the Savannah River Site, South Carolina. Water Resour Res 43:W04429. DOI: https://doi.org/10.1029/2005WR003996

Millonas M (1994) Swarms, phase transitions and collective intelligence in artificial life III. In: Langton C (ed). Addison-Wesley, Reading Mass, USA, pp 417–445

Mirjalili S, Gandomi AH, Mirjalili SZ, Saremi F, H., Mirjalili, S. M., (2017) Salp Swarm Algorithm: a bio-inspired optimizer for engineering design problems. Adv Eng Softw 114:163–191

Mirjalili S, Lewis A (2016) The whale optimization algorithm. Adv Eng Softw 95:51–67. https://doi.org/10.1016/j.advengsoft.2016.01.008

Mirjalili S, Mirjalili SM, Lewis A (2014) Grey wolf optimizer. Adv Eng Softw 69:46–61

Monteiro Santos FA (2010) Inversion of self-potential of idealized bodies' anomalies using particle swarm optimization. Comput Geosci 36:1185–1190

Murthy IVR, Sudhakar KS, Rao PR (2005) A new method of interpreting self-potential anomalies of two-dimensional inclined sheets. Comput Geosci 31:661–665

Murty SBV, Haricharen P (1985) Nomograms for the complete interpretation of spontaneous potential profiles over sheet-like and cylindrical two-dimensional sources. Geophys 50:1127–1135

Paine J (2007) Developments in geophysical inversion in the last decade. advances in In: Milkereit B (ed) Geophysical inversion and modeling, Proceedings of exploration 07: fifth decennial international conference on mineral exploration, pp 485–488

Parker RL (1977) Understanding inverse theory. Ann Rev Earth Plane Sci 5:35–64

Parker RL (1974) Best bounds on density and depth from gravity data. Geophys 39:644–649

Parker RL (1975) The theory of ideal bodies for gravity interpretation. Geophys J Roy Astron Soc 42:315–334

Patella D (1997) Introduction to ground surface self-potential tomography. Geophys Prospect 45:653–682

Petrowsky A (1928) Problem of hidden polarized sphere, Philosophical Magazine 5, 334, 914–927

Pric K (1994) Genetic annealing. Dr Dobb's J 127–132

Paul MK (1965) Direct interpretation of self-potential extension anomalies caused by inclined sheets of infinite horizontal extension. Geophys 30:418–423

Polya G (1945) How to solve It. Princeton University Press, Princeton, NJ

Rani P, Di Maio R, Piegari E (2015) High-resolution spectral analysis methods for self-potential data inversion. In: Expanded abstract volume of the 85th SEG annual meeting and exposition. New Orleans, pp 1596–1601

Rao AD, Babu RHV (1984) Quantitative interpretation of sclfpotential anomalies due to two dimensional sheet-like bodies. Geophys 48:1659–1664

Rao BSR, Murthy IVR, Reddy SJ (1970) Interpretation of self-potential anomalies of some geometric bodies. Pure Appl Geophys 78: 66–77

Rechenberg I (1965) Cybernetic solution path of an experimental problem. technical report, royal aircraft establishment library translation no 1112, Farnborough, UK

Reddi AGB, Madhusudan IC, Sarkar B, Sharma JK (1982) An album of geophysical responses from base metal belts of Rajasthan and Gujarat (Calcutta: geological survey of India). Miscellaneous Publication, no 51

Reddy SS (2017) Optimal reactive power scheduling using cuckoo search algorithm. Int J Electr Comput Eng (IJECE) 7(5):2349–2356. https://doi.org/10.11591/ijece.v7i5.pp2349-2356

Ramillien G, Mazzega P (1999) Non-linear altimetric geoid inversion for lithospheric elastic thickness and crustal density. Geophys J Int 138:667–678

Roy SVS, Mohan NL (1984) Spectral interpretation of self-potential anomalies of some simple geometric bodies. PAGEOPH 78:66–77

Russell S, Norvig P (1995) Artificial intelligence: a modern approach. Prentice-Hall

Sundararajan N, Kumar IA, Mohan NL, Rao SVS (1990) Use of the hilbert transform to interpret self-potential anomalies due to two-dimensional inclined sheets. Pure Appl Geophys 133:117–126

Sato M, Mooney HM (1960) The electrochemical mechanism of sulfide self-potentials. Geophys 25:226–249

Shaw P (1998) Using constraint programming and local search methods to solve vehicle routing problems. In Maher M, Puget J-F (eds), CP'98 principle and practice of constraint programming, LNCS, vol 1520. Springer, pp 417–431

Sharma SP, Biswas A (2013) Interpretation of self-potential anomaly over 2D inclined sheet structure using very fast simulated annealing global optimization—an insight about ambiguity. Geophys 78(3):WB3-WB15

Skianis GA, Hernandez MC (1999) Effects of transverse electric anisotropy on self-potential anomalies. J Appl Geophys 41:93–104

Skianis GA, Papadopoulos TD, Vaiopoulos DA (2000) A study of the SP field produced by a polarized sphere in an electrically homogeneous and transversely anisotropic ground. In: Development and application of computer techniques to environmental studies VII, CA. In: Brebbia P, Zannetti, Ibarra-Berastegi G (eds) © WIT Press. www.witpress.com, ISBN 1-85312-819-8

Shah H, Tairan N, Mashawani WK, Alsewari AA, Jan MA, Badshah G (2017) Hybrid global crossover bees algorithm for solving boolean function classification task. Lect Notes Comput Sci (international conference on intelligent computing). https://doi.org/10.1007/978-3-319-63315-2_41

Silva JBC, Hohmann GW (1983) Nonlinear magnetic inversion using a random search method. Geophys 48:1645–1658

Singh A, Biswas A (2016) Application of global particle swarm optimization for inversion of residual gravity anomalies over geological bodies with idealized geometries. Nat Resour Res 25(3):297–314

Srivastava S, Datta D, Agarwal BNP, Mehta S (2014) Applications of ant colony optimization in determination of source parameters from total gradient of potential fields. Near Surf Geophys 12:373–389

Srivastava S, Agarwal BNP (2009) Interprettaion of self-potential anomalies by enhanced local wave number technique. J Appl Geophys 68:259–268

Stern W (1945) Relation between spontaneous polarization curves and depth, size and dip of ore bodies: transactions of the american institute of mineralogy, metallurgy. PetUm Eng Min Eng 164:189–196

Storn R, Price K (1995) Differential evolution: a simple and efficient adaptive scheme for global optimization over continuous spaces. technical report TR-95–012, Int CS Institute, University of California, Mar 1995

Sweilam NH, Gobashy MM, Hashem T (2008) Using particle swarm optimization with function stretching (SPSO) for inverting gravity data: a visibility study. Proc Math Phys Soc Egypt 86(2):259–281. https://www.academia.edu/42806956/Using_Particle_Swarm_Optimization_with_Function_Stretching_SPSO_For_inverting_Gravity_Data_-A_Visibility_Study

Sweilam NH, El-Metwally K, Abdelazeen M (2007) Self-potential signal inversion to simple polarized bodies using the particle swarm optimization method: a visibility study. Appl Geophys 6(1):195–208. Egyptian society of applied petrophysics. https://www.academia.edu/42807118/self_potential_signal_inversion_to_simple_polarized_bodies_using_the_particle_swarm_optimization_method_a_visibility_study

Tereshko V, Loengarov A (2005) Collective decision-making in honey bee foraging dynamics. Comput Inf Sys J 9(3):1–7

Tereshko V, Lee T (2002) How information mapping patterns determine foraging behaviour of a honey bee colony. Open Syst Inf Dyn 9:181–193

Tereshko V (2000) Reaction-diffusion model of a honeybee colony's foraging behaviour. In: Schoenauer M (ed) parallel problem solving from nature VI. Lecture notes in computer science, vol 1917. Springer, Berlin, pp 807–816

Teodorovi´c D (2003) Transport modeling by multi-agent systems: a swarm intellgence approach. Transport Plan Technol 26(4):289–312

Telford WM, Geldart LP, Sheriff RE, and Keys DA (1976) Applied Geophysics. Cambridge University Press

Tlas M, Asfahani J (2013) An approach for interpretation of self-potential anomalies due to simple geometrical structures using flair function minimization. Pure Appl Geophys 170:895–905

Tlas M, Asfahani J (2008) Using the adaptive simulated annealing (ASA) for quantitative inter-
   pretation of self-potential anomalies due to simple geometrical structures. JKAU Earth Sci
   19:99–118
Tlas M, Asfahani J (2007) A best-estimate approach for determining self-potential parameters
   related to simple geometric shaped structures. Pure Appl Geophys 164:2313–2328
Voudouris C, Tsang E (1995) Guided local search. technical report CSM-247, University of Essex,
   UK
Voudouris C (1998) Guided local search: an illustrative example in function optimization. BT
   Technol J 16(3):46–50
Vasco DW, Johnson LR, Majer EL (1993) Ensemble inference in geophysical inverse problems.
   Geophys J Int 115:711–728
Voges KE, Pope N (2006) Business applications and computational intelligence. Idea Group
   Publishing
Wedde HF, Farooq M, Zhang Y (2004) Bee hive: an efficient fault-tolerant routing algorithm inspired
   by honey bee behavior, ant colony, optimization and swarm intelligence. In: Proceedings of the
   4th international workshop, ANTS, Brussels, Belgium
Wynn JC, Sherwood SI (1984) The self-potential (sp) method: an expensive reconnaissance and
   archaeological mapping tool. J Field Archaeol 11:195–204
Yang XS, Deb S (2009) Cuckoo search via L´evy flights. In: Proceedings of world congress on
   nature and biologically inspired computing (NaBIC 2009), Dec, India. IEEE Publications, USA,
   pp 210–214
Yang XS (2010) A new metaheuristic bat-inspired algorithm. In: nature inspired cooperative strategies for optimization (NISCO 2010). Gonzalez JR et al (Eds) studies in computational intelligence,
   vol 284. Springer, Berlin, pp 65–74
Yüngül S (1950) Interpretation of spontaneous polarization anomalies caused by spheroidal ore
   bodies. Geophysics 15:237–246
Zhang Y, Balochian S, Agarwal P, Bhatnagar V, Housheya OJ (2014) Artificial intelligence and its
   applications. Math Probl Eng Article ID 840491:10

Chapter 3
Self-potential Inversion and Uncertainty Analysis via the Particle Swarm Optimization (PSO) Family

Juan Luis Fernández-Martínez and Zulima Fernández-Muñiz

Abstract Water flow in the subsoil and pumping tests generate electrical currents measurable at the ground surface and terms spontaneous potential (SP) anomalies that are well correlated with the geometry of the water table. In this paper, we present the application of the Particle Swarm Optimization (PSO) family to estimate the water table elevation from SP measurements at the ground surface. The search is performed in a reduced space that is generated via Principal Component analysis performed in a set of models that are randomly generated taking into account the regularity of the SP anomaly. The PSO members used in this research perform a sampling of the water table coordinates in the PCA space, the electro-kinetic coupling constant and the reference hydraulic head, in a reduced dimensional space. Besides, based on the samples gathered on the low misfit area we are able to compute a fast approximation of the posterior distribution of the SP model parameters with very humble computational resources.

Keywords Streaming potential · Global optimization · GPSO · CC-PSO · RR-PSO · Uncertainty analysis

3.1 Introduction

The self-potential technique in near surface geophysics involves the passive measurement of the electrical potential distribution at the Earth's surface acquired by non-polarizable electrodes. SP anomalies are associated to a variety of situations such as: 1. Water flow in subsoil that generates electrical currents measured at surface named as streaming potential and correlate very well to the hydraulic head (Fournier

Right Running Head: PSO applied to the Streaming Potential inverse problem.

J. L. Fernández-Martínez (✉) · Z. Fernández-Muñiz
Group of Inverse Problems, Optimization and Machine Learning. Mathematics Department, University of Oviedo, C/Federico García Lorca 18, 33007 Oviedo, Spain
e-mail: jlfm@uniovi.es

Z. Fernández-Muñiz
e-mail: zulima@uniovi.es

© The Author(s), under exclusive license to Springer Nature Switzerland AG 2021   105
A. Biswas (ed.), *Self-Potential Method: Theoretical Modeling and Applications in Geosciences*, Springer Geophysics, https://doi.org/10.1007/978-3-030-79333-3_3

1989). 2. Pumping tests used to estimate the physical parameters of the aquifers also generate a SP signal that serves to monitor the shape of the cone of depression 3. The "electro-redox" effect associated with redox potential gradients (Corry 1985; Naudet et al. 2004; Naudet and Revil 2005).

The spontaneous potential inverse problem in hydrogeology is a non-intrusive method that uses the spontaneous potential measurements at different points of the surface to image the water table level. For that purpose, it is also needed to estimate the electro-kinetic constant that serves to take into account the coupling between the movement of the water molecules and its behavior as dipoles. This geophysical method has been used to obtain hydraulic information on subsurface flows from the analysis of SP data (Bogoslovsky and Ogilvy 1973; Morgan et al, 1989; Birch 1998; Sailhac and Marquis 2001; Revil et al. 2003; Darnet et al. 2004; Bolève et al. 2007; Minsley et al. 2007; Maineult et al. 2008), to monitor redox processes occurring in ore bodies (Bigalke and Grabner 1997) and contaminant plumes (Naudet et al. 2004), among other applications.

3.2   The Streaming Potential Forward and Inverse Problems

The electric flow (J_e in A/m^2) in a water saturated porous medium is related to the hydraulic pressure gradient (∇P) and to the electric potential gradient (∇V) as follows:

$$J_e = \sigma_r \nabla V - C_s \sigma_r \nabla P, \tag{3.1}$$

where σ_r is the rock conductivity (S/m) of the porous medium and C_s is called the electro-kinetic coupling coefficient (V/Pa) of the saturated zone. C_s can be measured in the laboratory as the ratio of the electric potential (∇V) induced by applying fluid flow (∇P) (Guichet et al. 2006).

In steady-state and without direct electric current source the conservation of flux:

$$\nabla \cdot J_e = \nabla \cdot (\sigma_r \nabla V - C_s \sigma_r \nabla P) = 0, \tag{3.2}$$

connects the electric streaming potential V to the hydraulic pressure P. Equation (3.2) can be written in integral form as follows (Fitterman 1978; Fournier 1989; Revil et al. 2003):

$$V(Q) = \frac{C'}{2\pi} \int_{\partial\Omega} (h - h_0) \frac{\mathbf{r} \cdot \mathbf{n}}{\|\mathbf{r}\|^3} d\Gamma + \frac{1}{4\pi} \int_{\Omega} \frac{\mathbf{E}}{\|\mathbf{r}\|} \cdot \frac{\nabla \rho}{\rho} dV. \tag{3.3}$$

Expression (3.3) provides the electrical potential (V in Volts) measured at an observation station Q located at the earth surface. In Eq. (3.3), h and h_0 are the hydraulic heads at any point of the water head and in a reference level, $\mathbf{n}$ is the outward normal to the water table and $\mathbf{r}$ is position vector from any source point in the water head to observation station Q. $\mathbf{E} = -\nabla\phi$ is the electrical field produced in the ground through the electro-kinetic coupling, ρ is the electrical resistivity of the rock (in Ω m), and C' (in Vm^{-1}) is the electro-kinetic coupling coefficient relating the hydraulic piezometric head Δh (m/s) to the electrical potential difference ΔV (mV).

In this model the electrical potential is the sum of two terms:

1. The first one is related to the current density induced by the water movement. The primary source term is such that each element of the water table acts as a dipole of strength $C'(h - h_0)$. The integral is the sum of all these dipoles that contribute to the SP signals recorded at the observation station Q with strength that depends on the distance between each dipole and the observation station Q.
2. The second one is related to electrical resistivity contrasts in the ground (secondary source of polarization). This second contribution can be neglected in a quasi-homogeneous resistivity earth. If this approximation is unfeasible, the electrical tomography can be used to determine the geoelectrical structure of the terrain and take this contribution into account (resistivity correction).

In this paper we will only consider the first contribution of the potential. The forward problem consists in, knowing the electro-kinetic constant and the water table, to compute the electric potential in any point of the surface. This problem involves the numerical approximation of the integral equation stated in (3.3).

3.3 The Inverse Problem and the Topology of the Cost Function

The inverse problem consists in recovering the depth of the water table and the coupling coefficient C' from the SP measurements performed at the ground surface. As it has been already commented the contribution related to electrical resistivity contrasts is either neglected or considered as a correction of the observed data (spontaneous potentials) before solving the inverse problem.

The inverse problem can be written as a set of equations:

$$V(\mathbf{r}_k) = f_k(h, C'), \quad k = 1, \ldots, m, \tag{3.4}$$

where $f_k(h, C')$ stands for the integral equation that provides the potential in $V(\mathbf{r}_k)$. For that purpose, we will write this problem in the form

$$\mathbf{F(m)} = \mathbf{d}^{obs}, \quad \mathbf{d}^{obs} \in R^s, \ \mathbf{m} \in R^n, \tag{3.5}$$

being $\mathbf{m} = (\mathbf{h}, C', h_0)$ the model parameters, $\mathbf{F} = (f_1(\mathbf{m}),\ f_2(\mathbf{m}),\ \dots,\ f_n(\mathbf{m}))$, the forward operator, and $\mathbf{d}^{obs} = (V(\mathbf{r}_1),\ V(\mathbf{r}_2),\ \dots,\ V(\mathbf{r}_n))$ contains the observed data (potentials) at different points of the surface. In this case $\mathbf{h}$ represents the coordinates of the water level in a given set of basis functions.

The inverse problem is typically cast into an optimization problem, consisting in minimizing the cost function:

$$c(\mathbf{m}) = w_1 \|\mathbf{d} - \mathbf{F}(\mathbf{m})\|_p + w_2 \|\mathbf{m} - \mathbf{m}_{ref}\|_{p'}, \tag{3.6}$$

which is a combination of the data misfit and the distance to a reference model $\mathbf{m}_{ref}$, being w_1, w_2 real weights used to provide the relative importance of both terms. In this paper the cost function is only composed by the data misfit.

The typology of the nonlinear system (3.5) depends on the number of observed data (s) with respect to the number of model parameters (n), that coincide with the number of discretization points of the water head plus one (C'). Very fine discretization of the water head (h) provokes the nonlinear system (3.5) to have a very high underdetermined character. This implies that when using linearization methods, the corresponding Jacobian matrix corresponds to an underdetermined (or rank-deficient) linear system with the corresponding difficulties associated to their solution.

The topography of the cost function in linear and nonlinear inverse problems and the effect of the noise and that of the regularization (Fernández-Martínez et al. 2012a, b; 2014a, b), showing that in the case of nonlinear problems the plausible solutions are located in one or several disconnected flat curvilinear valleys of the cost function landscape. As a consequence, the reference model and the prior information used to stabilize the inversion greatly influenced the results that have been obtained. This way of solving the inverse problem is not robust independently of the algorithm that has been used to find the solution.

Local optimization methods are not able to discriminate among the multiple choices consistent with the end criteria and may land quite unpredictably at any point of the nonlinear equivalence region $\mathbf{M}_{tol}$:

$$\mathbf{M}_{tol} = \frac{\|\mathbf{F}(\mathbf{m}) - \mathbf{d}^{obs}\|}{\|\mathbf{d}^{obs}\|} < tol, \tag{3.7}$$

which is composed by the geophysical models that fit the observed data within a given error tolerance. The case of linear inverse problems is simpler since the region of equivalence for a given error bound coincides with hyper-quadric of equivalence, that in the case of purely over determined systems is a hyper ellipsoid and degenerates to a hyper elliptical cylinder in the case of underdetermined and rank deficient problems (Fernández-Martínez et al. 2012). Besides the noise in data deforms the topography of the cost function, shifting the minimum to another model that belongs to the nonlinear equivalence region (Fernández-Martínez et al. 2014a, b). Tikhonov's regularization used in nonlinear least-squares methods only serve to obtain stable solutions, but it does not provoke the uncertainty of the solution to vanish. Furthermore, the uncertainty analysis of the corresponding linearized system does not provide a correct estimation of the nonlinear uncertainty region (Fernández-Martínez et al. 2013).

Uncertainty in the inverse problem solutions is always present and it is mainly due to several causes:

1. The forward model F is a simplification of reality: modelling hypothesis and numerical approximations provoke that the inverse solution is not unique and might differ drastically from the reality.
2. Data are noisy and only partially sample the domain of interest.
3. The data partially informs about the question that is going to be solved, that is, the inverse problem is intrinsically ill-posed. A simple example would be trying to identify two number whose sum is known. The ill-posed character of this problem is related to the question itself.

The only solution consists in taking into account the uncertainty of the solution, performing a correct quantification, which entitles the sampling of the nonlinear uncertainty region stated in (3.7). Bayesian approaches and global optimization methods when they are used in their exploratory forms are used to accomplish this task.

Bayesian approaches and Monte Carlo methods (Scales and Tenorio 2001; Mosegaard and Tarantola 1995) can be used to solve the inverse problem as a sampling problem to perform the model appraisal. These methods provide the posterior probability distribution of the model parameters that is obviously related to the topography of the cost function, since they sample many times in a random walk, with a bias towards increased sampling of areas lower data misfit and higher posterior probability. This procedure is called in the mathematical literature importance sampling, and tries to describe the topography of the cost function in the neighborhood of the low misfit regions from the collected samples. These methods are hampered by the curse of dimensionality (Fernández-Martínez and Fernández Muñiz, 2020).

Global optimization algorithms include among others, well known techniques such as Genetic Algorithms (Holland 1992), Simulated Annealing (Kirkpatrick et al. 1983), Particle Swarm Optimization (Kennedy and Eberhart 1995), Differential Evolution (Storn and Price 1997) and the Neighborhood Algorithm (Sambridge 1999a, b).

Posterior sampling techniques are closely related to global optimization algorithms, which can be used to provide a proxy for the true posterior distribution. In many practical situations, prior information is not available, and global optimization methods are a good alternative for avoiding the strong dependence of the solution upon noisy data.

Bayesian approaches are computationally expensive and they might not even be feasible in the case of high dimensional problems with very costly forward problems. Linde et al. (2007) applied a Bayesian estimation method using geostatistical techniques to integrate SP and piezometric data in order to estimate the water table throughout a catchment. Jardani et al. (2009) proposed an algorithm to solve the

SP inverse problem in a Bayesian framework. Darnet et al. (2003) used genetic algorithms and Fernández-Martínez et al. (2010) used the generalized PSO (GPSO).

In this paper we present the application of different members of the PSO family to estimate the depth of the water table from SP data, showing the application to field self-potential data collected in the vicinity of a pumping well (Bogoslovsky and Ogilvy 1973). The algorithm proposed here uses a dimensionality reduction of the water head based on a set of preliminary templates that are generated using the semi-empirical relationship proposed by Revil et al. (2003):

$$V(P) = (h - h_0)C'. \qquad (3.8)$$

Model reduction is achieved by PCA performed in this set of random models. The PSO family members sample the electro-kinetic constant, the reference water head, and the coordinates of water table in the PCA basis set. The use of exploratory PSO versions allows approximating the posterior distribution of the model parameters as it has been shown in different publications concerning the analysis an application of the PSO family members (Fernández Martínez et al. 2010b, 2012b; Pallero et al. 2015, 2017).

3.4   The PSO Family

Particle Swarm Optimization (PSO) is a stochastic evolutionary computation technique (Kennedy and Eberhart 1995) that was initially inspired by the social behavior of individuals (called particles) in nature, such as birds or fish. Let us suppose that we want to solve an optimization problem that consists in the minimization of a cost function $C(\mathbf{m})$ in a given subset $\mathbf{M}$ of the n-dimensional space. In the present case $\mathbf{m}$ are the geophysical model parameters, that is, the electro kinetic constant (C'), the reference water head (h_0) and the coordinates of the water table $\mathbf{h}$ in a basis set functions $\varphi_k(x)$:

$$h(x) = \sum_{k=1}^{n} h_k \varphi_k(x) = (h_1, h_2, \ldots, h_n)_{\{\varphi_k\}_{k=1}^n} = \mathbf{h}_{\{\varphi_k\}_{k=1}^n}. \qquad (3.9)$$

In PSO, each model $\mathbf{m}$, called a particle, samples the search space according to its own, $\mathbf{l}_i(k)$ and its companions, $\mathbf{g}(k)$ searching experience, that depend on the cost function, according to the following

The way the PSO algorithm is applied to solve an inverse problem is very intuitive:

1.   A prismatic space of admissible geophysical models, $\mathbf{M}$, is defined:

$$l_j \leq x_{ji} \leq u_j, \quad 1 \leq j \leq n, \quad 1 \leq i \leq N_{size},$$

where l_j, u_j, are the lower and upper limits for the j-th coordinate of each geophysical model (called i), n is the number of parameters in the inverse problem, and N_{size} is the swarm size. Each particle (or plausible geophysical model) has its own position on the search space and velocity, which stand for the parameter perturbations needed for these positions to find the solutions of the optimization problem. Without any lack of generality the velocities are initially set to zero.

2. PSO updates at each iteration the positions,$\mathbf{x}_i(k)$ and velocities,$\mathbf{v}_i(k)$ of each model in the swarm according to the following rule:

$$\mathbf{v}_i(k+1) = w\,\mathbf{v}_i(k) + \phi_1(\mathbf{g}(k) - \mathbf{x}_i(k)) + \phi_2(\mathbf{l}_i^k - \mathbf{x}_i(k))$$
$$\mathbf{x}_i(k+1) = \mathbf{x}_i(k) + \mathbf{v}_i(k+1). \tag{3.10}$$

The velocity of each particle, i, at each iteration, k, is a function of three major components: 1. The inertia term, which consists of the old velocity vector of the particle,$\mathbf{v}_i(k)$ weighted by a real constant, w called inertia. 2. The social learning term, which is the difference between the global best position found $\mathbf{g}(k)$ and the particle's current position $(\mathbf{x}_i(k))$. 3. The cognitive learning term, which is the difference between the particle's best position so far found, $\mathbf{l}_i(k)$, and the particle current position, $\mathbf{x}_i(k)$. w, a_g, a_l are the PSO parameters: the inertia and local and global acceleration constants; $\phi_1 = r_1 a_g$, $\phi_2 = r_2 a_l$ are the stochastic global and local accelerations, and r_1, r_2 are vectors of random numbers uniformly distributed in $(0, 1)$, to weight the global and local acceleration constants,a_g and a_l of any coordinate particle in the swarm.

The flowchart for the PSO algorithm is as follows:

1. A prismatic search space for the model parameters is given.
2. An initial swarm of N_{size} particles is uniformly distributed in the search space and their initial velocities are (usually) set to zero.
3. The misfit of the initial population is evaluated solving N_{size} forward problems, and the global best and the previous best of each particle are determined. This step is very important since no inversion is performed, only the solution of the forward models, one for each particle in the swarm, to establish their misfit. Therefore, to use this kind of algorithms the forward problem has to be fast to solve.
4. Drawing of the random numbers r_1, r_2, and updating of the velocities and positions of each particle of the swarm using formula 7.
5. Iterate to point 3 till the maximum number iterations is finished or some criteria are fulfilled. Typically, in this kind of sampling procedures, the algorithm finishes by iterations, when a correct sampling is performed, and/or the swarm has collapsed.

Although many PSO heuristic variants have been proposed in the literature, the stochastic convergence of this algorithm is related to the stability of the swarm, particularly, the first order stability (stability of the mean trajectories) depends on the total mean acceleration, $\overline{\phi} = \frac{a_g + a_l}{2}$ and on the inertia constant. The PSO algorithm can be physically interpreted as a stochastic damped mass-spring system (Fernández-Martínez et al. 2008). PSO corresponds to a particular finite-difference discretization of the differential equation:

$$x''(t) + (1 - w)x'(t) + (\phi_1 + \phi_2)x(t) = \phi_1 g(t) + \phi_2 l(t). \tag{3.11}$$

This equation describes the continuous movement of each particle in the swarm. PSO corresponds to a centered discretization in acceleration,

$$x''(t) = \frac{x(t + \Delta t) - 2x(t) + x(t - \Delta t)}{\Delta t^2}, \tag{3.12}$$

and a regressive schema in velocity,

$$x'(t) = \frac{x(t) - x(t - \Delta t)}{\Delta t}. \tag{3.13}$$

The GPSO algorithm (Fernández Martínez and García Gonzalo 2008), which is a PSO generalization for any time step, Δt, can be written as:

$$v(t + \Delta t) = (1 - (1 - w)\Delta t)v(t) + \phi_1 \Delta t(g(t) - x(t)) + \phi_2 \Delta t(l(t) - x(t)),$$
$$x(t + \Delta t) = x(t) + \Delta t\, v(t + \Delta t). \tag{3.14}$$

Using $\Delta t = 1$, this equation reduces to the PSO algorithm (Eq. 3.7). Due to the random effect introduced by the random numbers r_1, r_2, the particle trajectories have to be considered as stochastic processes whose first (mean) and second order moments (variance and temporal covariance) are important to understand the algorithm convergence. Other PSO family members (CC-PSO, CP-PSO, PP-PSO, RR-PSO and the 4 points PSO optimizers) were developed using different discretizations for the velocity and the accelerations (Fernández-Martínez and García-Gonzalo 2009, 2012; García-Gonzalo and Fernández-Martínez 2014).

The stochastic stability analysis of the PSO trajectories (Fernández-Martínez and García-Gonzalo 2008, 2009, 2011) served to establish the relationship between PSO convergence and the first- and second-order stability of the trajectories of the particles considered as stochastic processes. In general terms the good (or suitable) PSO parameters sets (w, a_g, a_l) are located in the neighborhood of the upper border of the second-order stability region for each member of the PSO family. This result was also generalized for any statistical distribution of the PSO parameters (see García-Gonzalo and Fernández-Martínez 2014). The cloud versions of these algorithms are based on the stochastic stability analysis of the trajectories. The advantage of the cloud versions is that no parameter tuning (inertia, local and global accelerations) is needed, since each particle in the swarm has its own PSO parameters that are randomly selected from a set of PSO parameters that are located in the neighborhood of the upper limit of their second-order stability regions. Additionally, in the cloud design, each particle has its corresponding time-step Δt, which is a numerical parameter that serves to achieve the exploration of the search space when this parameter increases and it is greater than 1.0. Conversely, the algorithm freezes the solution found when Δt is decreased to values lower than 1.0. Table 3.1 shows all the GPSO family members, with the corresponding expressions and the references where these algorithms have been published. Particularly, in the case of RR-PSO the optimum parameter sets are located along the line $\overline{\phi} = 3(w - 3/2)$, mainly for inertia values $w > 2$ (see Fernández-Martínez and García-Gonzalo 2012). This straight line remains invariant when the number of parameters increases and it is independent on the typology of the cost functions (multimodal or valley shape). Additionally this line is located in a region of medium attenuation and very high frequency for the particle trajectories. This feature confers to RR-PSO a good equilibrium between exploration and exploitation, allowing for an efficient and exploratory search. The numerical experiments using different benchmark functions have shown that the best-performing algorithm of the PSO family is RR-PSO. Among the rest of family members, CP-PSO is the most exploratory. Therefore, both, RR-PSO and CP-PSO constitute an interesting choice for performing nonlinear uncertainty analysis and exploring the cost-function topography efficiently. PP-PSO has the same velocity update as GPSO, but the positions of the particles are in time t instead of $t + 1$. PP-PSO has a more exploratory character than GPSO but a lower convergence rate. Finally, CC-PSO has showed in the numerical analysis the fastest convergence rate.

In general terms, all the PSO family members provide excellent results as long as the parameter tuning is correct done taking into account the corresponding second-order stability of the trajectories. Besides, no fancy mechanisms are needed to avoid the misunderstood phenomenon of premature convergence.

All these mathematical results make the PSO family to be a very unique algorithm, different from other heuristic approaches. Additionally, the PSO family members are able to provide a set of representative samples of the nonlinear region of equivalence, which can be used to infer an approximate posterior of the model parameters in nonlinear inverse problems much faster than Monte Carlo methods and much more realistically than linear analysis techniques combined with local optimization methods.

3.5 PSO Design in the SP Case

The successful application of the PSO family to real problems is based on three main features:

1. The understanding the physics involved in the forward problem and adapting the PSO algorithm to the inverse problem peculiarities.
2. The use of a robust family of PSO optimizers. Particularly, the PSO family members used in this paper (Fernández-Martínez and García-Gonzalo 2009,

Table 1 PSO family members

Algorithm	Expression
GPSO	$v(t + \Delta t) = (1 - (1 - w)\Delta t)v(t) + \phi_1 \Delta t(g(t) - x(t)) + \phi_2 \Delta t(l(t) - x(t)).$ $x(t + \Delta t) = x(t) + \Delta t \, v(t + \Delta t)$
CC-GPSO	$v(t + \Delta t) = \dfrac{2 + (w - 1)\Delta t}{2 + (1 - w)\Delta t}v(t) + \dfrac{\Delta t}{2 + (1 - w)\Delta t}\displaystyle\sum_{k=0}^{1}\left[\phi_1(l(t - t_0 + k\Delta t) - x(t + k\Delta t)) + \phi_2(g(t - t_0 + k\Delta t) - x(t + k\Delta t))\right],$ $x(t + \Delta t) = x(t) + \Delta t\left[\dfrac{2 + (w - 1)\Delta t}{2}v(t) + \dfrac{\Delta t}{2}\phi_1(l(t - t_0) - x(t)) + \dfrac{\Delta t}{2}\phi_2(g(t - t_0) - x(t))\right]$
CP-GPSO	$v(t + \Delta t) = \dfrac{1 - \phi \Delta t^2}{1 + (1 - w)\Delta t}v(t) + \dfrac{\phi_1 \Delta t}{1 + (1 - w)\Delta t}(g(t + \Delta t - t_0) - x(t)) + \dfrac{\phi_2 \Delta t}{1 + (1 - w)\Delta t}(l(t + \Delta t - t_0) - x(t)).$ $x(t + \Delta t) = x(t) + \Delta t \, v(t)$
PP-GPSO	$v(t + \Delta t) = (1 - (1 - w)\Delta t)v(t) + \phi_1 \Delta t(g(t) - x(t)) + \phi_2 \Delta t(l(t) - x(t)).$ $x(t + \Delta t) = x(t) + \Delta t \, v(t)$
RR-GPSO	$v(t + \Delta t) = \dfrac{v(t) + \phi_1 \Delta t(g(t) - x(t)) + \phi_2 \Delta t(l(t) - x(t))}{1 + (1 - w)\Delta t + \phi \Delta t^2},$ $x(t + \Delta t) = x(t) + \Delta t \, v(t + \Delta t)$
RC-GPSO	$v(t) = \dfrac{2}{(1 - w)\Delta t}(v(t - \Delta t) - \Delta t(x(t) - x(t - \Delta t))) + \dfrac{1}{(1 - w)}(\phi_1(g(t - t_0) - x(t)) + \phi_2(l(t - t_0) - x(t))).$ $x(t + \Delta t) = x(t - \Delta t) + 2\Delta t \, v(t)$

(continued)

Table 1 (continued)

Algorithm	Expression
RP-GPSO	$v(t) = \dfrac{1}{(1-w)\Delta t}(v(t-\Delta t) - v(t-2\Delta t) + \Delta t(\phi_1(g(t-t_0) - x(t)) + \phi_2(l(t-t_0) - x(t))))$. $x(t+\Delta t) = x(t) + \Delta t\, v(t)$
PR-GPSO	$v(t+2\Delta t) = v(t+\Delta t) - \Delta t(1-w)v(t) + \Delta t(\phi_1(g(t-t_0) - x(t)) + \phi_2(l(t-t_0) - x(t)))$. $x(t+2\Delta t) = x(t+\Delta t) + \Delta t\, v(t+2\Delta t)$
PC-GPSO	$v(t+\Delta t) = \dfrac{(w-1)\Delta t}{2}v(t) + x(t+\Delta t) + x(t) + \Delta t\left(\dfrac{1}{2}\phi_1(g(t-t_0) - x(t)) + \phi_2(l(t-t_0) - x(t))\right)$. $x(t+2\Delta t) = x(t) + 2\Delta t\, v(t+\Delta t)$

2012) belong to the category of free parameter tuning, since the optimum parameters in each case are selected taking into account second order stability criteria of the swarm trajectories.

3. The approach of the inverse problem as a sampling problem, since inverse problems have multiple plausible solutions that are compatible with the observed data.

In the SP case, the parameters optimized by the PSO algorithm are the piezometric heads (h_i), the electro kinetic coupling coefficient (C') using self-potential measurements performed at the ground surface, and the reference hydraulic head (h_0). The forward model assumes that the electro kinetic coupling coefficient is homogeneous in the saturated zone. The number of parameters is $nnodes + 2$, where $nnodes$ is the number of discretization nodes used to interpolate the water head. If the number of nodes increases to finely discretize the water head, the inverse problem becomes highly underdetermined. To avoid this fact, we use Principal Component Analysis to reduce the dimension.

The algorithm workflow is as follows (Fig. 3.1):

Fig. 3.1 Algorithm workflow

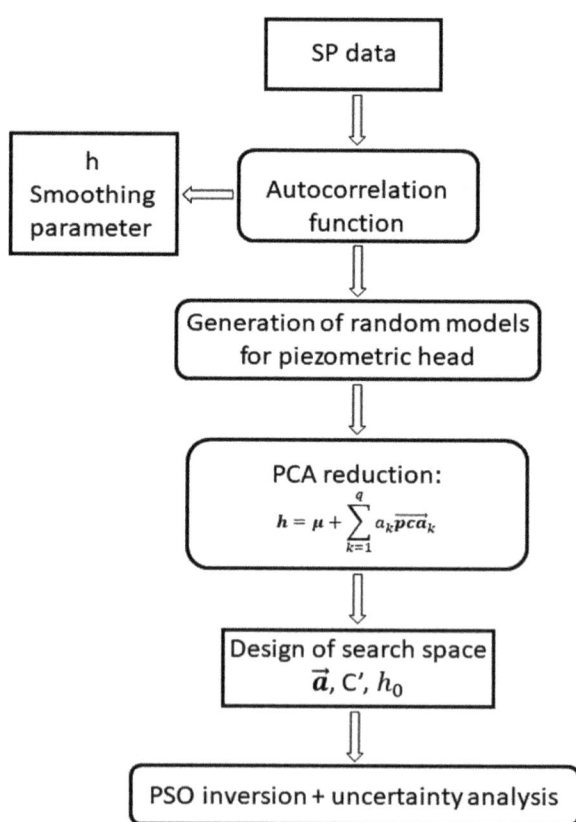

1. First, we establish a search space for reference hydraulic head (h_0).
2. The search space for the electro kinetic constant: $[C_{min}, C_{max}]$ is based on the relationship proposed by Revil et al. (2003): $V(P) = (h - h_0)C'$, taking into account the bounds provided to the water head.
3. Based on these values we randomly generate a set of models within the bounding limits, considering different realizations of h_0, C_{min} and C_{max} in the corresponding search space drawn with an uniform distribution:

$$h_{min}(P) = h_0 + \frac{V(P)}{C'_{max}},$$

$$h_{max}(P) = h_0 + \frac{V(P)}{C'_{min}}. \tag{3.15}$$

4. PCA is performed the diagonalization of the experimental Covariance matrix $\mathbf{C} = (\mathbf{X} - \boldsymbol{\mu})^T (\mathbf{X} - \boldsymbol{\mu})$ where $\mathbf{X}$ is the matrix that has as columns the random models of water tables and $\boldsymbol{\mu}$ is the average of all of them. The diagonalization of C provides a set of q eigenvectors (PCA) that serve to span most of the variability of the original set. A water model is expressed in the PCA basis set as follows:

$$\mathbf{h}_j = \boldsymbol{\mu} + \sum_{k=1}^{q} c_k \mathbf{pca}_k, \tag{3.16}$$

where $\mathbf{pca}_k$ are the q first PCA vectors of the expansion, and c_k the corresponding coordinates.

5. The search is performed in the PCA basis set and the search space is obtained by projecting all the examples (geophysical models) into the pca basis and finding their minimum and maximum.

The PSO methodology allows constraining of the search domain for the optimization of the parameters. Figure 3.2 shows the search space of the water table coordinates in the PCA basis set, composed of the first 15 vectors. In this case we have used the maximum and minimum values of the coordinates. For the upper and lower limits of the coupling coefficient C', we have used a broad interval at the exploratory stages constrained to the fact that the water table stays within the prior ranges. Beside, based on the SP anomaly, it is also convenient to establish the sign of the C' constant to improve the search efficiency. In this case we have used such as

$$\left[C'_{min}, C'_{max} \right] = [-20, \ 0.5] \, \text{mV/m}. \tag{3.17}$$

The search space for the water reference level was:

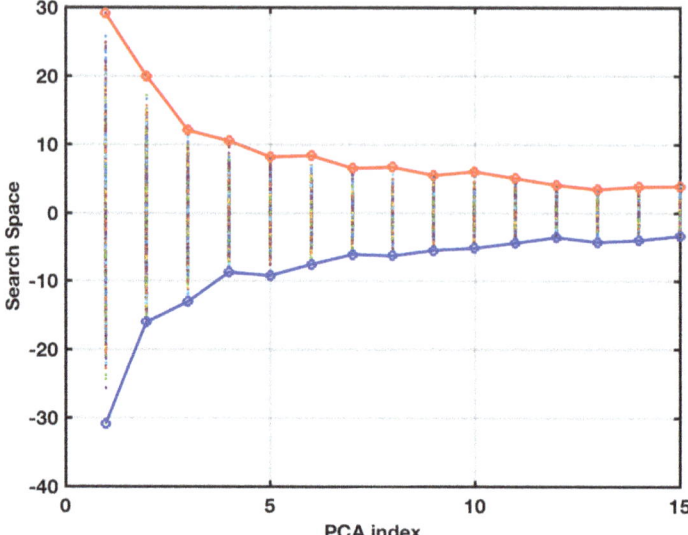

Fig. 3.2 Random generation of water head models and projection on to their PCA space of dimension 15. The search space has the shape of a funnel and it is obtained by finding the maximum and the minimum of each of the PCA coordinates of these random models. This projection reduces the dimension from n, being n the number of interpolation points of the water head, to 15

$$[h_{0\,min}, h_{0\,max}] = [15, \ 20]\,\text{m}. \tag{3.18}$$

Some important peculiarities of the PSO version used for SP optimization are:

1. The SP inverse problem is ill-posed, that is, very different geophysical models exhibiting very different spatial correlation lengths can account for the observed data within the same error tolerance. The water head parameters are not spatially independent, and to improve the analysis, the different inverted water models are smoothed after PCA reconstruction. The smoothed length is estimated from the autocorrelation function of the SP data. Figure 3.3 shows the autocorrelation function of the SP data. The spatial correlation vanishes at a distance of 10 lags. This is the length that was used to smooth the water table models after PCA reconstruction.

2. In previous research concerning the streaming potential inverse problem we have shown via synthetic modeling that the most important tradeoff between these parameters appears occurs between the drawdown amplitude and the coupling constant. This tradeoff is also observed on real cases (Fernández-Martínez et al. 2010b). Besides it has been shown that the topography of the cost function corresponds to a valley shape whose size increases under the presence of noise. These valleys exhibited sinkholes that might provoke entrapment of the geophysical when local optimization methods are used.

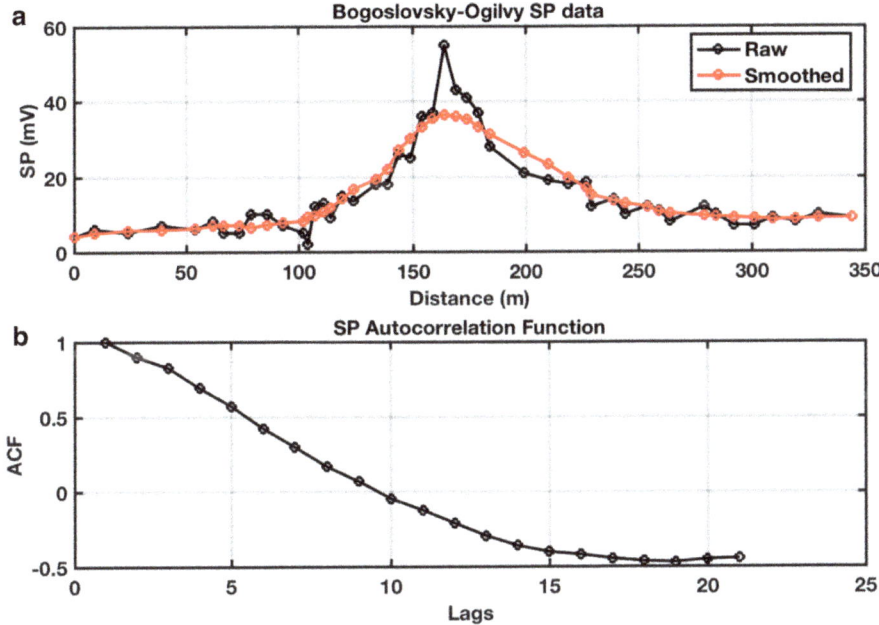

Fig. 3.3 Bogoslovsky-Ogilvy SP data. **a** Raw and Smoothed data using a length of 10 points. **b** SP autocorrelation function. The autocorrelation vanishes approximately at 10 lags

3.6 Modeling the Bogoslovsky and Ogilvy Dataset

To show the performance of the methodology in a real case we used f the SP data collected in the vicinity of a pumping well published by Bogoslovsky and Ogilvy (1973). This dataset has been also modeled by several other authors (Revil et al. 2003; Darnet et al. 2004; Minsley et al. 2007; Fernández-Martínez et al. 2010b). In this case we compared 3 main algorithms of the PSO family: GPSO, CP-PSO and RR-PSO. GPSO is the cloud version of PSO, while CP-PSO and RR-PSO are more exploratory versions.

Figure 3.4a shows the SP prediction and uncertainty analysis obtained via the GPSO algorithm compared to the water table elevations measured in several monitoring wells (inversed triangles). The borehole information was not used for the inversion. It can be observed that that the predicted (or inverted) SP anomaly is close to the smoothed signal with the lag given by the ACF shown in Fig. 3.3. In other words, the inversion can fit fairly well smoothed version of the anomaly but not its higher frequency peaks. To obtain the inversion shown in Fig. 3.4a we have used a correlation length of ten points, as suggested by the autocorrelation function of the SP data (Fig. 3.3a), supposing that the water head and the SP signal have a similar correlation length. This smoothing parameter is needed to introduce some regularity requirements into the inverse solution which is performed in the PSO

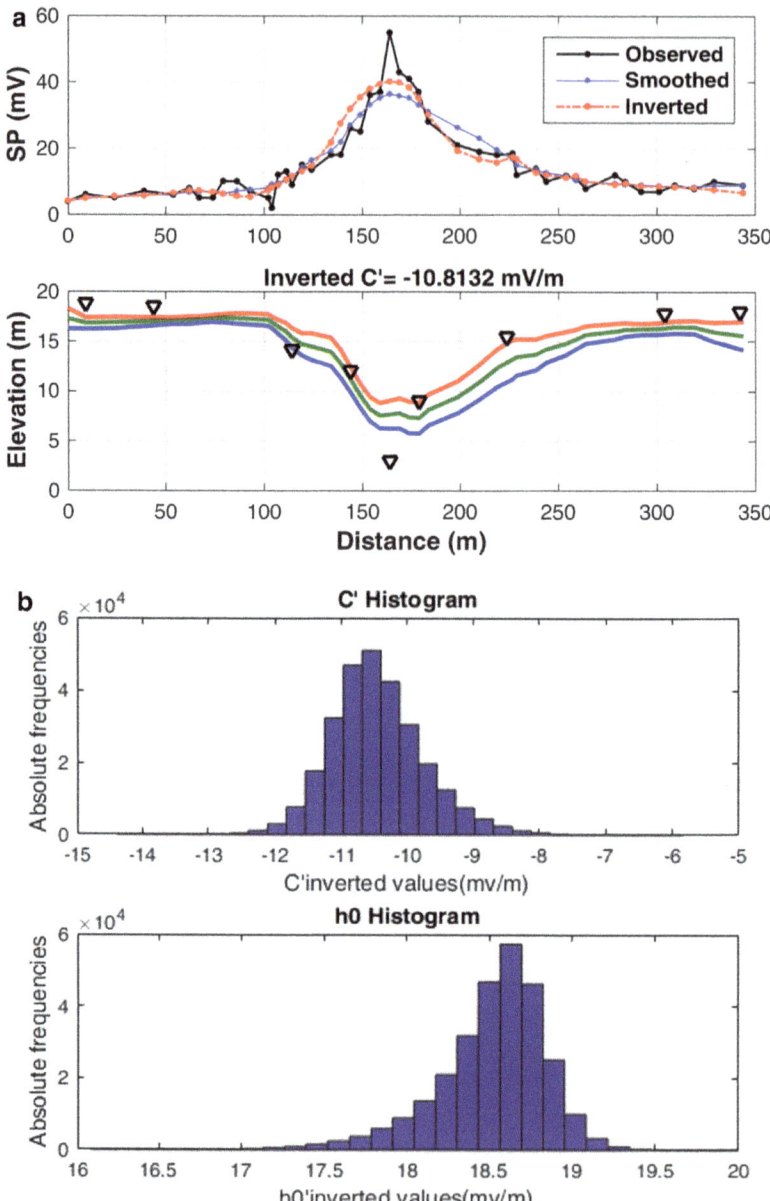

Fig. 3.4 Bogoslovsky-Ogilvy SP data-PSO. **a** Uncertainty analysis of the water table prediction. **b** Posterior histograms of C' and h_0

sampling procedure before solving the forward problem for each of the particles in the swarm. Figure 3.4a also shows the median inverted water head and the lower and upper bound given by the interquartile range obtained via the uncertainty analysis of the water head models sampled by GPSO in the region of relative misfit lower than 20%. There is a good agreement between the prediction of the GPSO algorithm and the water table measured at the monitoring wells except in the central borehole corresponding to the deepest water table coordinate which is closer to the lower bound. The rest are closer to the median and upper bound (lower interquartile). Expanding the lower and upper percentiles to 10 and 90% will make a closer solution to these points. The coupling coefficient shows and histogram with values between -13 and -8, having the mode approximately in -10.5 mV/m. The reference water level has the mode in 18.5 with a variability between 17.5 and 19.5 m (Fig. 3.4b).

Fig. 3.5 GPSO. **a** Error curve. **b** Topography of the cost function in the PCA space

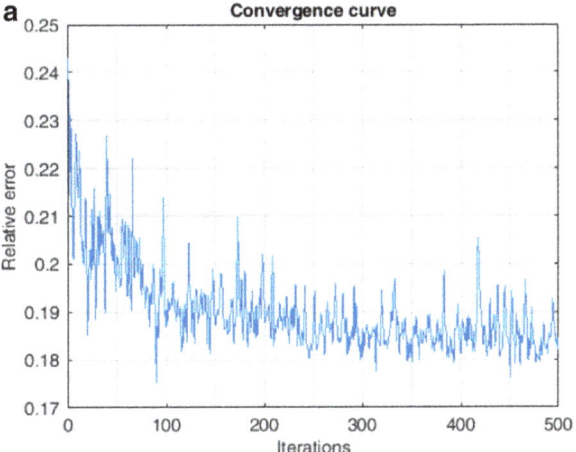

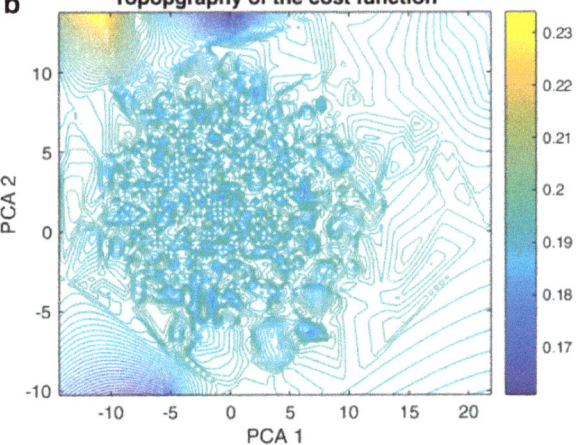

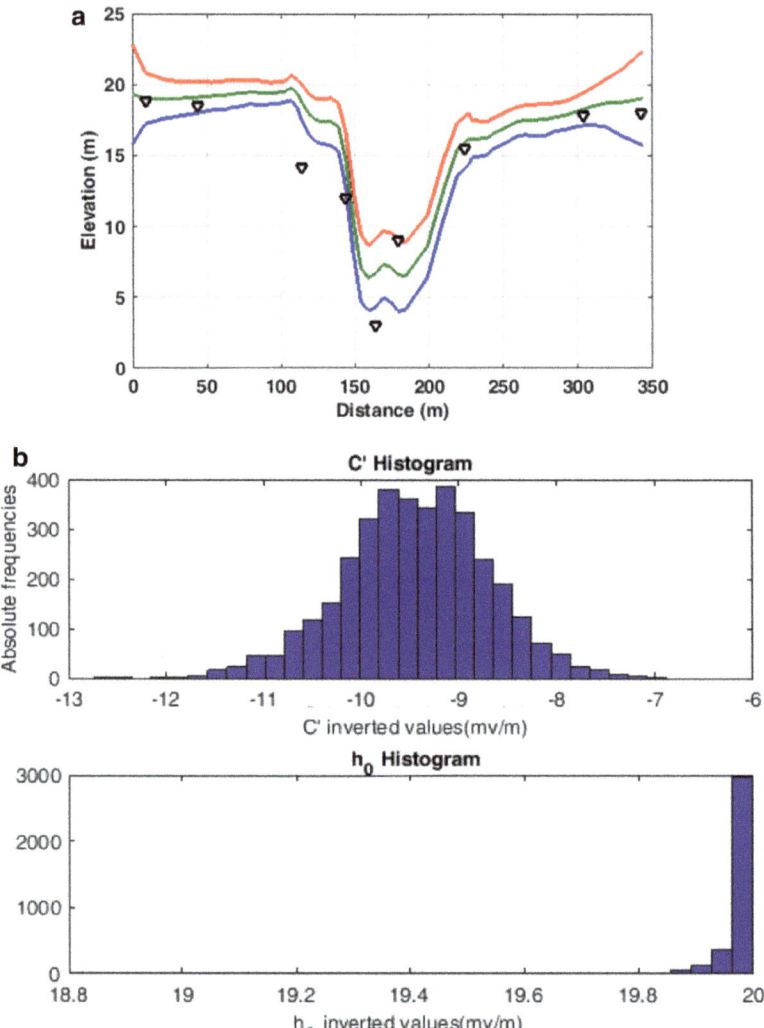

Fig. 3.6 CP-PSO. **a** Uncertainty analysis of the water table. **b** Posterior histograms of C' and h_0

Figure 3.5a shows the convergence curve for GPSO. It can be observed that the relative error is highly variable since the global best was updated in each iteration, although the misfit error might increase. This non-elitist feature makes the error curve to be very variable. Figure 3.5b shows the topography function in the 2D PCA space interpolated from the models that have been sampled by GPSO. It can be observed the complex topography of the low misfit region with sinkholes. In this it can not be clearly seen the valley shape.

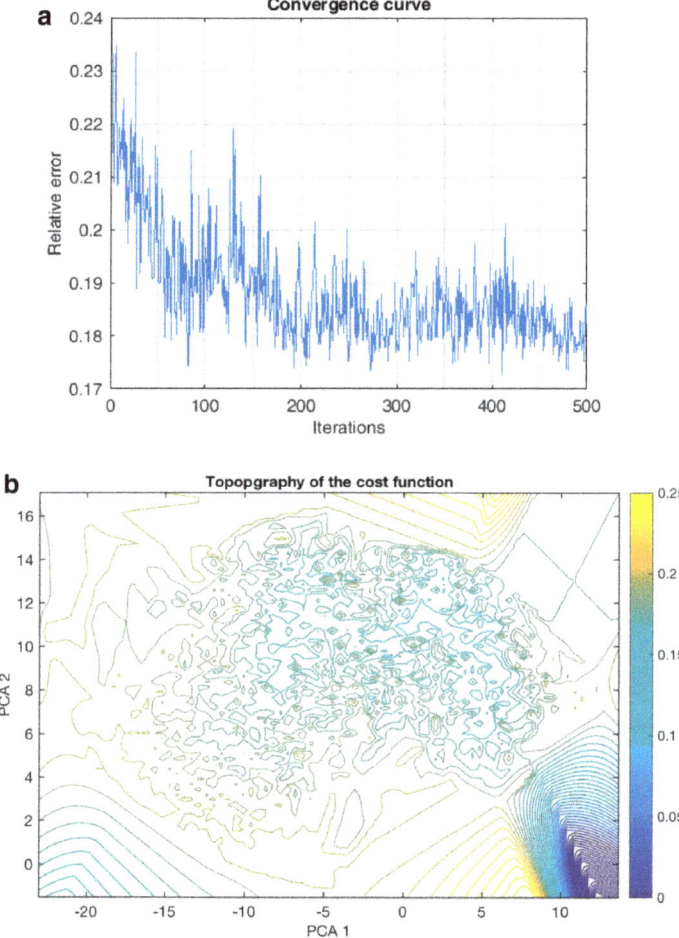

Fig. 3.7 CP-PPSO. **a** Error curve. **b** Topography of the cost function in the PCA space

Figures 3.6 and 3.7 show the same results for CP-PSO, and Figs. 3.8 and 3.9 for the RR-PSO. These algorithms are more exploratory than GPSO. It can be observed the following:

1. CP-PSO estimated the mode of the electro-kinetic constant between -9 and -10 and the reference level in 20 (Fig. 3.6b). The shape of the convergence curve is also very noisy (Fig. 3.7a). The shape of the water head in sharper than in the GPSO case, particularly close to the well (Fig. 3.6a). The bounds also include most the observed data at the wells. This interpretation suggests the presence of a normal fault around 100 meters. In this case the curvilinear valley shape of the cost landscape is clearer seen than in the GPSO case (Fig. 3.7b).

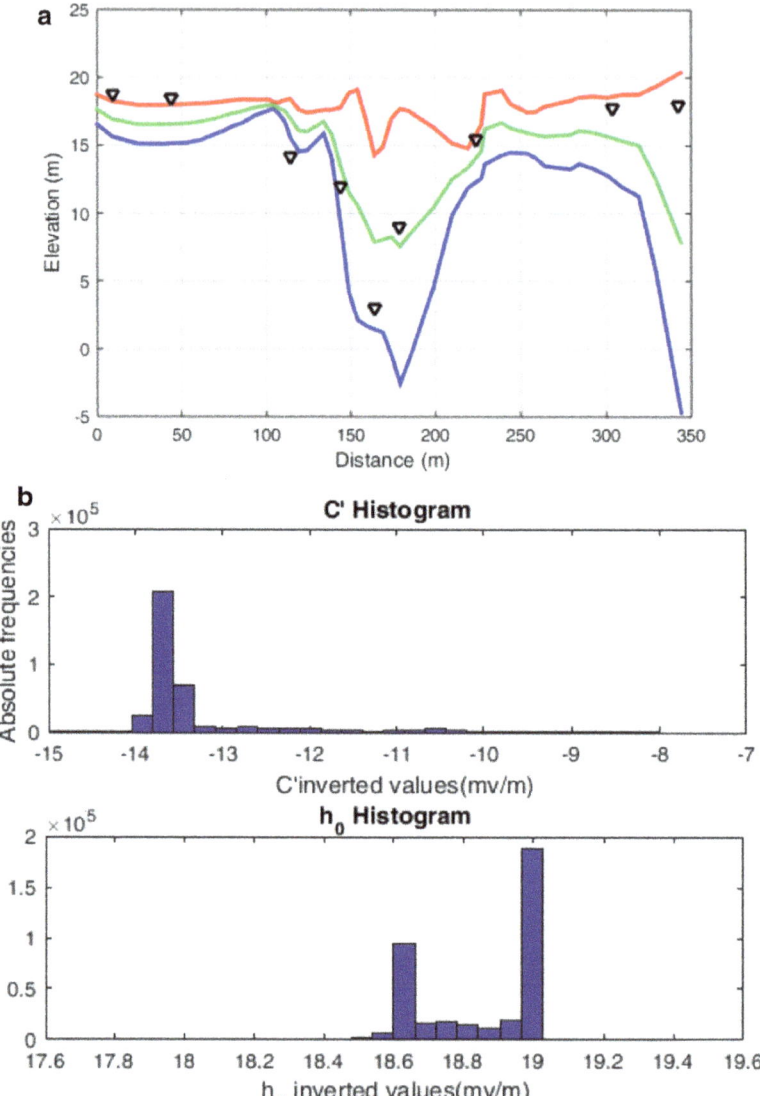

Fig. 3.8 RR-PSO. **a** Uncertainty analysis of the water table. **b** Posterior histograms of C' and h_0

2. RR-PSO estimated the mode of the Electro kinetic constant at -14 and the reference level in 19 meters (Fig. 3.8b). The convergence curve (Fig. 3.8a) is almost monotonous, showing a greater convergence speed that the other two algorithms, reaching the smallest misfit within the first 50 iterations, and the valley shape (Fig. 3.9b) is in this case very well delineated embedded in a broader area of higher misfits. The water head model and its uncertainty analysis (Fig. 3.9a) shows an interpretation closer to CP-PSO. As it has been mentioned

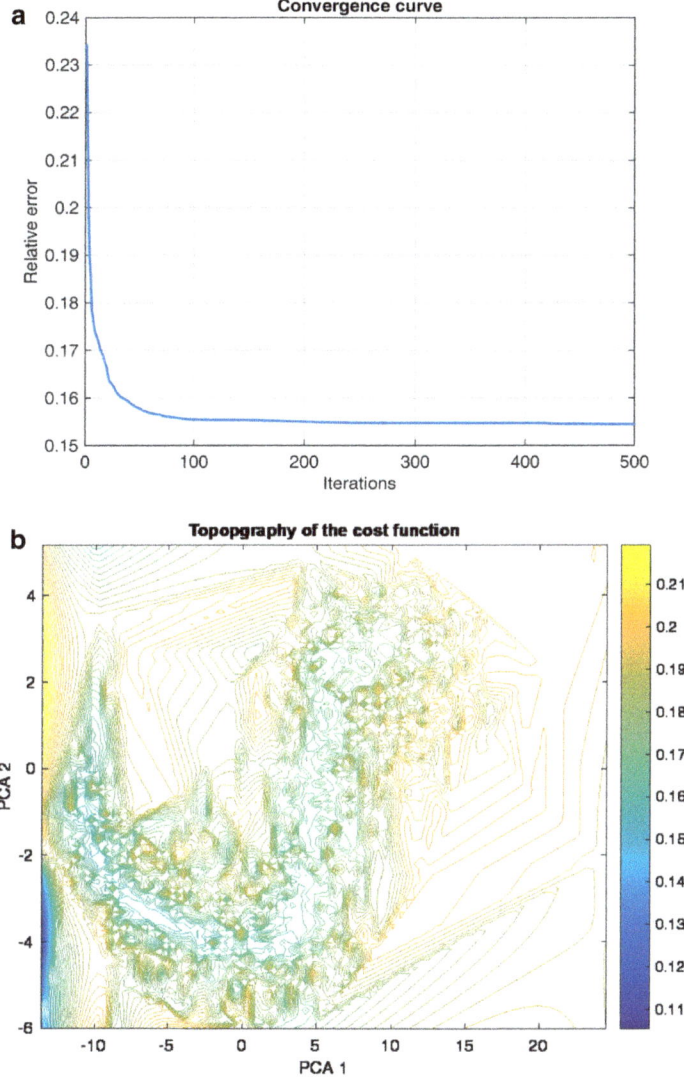

Fig. 3.9 RR-PPSO. **a** Error curve. **b** Topography of the cost function in the PCA space

RR-PSO is a very unique algorithm due to its exploratory and convergence capabilities.

3. Figures. 3.10 and 3.11 show the results for PP-PSO, CC-PSO and the 4 points algorithms. Although some results are different all the algorithms provide in terms of the water head inverted levels a similar interpretation. The highest variability is observed in the electro-kinetic constant, confirming the existing trade-off with the water level. Also the interpolated topography of the cost

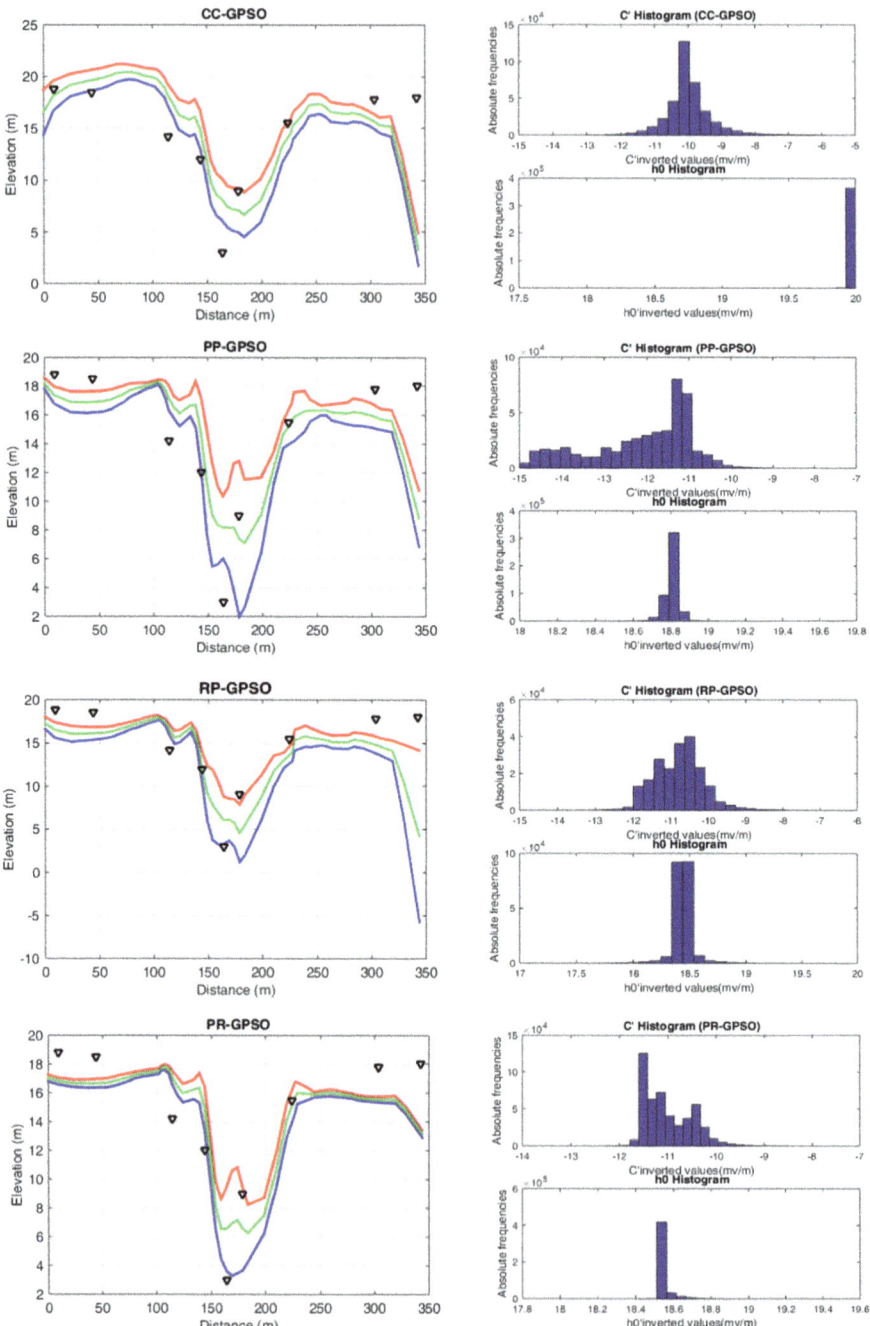

Fig. 3.10 Water head and histograms of the Electrokinectic constant and water reference level inverted by different PSO members: CC-GPSO, PP-GPSO, RP-GPSO, PR-GPSO, PC-GPSO and RC-GPSO

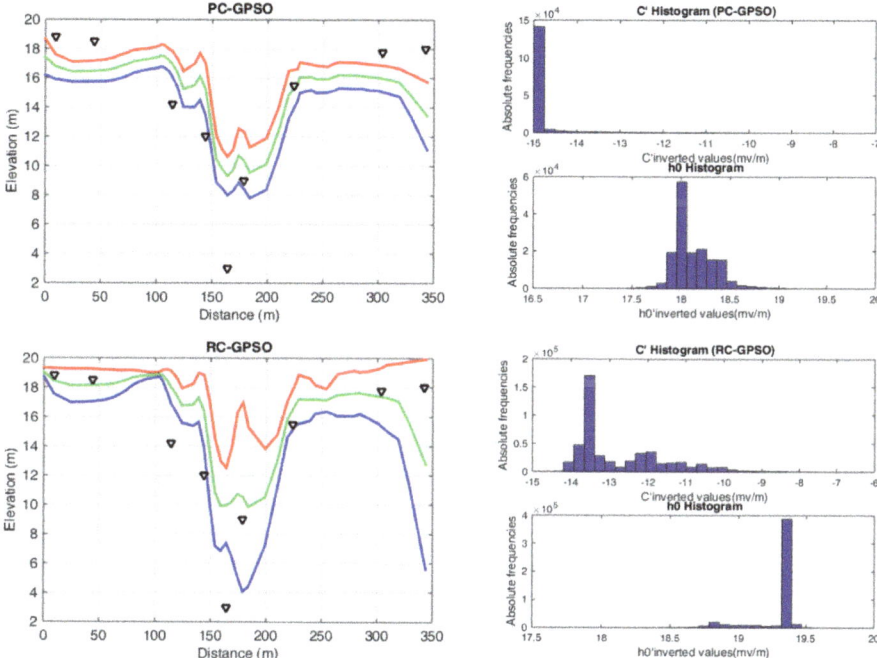

Fig. 3.10 (continued)

function obtained from the sampled models vary, but in all the cases we can observe the valley shape, in some cases better outlined (PP-GSPO and RC-GPS0) than the others.

3.7 Conclusions

The streaming potential inverse problem as many other geophysical problems is recognized to be ill-posed. In absence of prior information, local methods provide unstable solutions, which are greatly dependent on the initial guess used. We have shown that the topography of the prediction error (cost function) for the streaming potential inverse problem corresponds to elongated valleys with localized sinkholes. This flat topography is also indicative of the existing trade-offs between the model parameters. Therefore, the SP inverse problem is ill-posed as many other inverse problems, therefore uncertainty assessment of the inverse solution must be a compulsory step. In this case we have performed this task via different members of the PSO family whose stochastic stability analysis has been performed in different publications from 2008 to 2014. The sampling was performed in a reduced PCA basis generated by a set of models that have been randomly generated between some prior bounds. The results are very consistent. We can conclude that PSO is a very simple algorithm

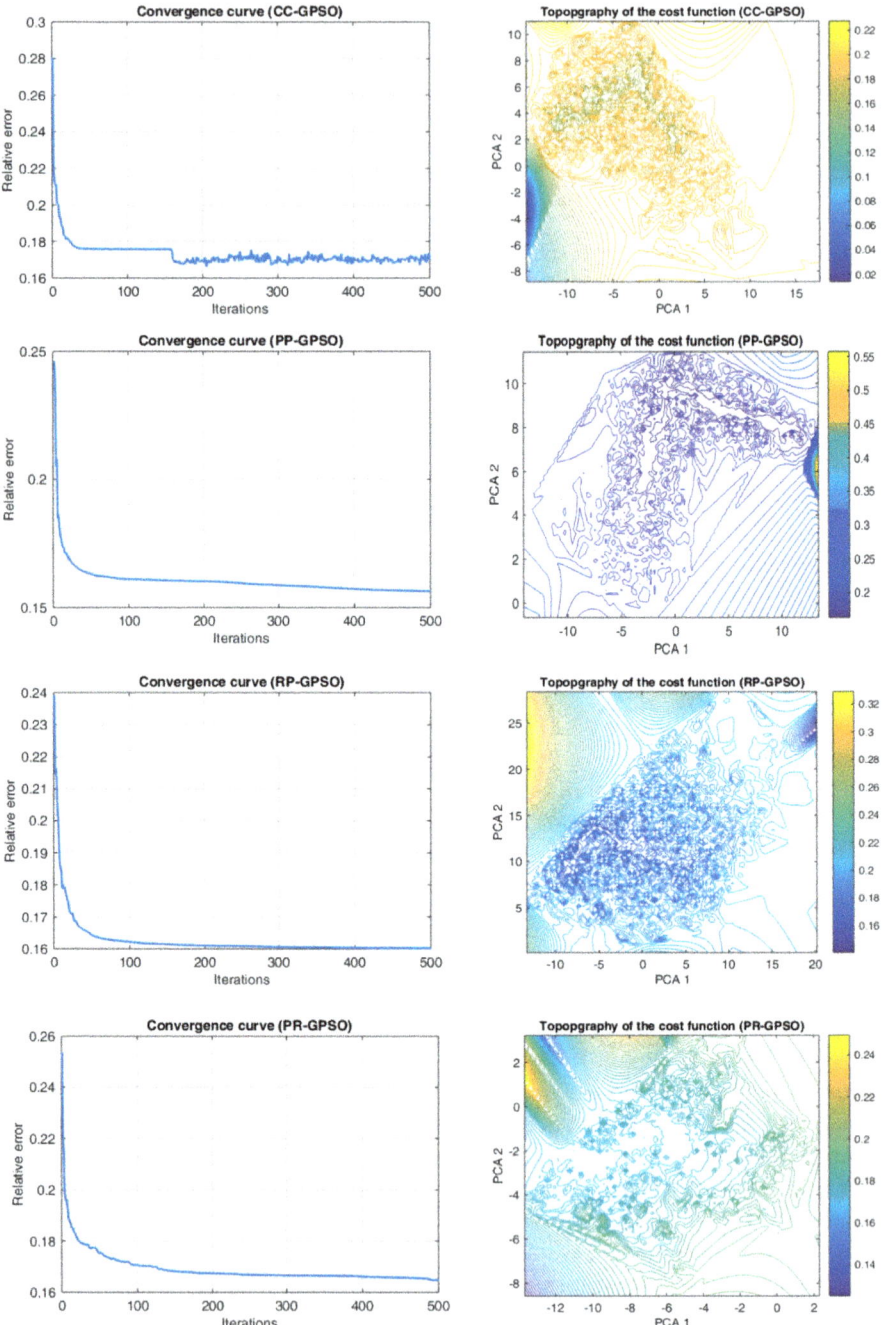

Fig. 3.11 Convergence curve and Topography of the cost function for different PSO members: CC-GPSO, PP-GPSO, RP-GPSO, PR-GPSO, PC-GPSO and RC-GPSO

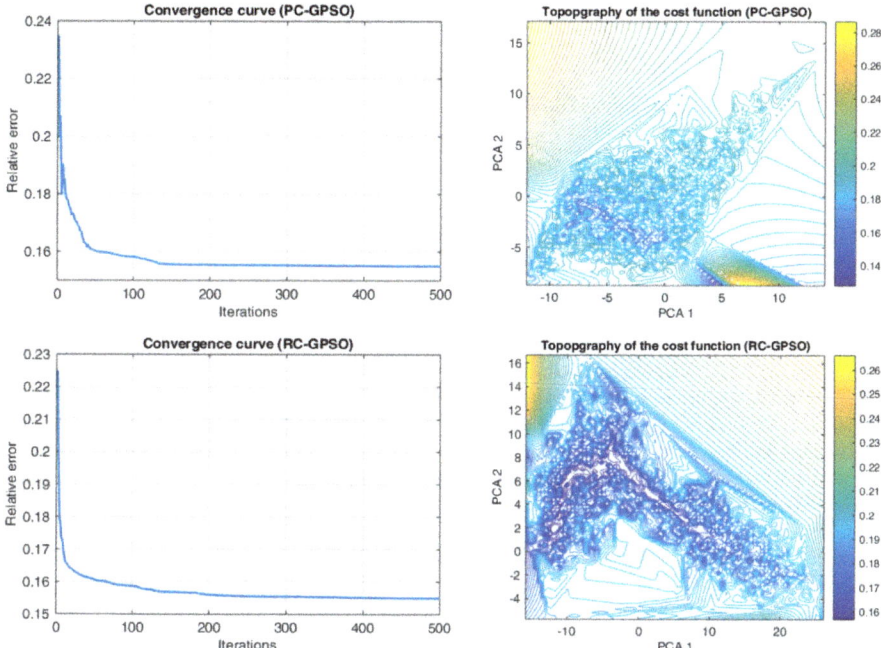

Fig. 3.11 (continued)

able to find plausible solutions and sampling the model space according to some minimum prior requirements: the search space and the regularity required for the water head model. The uncertainty analysis provided by these algorithms help to adopt the right geological decisions.

Acknowledgements We acknowledge Véronique Naudet, formerly at the Université Bordeaux 1, who introduced us to this interesting geophysical inverse problem.

References

Bigalke J, Grabner EW (1997) The geobattery model: a contribution to large-scale electrochemistry. Electrochim Acta 42:3443–3452

Birch FS (1998) Imaging the water table by filtering self-potential profiles. Ground Water 36:779–782

Bogoslovsky VV, Ogilvy AA (1973) Deformations of natural electric fields near drainage structures. Geophys Prospect 21:716–723

Bolève A, Revil A, Janod F, Mattiuzzo JL, Jardani A (2007) Forward Modeling and validation of a new formulation to compute self-potential signals associated with ground water flow. Hydrol Earth Syst Sci 11:1661–1671

Corry CE (1985) Spontaneous polarization associated with porphyry sulfide mineralization. Geophysics 50:1020–1034

Darnet M, Marquis G, Sailhac P (2003) Estimating aquifer hydraulic properties from the inversion of surface Streaming Potential (SP) anomalies. Geophys Res Lett 30(13), 1679. https://doi.org/10.1029/2003GL017631

Darnet M, Maineult A, Marquis G (2004) On the origins of self potential (SP) anomalies induced by water injections into geothermal reservoirs. Geophys Res Lette, 31. https://doi.org/10.1029/2004GL020922

Fernández-Martínez JL, García-Gonzalo E, Fernández-Alvarez JP (2008) Theoretical analysis of Particle Swarm trajectories through a mechanical analogy. Int J Comput Intell Res 4:93–104

Fernández Martínez JL, García Gonzalo E (2008) The generalized PSO: a new door to PSO evolution. J Artif Evol Applications, 1–15

Fernández-Martínez J., García Gonzalo E (2009) The PSO family: deduction, stochastic analysis and comparison. Swarm Intelligence 3:245–273

Fernández Martínez JL, García-Gonzalo E, Fernández-Álvarez JP, Kuzma HA, Menéndez-Pérez CO (2010a) PSO: A powerful algorithm to solve geophysical inverse problems. application to a 1D-DC resistivity case. J Appl Geophys 71:13–25

Fernández Martínez JL, García-Gonzalo E, Naudet V (2010b) Particle swarm optimization applied to solving and appraising the streaming-potential inverse problem. Geophysics 75(4):WA3–WA15

Fernández Martínez JL, García-Gonzalo E (2011) stochastic stability analysis of the linear continuous and discrete PSO models. IEEE Trans Evol Comput 15:405–423

Fernández Martínez JL, García-Gonzalo E (2012) Stochastic Stability and Numerical analysis of two novel algorithms of PSO family: PP-PSO and RR-PSO. Int J Artif Intell Tools 21:1240011–1240029

Fernández-Martínez JL, Fernández-Muñiz Z, Tompkins MJ (2012a) On the topography of the cost functional in linear and nonlinear inverse problems. Geophysics 77(1):W1–W15

Fernández Martínez JL, Mukerji T, García-Gonzalo E, Suman A (2012b) Reservoir characterization and inversion uncertainty via a family of particle swarm optimizers. Geophysics 77:M1–M16

Fernández Martínez JL, Fernández-Muñiz Z, Pallero JLG, Pedruelo-González LM (2013) From Bayes to Tarantola: New insights to understand uncertainty in inverse problems. J Appl Geophys 98:62–72

Fernández Martínez JL, Pallero JLG, Fernández-Muñiz Z, Pedruelo-González LM (2014a) The effect of noise and Tikhonov's regularization in inverse problems. Part I: The linear case. J Appl Geophys 108:176–185

Fernández Martínez JL, Pallero JLG, Fernández-Muñiz Z, Pedruelo-González LM (2014b) The effect of noise and Tikhonov's regularization in inverse problems. Part I: The linear case. J Appl Geophys 108:186–193

Fernández Martínez JL, Fernández-Muñiz Z (2020) The curse of dimensionality in inverse problems. J Comput Appl Math 369:112571–112584

Fitterman DV (1978) Electrokinetic and magnetic anomalies associated with dilatant regions in a layered Earth. J Geophys Res 83:5923–5928

Fournier C (1989) Spontaneous potentials and resistivity surveys applied to hydrogeology in a volcanic area: case history of the Chaine des Puys (Puy-De-Dôme, France). Geophys Prospect 37:647–668

García-Gonzalo E, Fernández Martínez JL (2014) Convergence and stochastic stability analysis of particle swarm optimization variants with generic parameter distributions. Appl Math Comput 249:286–302

Guichet X, Jouniaux L, Catel N (2006) Modification of streaming potential by precipitation of calcite in a sand-water system: laboratory measurements in the pH range from 4 to 12. Geophys J Int 166:445–460

Holland JH (1992) Adaptation in natural and artificial systems. MIT Press

Jardani A, Revil A, Barrash W, Crespy A, Rizzo E, Straface S, Cardiff M, Malama B, Miller C, Johnson T (2009) Reconstruction of the water table from self-potential data: a bayesian approach. Ground Water 47:213–227

Kennedy J, Eberhart RC (1995) Particle swarm optimization: proceedings IEEE International Conference on. Neural Networks 4:1942–1948

Kirkpatrick S, Gelatt CD, Vecchi MP (1983) Optimization by simulated annealing: science 220:671–680

Linde N, Revil A, Bolève A, Dagès C, Castermant J, Suski B, Voltz M (2007) Estimation of the water table throughout a catchment using self-potential and piezometric data in a Bayesian framework. J Hydrol 334:88–98

Maineult A, Strobach E, Renner J (2008) Self-potential signals induced by periodic pumping tests, J Geophys Res 113:B01203. https://doi.org/10.1029/2007JB005193

Minsley BJ, Sogade J, Morgan FD (2007) Three-dimensional source inversion of self-potential data. J Geophys Res 112:B02202. https://doi.org/10.1029/2006JB004262

Morgan FD, Williams ER, Madden TR (1989) Streaming potential properties of westerly granite with applications. J Geophys Res 94:12449–12461

Mosegaard K, Tarantola A (1995) Monte Carlo sampling of solutions to inverse problems. J Geophys Res 100:12431–12447

Naudet V, Revil A, Rizzo E, Bottero JY, Bégassat P (2004) Groundwater redox conditions and conductivity in a contaminant plume from geoelectrical investigations. Hydrol Earth Syst Sci 8:8–22

Naudet V, Revil A (2005) A sandbox experiment to investigate bacteria-mediated redox processes on self-potential signals. Geophys Res Lette 32(11). https://doi.org/10.1029/2005gl022735

Pallero JLG, Fernández-Martínez JL, Bonvalot S, Fudym O (2015) Gravity inversion and uncertainty assessment of basement relief via Particle swarm optimization. J Appl Geophys 116:180–191

Pallero JLG, Fernández-Martínez JL, Bonvalot S, Fudym O (2017) 3D gravity inversion and uncertainty assessment of basement relief via Particle Swarm Optimization. J Appl Geophys 139:338–350

Revil A, Naudet V, Nouzaret J, Pessel M (2003) Principles of electrography applied to self-potential electro-kinetic sources and hydrogeological applications. Water Resour Res 39:1114. https://doi.org/10.1029/2001WR000916

Sailhac P, Marquis G (2001) Analytic potentials for the forward and inverse modeling of SP anomalies caused by subsurface fluid flow. Geophys Res Lett 28:1851–1854

Sambridge M (1999a) Geophysical inversion with a neighbourhood algorithm–I: searching a parameter space. Geophys J Int 138:479–494

Sambridge M (1999b) Geophysical Inversion with a Neighbourhood Algorithm–II: appraising the ensemble. Geophys J Int 138:727–746

Scales JA, Tenorio L (2001) Prior information and uncertainty in inverse problems. Geophysics 66:389–397

Storn R, Price K (1997) Differential evolution—a simple and efficient heuristic for global optimization over continuous spaces. J Global Optim 11:341–359

Chapter 4
A Comparison of the Model Parameter Estimations from Self-Potential Anomalies by Levenberg-Marquardt (LM), Differential Evolution (DE) and Particle Swarm Optimization (PSO) Algorithms: An Example from Tamış-Çanakkale, Turkey

Petek Sindirgi and Şenol Özyalin

Abstract In geophysics, it is particularly important to choose an adequate optimization algorithm for parameter estimation. In this study, the success of Levenberg-Marquardt (LM), Differential Evolution (DE) and Particle Swarm Optimization (PSO) inversion algorithms has been tested by applying to the synthetic and field self-potential (SP) anomalies. Even though it is not preferred to compare derivative-based algorithms with metaheuristics, thanks to a LM-based limitation procedure first proposed in this study, a comparison could be realized. First, a synthetic SP data have been inverted by LM, DE and PSO algorithms. Then, SP field data set collected from Tamış-Çanakkale, Turkey was evaluated by the same algorithms. The estimated model parameters by these algorithms were compared with each other. We also inverted vertical electrical sounding (VES) data set collected from the same region, and an earth model was constructed by using both SP and VES methods. The results from each geophysical method point out the same location for a fault. Based on these studies, it can be concluded that DE, PSO, and LM algorithms may be confidently used in SP modelling studies.

Keywords Differential evolution · Levenberg-Marquardt · Particle swarm optimization · Self-potential · Vertical electrical sounding

P. Sindirgi (✉) · Ş. Özyalin
Faculty of Engineering, Department of Geophysical Engineering, Dokuz Eylül University, İzmir, Turkey
e-mail: petek.sindirgi@deu.edu.tr

Ş. Özyalin
e-mail: senol.ozyalin@deu.edu.tr

4.1 Introduction

Electrical methods are frequently used to detect the location of systems including the groundwater. Self-potential (SP) and vertical electrical sounding (VES) are proven methods to be successful in groundwater explorations such as groundwater pollution studies, fresh, saltwater interference problems, and geothermal exploration (Ogilvy et al. 1969; Corwin and Hoover 1979; Schiavone and Quarto 1984; Hamzah et al. 2007; Karlık and Kaya 2001).

The VES technique is used to determine the resistivity changes from the surface to the depth. It is mainly based on the principle of measuring the response of the earth to an electric current applied to the ground. The VES method is useful in determining the depth, geometry and resistivity of the layers (Hamzah et al. 2007; Kaya et al. 2015).

Self-potential is an electrical phenomenon that is so easy to measure but it is also so hard to determine the source mechanism. These mechanisms can be specified as electro-kinetic (streaming), thermo-electric, diffusion, and electro-chemical potential. Self-potential method can be applied for determining the possible faults containing fluid in the study area (Yüngül 1950; Fitterman and Corwin 1982; Corwin 1990; Monteiro Santos et al. 2002; Revil et al. 2003). Potential anomalies created by fluid-containing faults are generally generated by electro-chemical sources.

SP anomalies can be analysed by different approaches. Since the use of the graphic-based evaluation methods (Yüngül 1950; Paul 1965; Rao et al. 1970), a new generation numerical methods have been developed for the evaluation of SP data in parallel with developing computer technology: The Fourier, Hartley, Hilbert Transforms and Wavelet analysis (Sundararajan et al. 1990; Asfahani et al. 2001; Gilbert and Pessel 2001; Al-Garni and Sundararajan 2011; Di Maio et al. 2016), Euler Deconvolution (Agarwal and Srivastava 2009; Sındırgı and Özyalın 2019), Gradient and Derivative Analysis (Abdelrahman et al. 1997, 1998, 2006; El-Araby 2004; Essa et al. 2008; Sındırgı et al. 2008; Abedi et al. 2012; Mehanee 2015), tomographic approach (Di Maio and Patella 1994; Patella 1997; Revil et al. 2001; Juliano et al.,2002), Artificial Neural Network algorithms (El-Kaliouby and Al-Garni 2009; Kaftan et al. 2014), and metaheuristic algorithms including Particle Swarm Optimization (PSO) (Juan et al. 2010, Monteiro Santos 2010; Göktürkler et al. 2016; Ekinci et al. 2019; Pekşen et al. 2011), Simulated annealing (SA) (Sharma 2012; Biswas and Sharma 2014, 2015), Genetic Algorithm (GA) and Differential Evaluation (DE) (Abdelazeem and Gobashy 2006; Fernández-Martínez et al. 2010; Göktürkler and Balkaya 2012; Di Maio et al. 2017; Ekinci et al. 2019).

In this study, a synthetic (noise-free and noisy) and a field SP data set (collected from Tamış-Çanakkale, Turkey) have been evaluated by three algorithms including the Levenberg-Marquardt (LM), PSO and DE. Also, to be able to compare LM to the metaheuristics, a new initial model selection process for LM was developed. Then, the estimated parameters have been compared with each other. The VES data set, collected from the same location, is also inverted, and the subsurface model for Tamış-Çanakkale anomaly has been constructed by combining the results from both VES and SP data.

4.2   Materials and Methods

4.2.1   Formulation of the SP Anomaly

Let $V(x, x_0, z_0, K, \theta, q)$ be the SP anomaly produced by a simple polarized causative body observed at any point on the earth's surface (Fig. 4.1). Formulation of the SP anomaly (Yüngül 1950; Murty and Haricharan 1985) can be written as;

$$V(x, x_0, z_0, K, \theta, q) = K \frac{(x - x_0)cos\theta + z_0 sin\theta}{\left[(x - x_0)^2 + z_0^2\right]^q} \tag{4.1}$$

where K is the electric dipole moment, x is the horizontal distance, x_0 is the distance from the origin, z_0 is the depth of the centre of the body, θ is the polarization angle, and q is the shape factor. The shape factor is dimensionless and its value for a sphere, horizontal cylinder, and semi-infinite vertical cylinder are 1.5, 1.0, and 0.5, respectively. The shape factor becomes near to zero as the structure approaches a horizontal sheet.

Fig. 4.1 An infinitely long horizontal cylinder model and its noise-free and noisy anomalies (The model parameters and their corresponding values are listed on the figure)

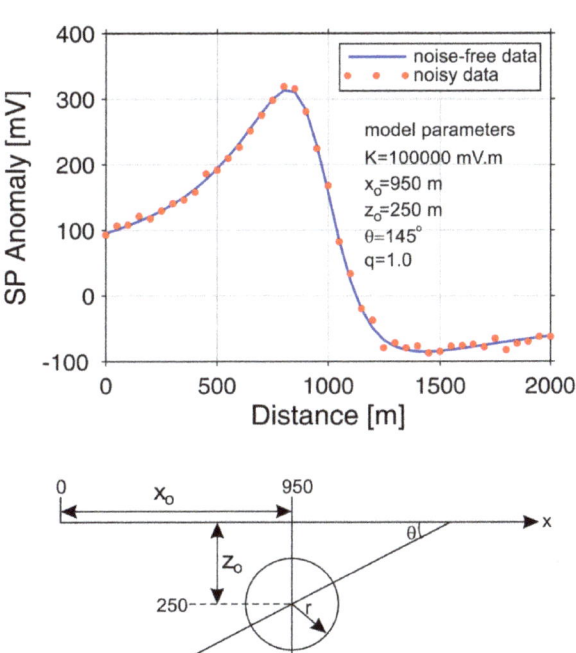

4.2.2   Algorithms

4.2.2.1   Levenberg-Marquardt (LM) Inversion Algorithm

Nonlinear least squares problems can be solved using the LM algorithm. The solution from the LM algorithm is not necessarily to be a global minimum. Generally, the LM algorithm is often preferred to Gauss-Newton and Steepest-Descent methods because it guarantees good convergence and non-singularity of the solution. Kenneth Levenberg introduced this algorithm in (1944), and Donald Marquardt (1963) improved it subsequently.

Generalized formulation of forward modelling problems can be written as,

$$d = G(m) \tag{4.2}$$

where d is the M-dimensional vector of observations and, m is a vector of model parameters (in previous study model parameters are x_0, z_0, K, θ, q) with the size of N×1. $G(m)$ is a nonlinear function predicted by the model. Model parameter m can be written by

$$m = m_0 + \Delta m \tag{4.3}$$

where m_0 is the initial model and Δm is the model parameters update.

Minimizing the model perturbation to the Gauss-Newton solution can be fulfil via minimizing the objective F,

$$F = (d - G(m_0 + \Delta m)) + \lambda \Delta m^2 \tag{4.4}$$

The sensitivity (Jacobian) matrix $J(M \times N)$ can be written as,

$$J = \begin{bmatrix} \frac{\partial G_i(m)}{\partial m_1} & \cdots & \frac{\partial G_i(m)}{\partial m_N} \\ \vdots & \ddots & \vdots \\ \frac{\partial G_M(m)}{\partial m_1} & \cdots & \frac{\partial G_M(m)}{\partial m_N} \end{bmatrix}$$

and using sensitivity matrix, Δm can be defined as

$$\Delta m = \left[J^T W J + I \right]^{-1} J^T W (d - G(m_0)) \tag{4.5}$$

where I is the identity matrix and λ is a damping factor shows the effect of model perturbation. If λ is small, Eq. (4.4) will become equal to Gauss-Newton solution equation. Generally, the initial value of λ is chosen large. If misfit is smaller than previous iteration λ is reduced, if not it is increased. W is a positive definite matrix and defined as (Jupp and Vozoff 1975)

$$W = \frac{1}{M}d_i^2, \quad i = 1, 2, \ldots, M \tag{4.6}$$

RMSE, which means the standard deviation of the residuals, is calculated as follows (Barnston 1992):

$$RMSE = \sqrt{\frac{1}{M}\sum_{i=1}^{M}(d_i - G(m)_i)^2} \tag{4.7}$$

4.2.2.2 Particle Swarm Optimization (PSO) Algorithm

The PSO was proposed in 1995 by the authors Kennedy and Eberhart. It is a population-based metaheuristic technique and is based on the social behaviour of animals (birds, fishes). While each individual searching for the solution in PSO is called a *particle*, the population of the particles is called a *swarm*. Particles move according to two important parameters in the search space. *Pbest* is the particle's best position found so far and *Gbest* is global best position found thus far in the entire swarm. According to these definitions, basic steps of the PSO algorithm can be listed as follows: (1) The algorithm is initialized by placing the particles with random velocities (v) and positions (x) in the search space. (2) *The fitness value* is used to understand how close a particle is to the solution. It is calculated for each particle. (3) Individual and global bests are updated by comparing them with the previous ones (pbest$_i$) and equalized to the current value of the fitness. Then the particle's prior best position (p_i) is assumed to be as the current position (x_i). The determined position of the particle with the best fitness value so far is assigned as the global best (g_i). (4.4) New velocity and position values are updated for each particle. (4.5) Stopping criterion is checked, if could not reach the threshold values, it is continued with step (4.2) (Fig. 4.2).

The position and velocity of a particle i can be updated as follows ($i = 1, 2, 3, \ldots, N$);

$$v_i = \omega v_i + c_1 rand()(p_i - x_i) + c_2 rand()(g_i - x_i) \tag{4.8}$$

$$x_i = x_i + v_i \tag{4.9}$$

where w is a weighting factor ($0 < \omega < 1$) known as inertia weight; c_1 and c_2 are individual and social behaviour coefficients, respectively. *rand()* is a function to generate pseudo-random numbers within [0, 1]. The updates of position and velocity of each particle end after reaching the stopping criterion (Kennedy and Eberhart 1995; Shi and Eberhart 1998; Poli et al. 2007; Luke 2009; Salmon 2011). In the light

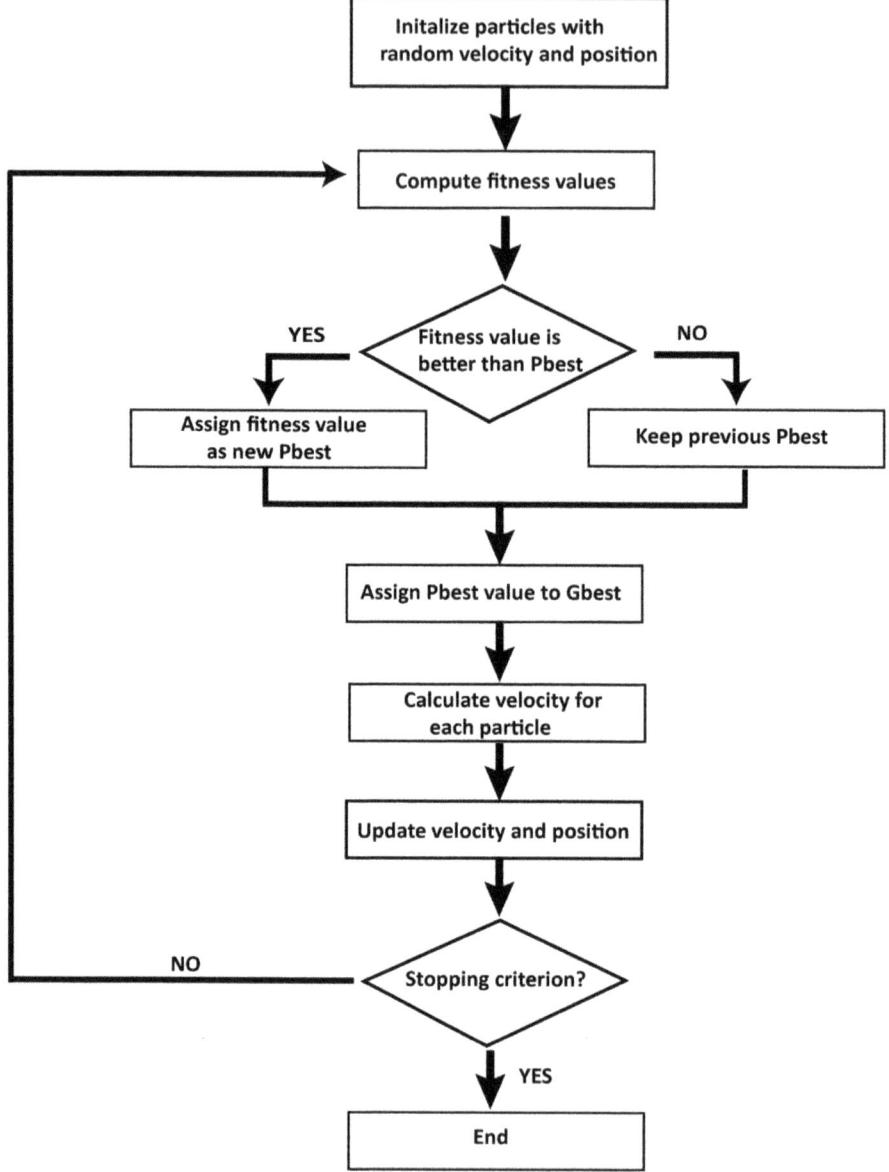

Fig. 4.2 The flow chart of the particle swarm optimization (Adopted from http://mnemstudio.org/particle-swarm-introduction.htm)

Table 4.1 The values of the parameters used in PSO and DE algorithms

DE parameters	Values	PSO parameters	Values
Number of parameter (D)	5	Number of parameter (D)	5
Population size and Weighting factor (F)	100 0.7	Particle number Inertia weight (ω)	100 1
Max. generation number (G)	100	Max. generation number (G)	100
Max. number of run	10	Max. number of run	10
Value to reach (VTR) (mV)	1e-11	Value to reach (VTR) (mV)	1e-11
Crossover probability (Cr)	0.9	Cognitive and social scaling factors (c1 and c2)	c1 = 2 c2 = 2

of this information, the values of the parameters used in PSO algorithm in this study are listed in Table 4.1.

4.2.3 Differential Evolution (DE) Algorithm

DE algorithm (Storn and Price 1995, 1997) is a population-based optimization algorithm and its applications in geophysics have increased in recent years. Different from the conventional gradient-based inversion methods, a good starting model is not a requirement for the DE algorithm to reach the global minimum. Three control parameters are the only requirements: number of population (Np), weighting factor (mutation constant, F) and crossover probability (Cr). The initial population is generated randomly in the initialization stage of the algorithm, then in the evolution stage population evolves from one generation to the next through mutation, crossover and selection operations until the termination criterion is satisfied (Fig. 4.3) (Li and Yin 2012; Ekinci et al. 2016).

The target vectors can be defined as $x_{i,G} = \left(x_{i,G}^1, x_{i,G}^2, \ldots, x_{i,G}^D\right), i = 1, 2, \ldots, Np$, where G is the current generation, and D is the number of parameters ($j = 1, 2, \ldots, D$). The jth component of the ith vector can be generated as follows:

$$x_{i,G}^j = x_l^j + rand().\left(x_u^j - x_l^j\right) \tag{4.10}$$

where $rand()$ symbolizes pseudo-random number between [0,1), also l and u are the lower and upper limits for each parameter.

The evolution cycle includes mutation, crossover and selection operations (Fig. 4.3). Mutation operation is run to form a donor (mutant) vector, $v_{i,G} = \left(v_{i,G}^1, v_{i,G}^2, \ldots, v_{i,G}^D\right), i = 1, 2, \ldots, Np$, for each target vector. Generally, there are five differential mutation strategies (Li and Yin 2012). Previous studies (Balkaya 2013; Ekinci 2016; Ekinci et al. 2017, 2019) are indicated that DE/best/1/bin supplies better solutions with a good estimation accuracy and less computing time for the

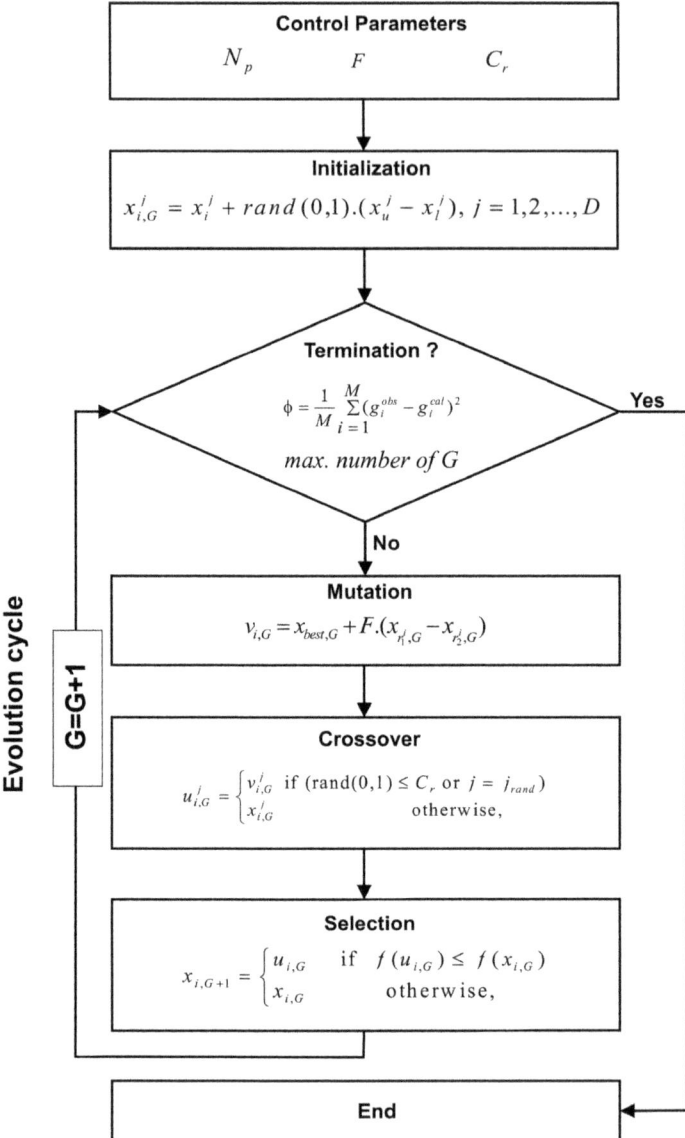

Fig. 4.3 The flow chart of the DE optimization algorithm (from Ekinci et al. 2016)

inversion of geophysical data sets. This strategy is preferred in the DE optimizations of the synthetic and field SP data in this work. Mutation operation for this strategy can be defined as below:

$$v_{i,G} = x_{best,G} + F.\left(x_{r^i_{1,G}} - x_{r^i_{2,G}}\right) \qquad (4.11)$$

Here, $x_{best,G}$ is the best individual vector in the population at generation G, and (x_{r_1}, x_{r_2}) is a pair of differential vectors.

Then, the trial vector $(u_{i,G})$ is produced by a recombination of the donor vector $(v_{i,G})$ and the target vector $(x_{i,G})$. The trail vector of the jth particle in the ith dimension at the G th iteration can be written as:

$$u_{i,G}^j = \begin{cases} v_{i,G}^j \ if \ (rand(0,1) \le Cr \ or \ j = j_{rand}) \\ x_{i,G}^{jk} \qquad\qquad otherwise. \end{cases}, \quad j = 1, 2, \ldots, D \quad (4.12)$$

where Cr is a crossover rate in the range $[0,1]$ and j_{rand} is a randomly chosen integer in the range $[1, D]$.

Selection operator is employed to select the next generation between the trial and target vectors.

$$x_{i,G+1} = \begin{cases} u_{i,G} if f(u_{i,G} \le f(x_{i,G}) \\ x_{i,G} otherwise. \end{cases} \qquad (4.13)$$

If the new generated trial vector gives a better fitness value than its previous one, the target vector is updated by using Eq. 13, else it is kept in the present population. The fitness value is calculated for each particle from the objective function, and the particle with the best value is selected as the solution in the current generation.

Evolution cycle ends when a predefined termination criterion is met. This criterion can be error energy, and/or maximum number of G. So, the vector yielding the lower error energy value is chosen as an optimum solution for the optimization problem. In this study both termination criteria were used.

For the number of M data, the objective function (Relative Error) can be calculated as follows:

$$\phi = \frac{1}{M} \sum_{i=1}^{M} \left(g_i^{obs} - g_i^{cal}\right)^2 \qquad (4.14)$$

where g^{obs} and g^{cal} are the observed and calculated data, respectively, and i indicates the observations. The square root of the Eq. (4.14) gives the Root Mean Square (RMS) value.

4.2.4 Parameter Estimation Studies

Synthetic noise-free and noisy (5%) infinitely long horizontal cylinder-shaped SP model anomalies are generated to test the parameter solution quality of the proposed algorithms. Then, to better analyse the pertinence of the suggested algorithms on real data, they applied to four SP profile data, which are selected from Tamış-Çanakkale SP anomaly.

The values of the DE and PSO parameters used during the test and field studies are summarized in Table 4.1. The codes for all algorithms are written in MATLAB® (ver.R2019a) software with a 3.10 GHz compatible computer with 6 GB memory.

4.2.5 Synthetic Examples

First, to test the parameter solution quality of the proposed algorithms, synthetic noise-free and noisy anomalies based on an infinitely long horizontal cylinder-shaped model were generated. The parameters used for this model were selected as $K = 100\,000$ mV.m, $z_0 = 500$ m, $x_0 = 950$ m, $x_0 = 145°$, $q = 1$, and profile length is 2000 m (assuming 50 m sampling interval) (Fig. 4.1). To calculate the noisy synthetic model, 5% Gaussian noise, were added to the synthetic data (Fig. 4.1). Thereafter the proposed algorithms have been applied for estimating the model parameters of the SP source body.

Local optimization (gradient-based) algorithms requires choose the initial parameter values close to the true solution, otherwise the algorithm may end up with a local minimum instead of a global one. To cope with this problem, a new approach to assign the initial values to the LM inversion algorithm has been introduced in this study. For this purpose, similar to the population-based metaheuristic methods, a set including 100 different models for SP have been generated randomly within certain ranges (Table 4.2), then objective function values for each model have been calculated by forward solution. Among these models the one with the lowest error energy has been taken as the initial model for LM. Finally, a LM inversion has been carried based on this initial model. Optionally this procedure may be repeated several times (Göktürkler and Balkaya 2012; Li and Yin 2012; Balkaya 2013), the one with the lowest error energy can be assigned as the solution.

Table 4.3 illustrates the initial models by the above mentioned routine for LM algorithm for noise-free and noisy SP data sets. As can be seen from the table the noisy data set produced larger RMS value as expected. The Tables 4.4 and 4.5 give the results of the parameter estimations by the LM, DE and PSO with both the noise-free

Table 4.2 Parameter ranges used to select LM algorithm initial parameters and generate the initial models by PSO and DE of noise-free and noisy synthetic SP anomalies	Parameters	True	Search Space	
			Minimum	Maximum
	x_0 (m)	950	500	1000
	z_0 (m)	250	100	500
	θ (°)	145	0	180
	K (mV. m$^{(2q-1)}$)	100000	10000	250000
	q	1	0.5	1.5

Table 4.3 Estimated initial SP model parameters for LM algorithm by the proposed approach. This approach has been repeated 10 times, and the model having the lowest error energy has been taken as the initial model

Anomaly	RUN	Parameters					
		x_0(m)	z_0(m)	θ (°)	K (mV. m$^{(2q-1)}$)	q	RMS (mV)
Noise-free	2	904.82	465.38	142.67	167640.86	1.01	2197.28895
Noisy (5%)	2	721.55	341.32	84.28	120239.64	1.08	11271.51951

Table 4.4 The best solutions from noise-free synthetic data set by three algorithms at the end of 10 independent runs

Algorithm	RUN	Parameters					
		x_0(m)	z_0(m)	θ (°)	K (mV. m$^{(2q-1)}$)	q	RMS (mV)
DE	4	950.00	250.00	145.00	100020.65	1	0.00251022
PSO	6	949.72	248.69	144.84	95110.06	1	0.13
LM	2	950.00	250.00	144.77	100015.32	1	0.67

Table 4.5 The best solutions from noisy synthetic data set by three algorithms at the end of 10 independent runs

Algorithm	RUN	Parameters					
		x_0(m)	z_0(m)	θ (°)	K (mV. m$^{(2q-1)}$)	q	RMS (mV)
DE	7	949.34	241.83	143.51	67884.33	0.97	5.31
PSO	10	950.32	246.03	144.04	78618.48	0.98	5.32
LM	2	949.33	241.78	143.97	67762.78	0.97	5.47

and noisy SP anomalies. The comparisons of the synthetic and calculated anomalies are illustrated in Fig. 4.4. When Tables 4.4 and 4.5 are compared for noise-free data (Fig. 4.4a–c), it is observed that the algorithms generated similar results in the vicinity of true model parameters. On the other hand, the results for noisy data sets (Fig. 4.4d–f) are deviated from the true model parameters. Based on Fig. 4.4 and Tables 4.4 and 4.5, it can be said that the DE algorithm is relatively better than the others for both noise-free and noisy data sets. The behaviour of parameters and error energy variations of DE solutions are only presented in figure form (Fig. 4.5), in order to save some space in the text.

4.2.6   Field Example

Çanakkale is tectonically active region on the Alpine-Himalayan Mountain Belt that corresponds to the northward movement of the Arabian plate and located in the middle segment of the NAF zone (Altınok et al. 2012). The main fault systems of the region

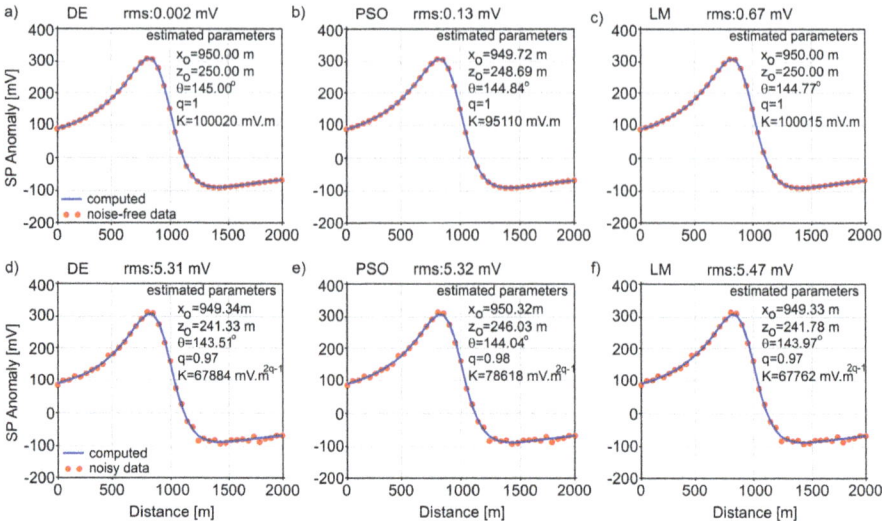

Fig. 4.4 Synthetic data: **a–c** Noise-free, **d–f** noisy data. Calculated anomalies from DE, PSO and LM algorithms (The estimated best-fitting parameters are listed on the figures)

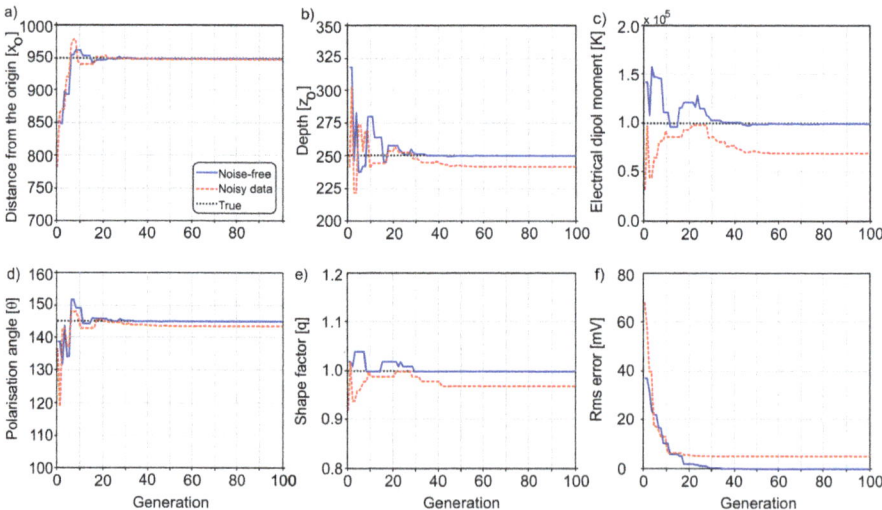

Fig. 4.5 The convergence characteristics of the DE algorithm. **a** Amplitude K, **b** the distance from the origin x_0, **c** depth z_0, **d** polarization angle θ, **e** Shape factor q

are Balabanlı, Kestanbol, Tuzla and Edremit Faults. There are a number of geothermal fields (Tuzla, Palamutova, Kestanbol, Küçükçetmi geothermal fields etc.) related with these faults in the study area. The field data sets including SP and VES in this study were collected near a segment of the Tuzla fault system. It represents the transition

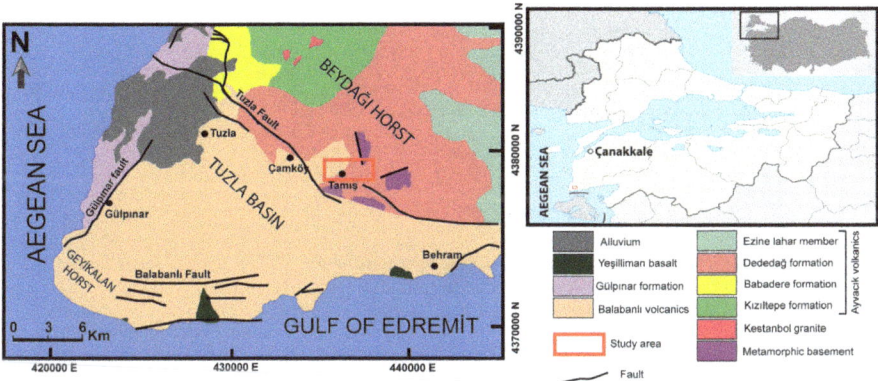

Fig. 4.6 Geological map of the study area (adapted and modified from Karacık and Yılmaz 1998; Sözbilir et al. 2018)

zone between the Beydağı Horst and Tuzla Basin. Geological units of the study area are the Balabanlı volcanics, Dededağ formations, and Karadağ metamorphics. The Balabanlı volcanics consist of pyroclastic rocks such as rhyodacitic ignimbrites and lavas. The Dededağ formation contains andesitic and trachyandesitic lavas and flow-breccias. The Balabanlı volcanics and Dededağ formation lie over the metamorphic basement (Karacık and Yılmaz 1998; Sözbilir et al. 2018) (Fig. 4.6).

The SP contour map and the superimposed locations of the VES measurements are shown in Fig. 4.7a. The VES method was carried out at five stations using the Schlumberger array. They have been inverted by a software based on a least-squares approach (IPI2WIN), and the inverted resistivity values can be seen in Fig. 4.7b. They indicate two distinct units. The first one is the surface volcanics characterized by low resistivities (10–50 Ωm), and the second one is the metamorphic units (having resistivities of 50–200 Ωm) forming the basement. It is seen that the depth to the basement ranges between approximately 350-600 m from the station VES-1 to VES-4, and the depth to the basement rock is approximately 180 m at the station VES-5. The difference between the depths may be explained as the effect of the Tuzla Fault System.

Four different profiles (P1, P2, P3, and P4) were selected for inversion (Fig. 4.7a). They have been evaluated by LM, PSO, and DE algorithms. Search spaces for these algorithms are given in Table 4.6. The procedure of assigning initial values for the LM inversion algorithm introduced in the present study (see Sect. 3.2) has also been applied to the Tamış-Çanakkale data set (Table 4.7). The same values for the algorithm-based parameters as the synthetic data evaluation were also used for the field data set. The measured and calculated data from SP profiles are given in Figs. 4.8, 4.9, 4.10 and 4.11. Tables 4.8, 4.9, 4.10 and 4.11 show the results of the model parameter estimations from the field data sets. Similar to the synthetic data, LM, PSO, and DE algorithms have been executed 10 times and the one has the minimum RMS value has been selected as the best-fitting model (Tables 4.8, 4.9, 4.10 and

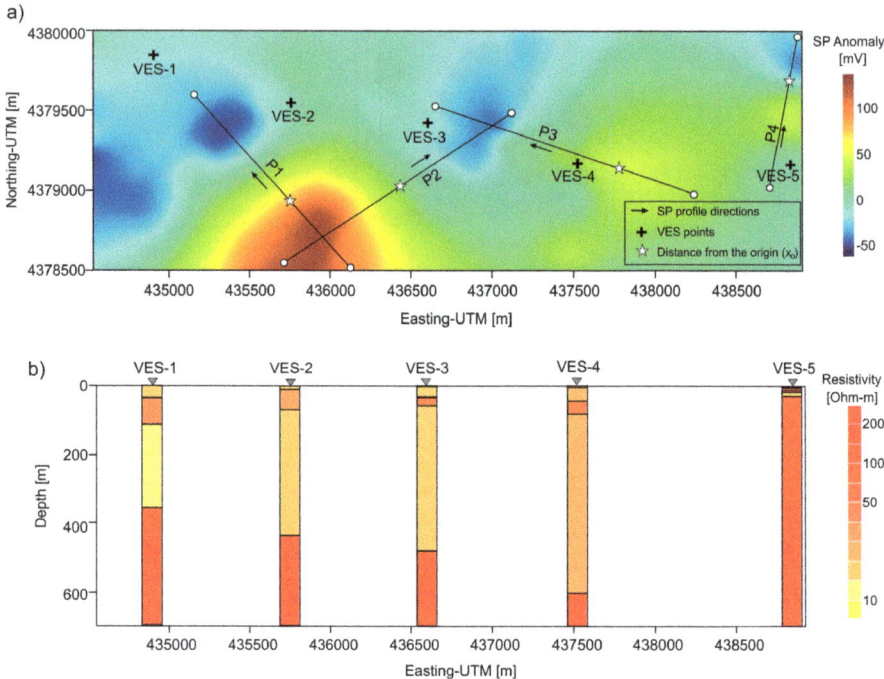

Fig. 4.7 **a** SP anomaly map, location of selected SP profiles (P1, P2, P3, and P4), VES stations (VES-1, VES-2, VES-3, VES-4, and VES-5), and estimated SP body locations (stars). **b** Layered earth models obtained from VES studies

Table 4.6 Parameter ranges used in LM, PSO and DE optimizations of the Tamış-Çanakkale anomalies

Parameters	Search Space	
	Minimum	Maximum
x_0 (m)	100	1500
z_0 (m)	100	1000
θ (°)	0	180
K (mV.m$^{(2q-1)}$)	10000	750000
Q	0.5	1.5

4.11). Although DE and PSO algorithms have smaller RMS errors than does LM with the help of the initial model determination procedure developed for the LM algorithm in this study, it is seen that the parameters are also successfully predicted with LM. When the tables are examined, it can be seen that all algorithms provided similar z_0 values (~500–700 m) for the SP profiles, except profile P4. On the other hand, the algorithms have determined a smaller z_0 values (~185 m) for P4. The calculated average depths (z_0) and origin to distances (x_0^{P1}, x_0^{P2}, x_0^{P3} and x_0^{P4}) of the SP body using by the algorithms are shown in Fig. 4.12. It is seen that there is a

Table 4.7 The initial SP parameters by the proposed approach in the present study for LM inversion of the Tamış-Çanakkale data set. This approach has been repeated 10 times, and the model having the lowest error energy has been taken as the initial model

Profile	RUN	Parameters					
		x_0(m)	z_0(m)	θ (°)	K (mV. m$^{(2q-1)}$)	q	RMS (mV)
P1	10	445.77	820.21	147.64	635087.96	1.12	272.87181
P2	9	1038.32	323.03	156.4	390919.62	1.32	551.5644
P3	4	665.56	261.31	174.41	311517.25	1.34	294.73764
P4	10	629.73	783.03	144.68	605134.85	1.1	1344.37074
Algorithm	RUN	Parameters					
		x_0(m)	z_0(m)	θ (°)	K (mV. m$^{(2q-1)}$)	q	RMS (mV)
DE	4	557.3	504.52	145.81	750000	1.18	3.22
PSO	4	556.71	500.43	145.55	634540.45	1.16	3.23
LM	10	557.43	505.62	146.37	786066.39	1.18	3.32

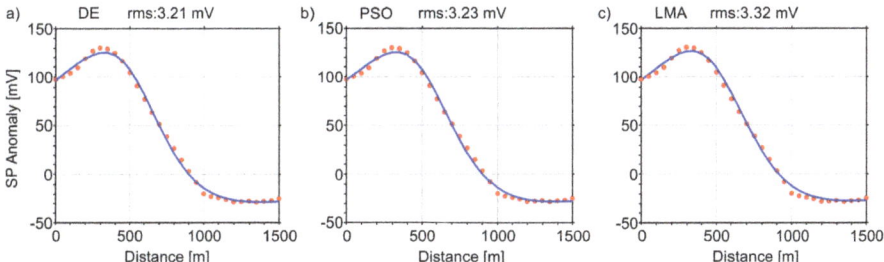

Fig. 4.8 **a** DE, **b** PSO and, **c** LM inversions of P1-profile

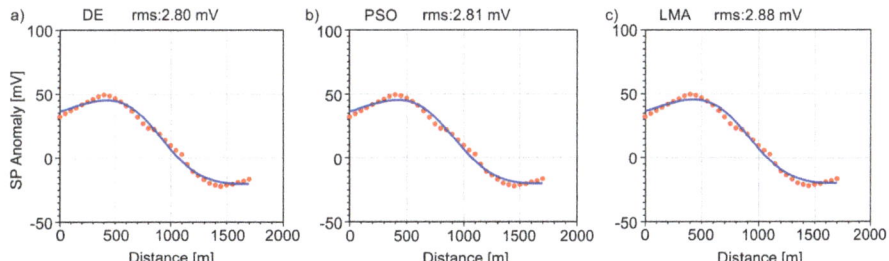

Fig. 4.9 **a** DE, **b** PSO and, **c** LM inversions of P2-profile

depth difference between the points x_0^{P3} and x_0^{P4}. Considering that the study area is in a horst-graben transition boundary, this difference may be related with the Tuzla Fault System. These findings are accordance with those of VES studies.

When we combine the geological units of the study area (Fig. 4.6) with the SP and VES findings, we can say that the surface volcanics become thinner and the

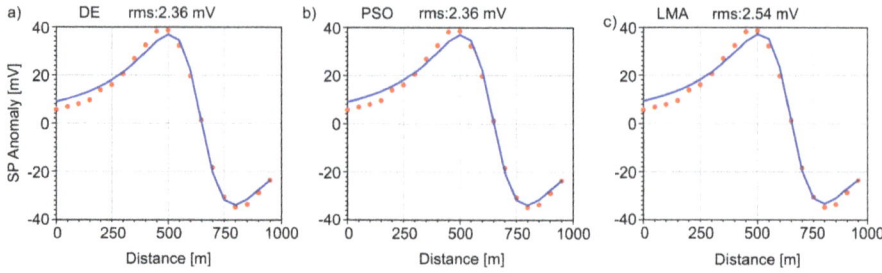

Fig. 4.10 **a** DE, **b** PSO and, **c** LM inversions of P3-profile

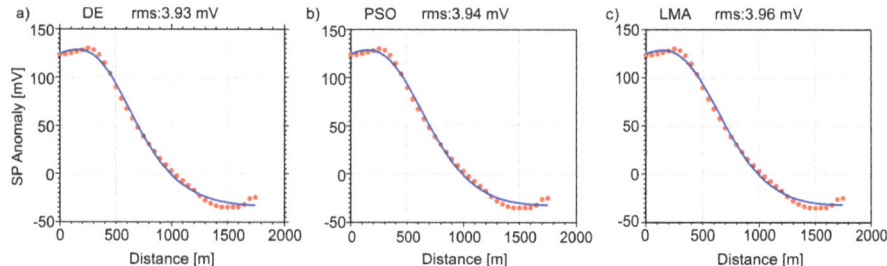

Fig. 4.11 **a** DE, **b** PSO and, **c** LM inversions of P4-profile

Table 4.8 The best solutions from Tamış-Çanakkale P1 data set by three algorithms at the end of 10 independent runs

Algorithm	RUN	Parameters					
		x_0(m)	z_0(m)	θ (°)	K (mV. m$^{(2q-1)}$)	q	RMS (mV)
DE	4	557.3	504.52	145.81	750000	1.18	3.22
PSO	4	556.71	500.43	145.55	634540.45	1.16	3.23
LM	10	557.43	505.62	146.37	786066.39	1.18	3.32

Table 4.9 The best solutions from Tamış-Çanakkale P2 data set by three algorithms at the end of 10 independent runs

Algorithm	RUN	Parameters					
		x_0(m)	z_0(m)	θ (°)	K (mV. m$^{(2q-1)}$)	q	RMS (mV)
DE	1	836.57	686.04	161.14	749999.99	1.21	2.8
PSO	5	836.51	685.93	161.14	750000	1.21	2.8
LM	9	834.49	650.39	159.91	243277.16	1.13	2.88

metamorphic basement units reach the shallower depths in the east and northeast of the study area. In the light of comparison of two geophysical methods we can said

Table 4.10 The best solutions from Tamış-Çanakkale P3 data set by three algorithms at the end of 10 independent runs

Algorithm	RUN	Parameters					
		x_0(m)	z_0(m)	θ (°)	K (mV. m$^{(2q-1)}$)	q	RMS (mV)
DE	4	480.21	633.49	140.98	37624.04	0.93	3.93
PSO	1	488.31	695.96	144.07	199675.88	1.04	3.94
LM	6	489.89	709.06	144.85	281692.89	1.07	3.96

Table 4.11 The best solutions from Tamış-Çanakkale P4 data set by three algorithms at the end of 10 independent runs

Algorithm	RUN	Parameters					
		x_0(m)	z_0(m)	θ (°)	K (mV. m$^{(2q-1)}$)	q	RMS (mV)
DE	1	644.98	185.99	177.91	750000	1.37	2.36
PSO	1	644.96	185.99	177.9	750000	1.37	2.36
LM	4	645.15	180.37	177.34	476253.67	1.33	2.54

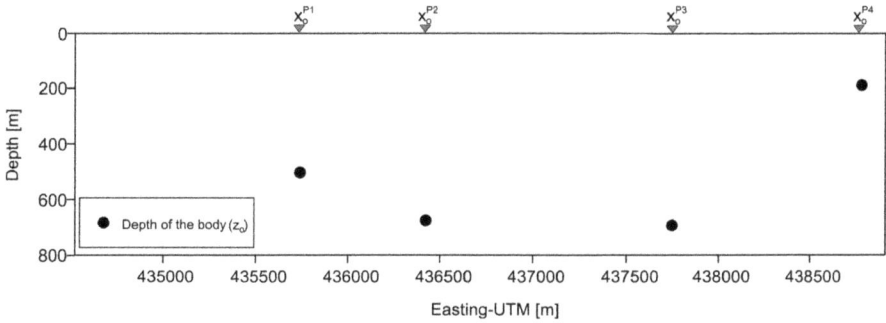

Fig. 4.12 Estimated source depths (dots) from the inversion of the SP profile data

that the depth values estimated from SP and VES methods are in good agreements with each other.

4.3   Conclusions

In this study, the model parameters of a polarized body have been determined by a derivative-based (LM), and two population-based optimization algorithms (DE and PSO), and the results are compared. Even though it is not preferred to compare derivative-based algorithms with metaheuristics, a comparison could be realized by a LM-based limitation procedure introduced in this study. By this limitation procedure, the misfit values from the LM algorithm have been observed as being close to those from DE and PSO for both synthetic and field data sets.

In this study, Tamış-Çanakkale SP anomaly from Turkey was also evaluated with the mentioned algorithms and the solutions of them compared to each other. RMS value of the LM solution is relatively higher than the others. Comparison of the estimated SP model parameters to the VES sections has indicated that the surface volcanics become thinner and the metamorphic basement units reach the shallower depths in the east and northeast of the study area.

As a result, the solutions by DE, PSO, and LM (with limitation procedure introduced by the present study) are represented by being in good agreement with each other and they have the ability to converge from local best to the general best, can be successfully applied in determining SP model parameters. The LM algorithm, after the process introduced by the present study, has yielded results comparable with the other algorithms PSO and DE. It also displayed better convergence characteristics after the proposed process.

References

Abdelazeem M, Gobashy M (2006) Self-potential inversion using genetic algorithm. J King Abdulaziz University Earth Sci 17:83–101

Abdelrahman EM, Ammar AA, Sharafeldin SM, Hassanein HI (1997) Shape and depth solutions from numerical horizontal self-potential gradients. Appl Geophys 36:31–43

Abdelrahman EM, Ammar AA, Hassanein HI, Hafez MA (1998) Derivative analysis of SP anomalies. Geophysics 63:890–897

Abdelrahman EM, Essa KS, Abo-Ezz ER, Soliman KS, El-Araby TM (2006) A least-squares depth–horizontal position curves method to interpret residual SP anomaly profiles. J Geophys Eng 3:252–259

Abedi M, Hafizi MK, Norouzi GH (2012) 2D interpretation of self-potential data using Normalized Full Gradient, a case study: galena deposit. Bollettino di Geofisica Teorica ed Applicata 53:213–230

Agarwal BNP, Srivastava S (2009) Analyses of self-potential anomalies by conventional and extended Euler deconvolution techniques. Comput Geosci 35:2231–2238

Al-Garni M, Sundararajan, N (2011). Hartley spectral analysis of self-potential anomalies caused by a 2-D horizontal circular cylinder. Arabian J Geosci 5(6). https://doi.org/10.1007/s12517-011-0285-8

Altınok Y, Alpar B, Yaltırak C, Pınar A, Özer N (2012) The earthquakes and related tsunamis of October 6, 1944 and March 7, 1867. NE Aegean Sea Nat Hazards 60(1):3–25

Asfahani J, Tlas M, Hammadi M (2001) Fourier analysis for quantitative interpretation of self-potential anomalies caused by horizontal cylinder and sphere. J King Abdulaziz University-Earth Sci 13:41–53

Balkaya Ç (2013) An implementation of differential evolution algorithm for inversion of geoelectrical data. J Appl Geophys 98:160–175

Barnston AG (1992) Correspondence among the correlation, RMSE, and Heidke forecast verification measures; refinement of the Heidke score. Wea Forecasting 7(4):699–709

Biswas A, Sharma SP (2014) Optimization of Self-Potential interpretation of 2-D inclined sheet-type structures based on Very Fast Simulated Annealing and analysis of ambiguity. J Appl Geophys 105:235–247

Biswas A, Sharma SP (2015) Interpretation of self-potential anomaly over idealized body and analysis of ambiguity using very fast simulated annealing global optimization. Near Surface Geophysics 13:179–195

Corwin RF (1990) The self-potential method for environmental and engineering applications. In: Ward SW (ed) Geotechnical and environmental geophysics I: 127–145

Corwin RF, Hoover DB (1979) The self-potential method in geothermal exploration. Geophysics 44(2):226–245

Di Maio R, Patella D (1994) Self-potential anomaly generation in volcanic areas. Acta Vulcanol 4:119–124

Di Maio R, Piegari E, Rani P, Avella A (2016) Self-potential data inversion through the integration of spectral analysis and tomographic approaches. Geophys J Int 206:1204–1220

Di Maio R, Rani P, Piegari E, Milano L (2017) Self-potential data inversion through a Genetic-Price Algorithm. Comput Geosci 94:86–95

El-Kaliouby HM, Al-Garni MA (2009) Inversion of self-potential anomalies caused by 2D inclined sheets using neural networks. J Geophys Eng 6:29–34

Ekinci YL, Balkaya Ç, Göktürkler G, Turan S (2016) Model parameter estimations from residual gravity anomalies due to simple-shaped sources using differential evolution algorithm. J Appl Geophys 129:133–147

Ekinci YL, Özyalın Ş, Sındırgı P, Balkaya Ç, Göktürkler G (2017) Amplitude inversion of the 2D analytic signal of magnetic anomalies through the differential evolution algorithm. J Geophys Eng 14:1492–1508

Ekinci YL, Balkaya Ç, Göktürkler G (2019) Parameter estimations from gravity and magnetic anomalies due to deep-seated faults: differential evolution versus particle swarm optimization. Turk J Earth Sci 28:860–881

Essa K, Mehanee S, Smith PD (2008) A new inversion algorithm for estimating the best fitting parameters of some geometrically simple body to measured self-potential anomalies. Explor Geophys 39:155–163

Fernández-Martínez JL, García-Gonzalo E, Naudet V (2010). Particle swarm optimization applied to solving and appraising the streaming-potential inverse problem. Geophysics, 75:WA3–WA15

Fitterman DV, Corwin RF (1982) Inversion of self-potential data from the Cerro-Prieto geothermal field Mexico. Geophysics 47:938–945

Gilbert D, Pessel M (2001) Identification of sources of potential fields with the continuous wavelet transform: application to self-potential profiles. Geophys Res Lett 28:1863–1866

Göktürkler G, Balkaya Ç (2012) Inversion of self-potential anomalies caused by simple-geometry bodies using global optimization algorithms. J Geophys Eng 9(5):498–507

Göktürkler G, Balkaya Ç, Ekinci, YL Turan S (2016) Metaheuristics in applied geophysics (in Turkish). Pamukkale Univ Muh Bilim Derg., 22(6):563–580. https://doi.org/10.5505/pajes.2015.81904

Hamzah U, Samsudin AR, Malim AP (2007) Groundwater investigation in Kuala Selangor using vertical electrical sounding (VES) surveys. Environ Geol 51(8):1349–1359

IPI2 WIN Free Version 3.0.1 (2000). Program for vertical electrical sounding curves 1-D interpreting along a simple profile. Department of Geophysics, Geological Faculty, Moscow State University, Russia. http://geophys.geol.msu.ru/ipi2win.htm Accessed 16 August 2020

Juan LFM, Esperanza GG, José PFÁ, Heidi AK, César OMP (2010) PSO, a powerful algorithm to solve geophysical inverse problems, application to a 1D-DC resistivity case. J Appl Geophys 71:13–25

Juliano T, Mauriello P, Patella D (2002) Looking inside Mount Vesuvius by potential fields integrated probability tomographies. J Volcanol Geotherm Res 113:363–378

Jupp DLB, Vozoff K (1975) Stable iterative methods for the inversion of geophysical data. Geophys J Roy Astron Soc 42(3):957–976. https://doi.org/10.1111/j.1365-246x.1975.tb06461.x

Kaftan I, Sındırgı P, Akdemir Ö (2014) Inversion of self potential anomalies with multilayer perceptron neural networks. Pure appl Geophys 171:1939–1949

Karacık Z, Yılmaz Y (1998) Geology of the Ignimbrites and and the associated volcano–plutonic complex of the Ezine area, northwestern Anatolia. J Volcanol Geoth Res 85:251–264

Karlık G, Kaya MA (2001) Investigation of groundwater contamination using electric and electro-magnetic methods at an open waste-disposal site: a case study from Isparta Turkey. Environ Geol 40(6):725–731

Kaya MA, Özürlan G, Balkaya Ç (2015) Geoelectrical investigation of seawater intrusion in the coastal urban area of Çanakkale. NW Turkey. Environmental Earth Sciences 73(3):1151–1160

Kennedy J, d Eberhart R (1995). Particle swarm optimisation Proceedings IEEE international conference on neural networks (Piscataway, NJ), p 1942

Levenberg K (1944) A method for the solution of certain non-linear problems in least squares. Q Appl Math 2(2):164–168. https://doi.org/10.1090/qam/10666

Li X, Yin M (2012) Application of differential evolution algorithm on self-potential data. PLoS ONE 7(12): https://doi.org/10.1371/journal.pone.0051199

Luke S (2009) Essentials of Metaheuristics (Lulu) p 233 Available free at http://cs.gmu.edu/~sean/book/metaheuristics/

Mehanee SA (2015) Tracing of paleo-shear zones by self-potential data inversion: case studies from the KTB, Rittsteig, and Grossensees graphite-bearing fault planes. Earth Planets Space 67:14. https://doi.org/10.1186/s40623-014-0174-y

Marquardt D (1963) An algorithm for least-squares estimation of nonlinear parameters. SIAM J Appl Math 11(2):431–441

Monteiro Santos FA (2010) Inversion of self-potential of Idealized bodies' anomalies using particle swarm optimization. Comput Geosci 36:1185–1190

Monteiro Santos FA, Almeida EP, Castro R, Nolasco R, Victor LM (2002) A hydrogeological investigation using EM34 and SP surveys. Earth Planets Space 54:655–662

Ogilvy AA, Ayed MA, Bogoslovsky VA (1969) Geophysical studies of water leakage from reservoirs. Geophys Prospect 17(1):36–62

Particle Swarm optimization. Introduction to Particle Swarm Optimization. http://mnemstudio.org/particle-swarm-introduction.htm. Accessed 10 Sep 2020

Patella D (1997) Introduction to ground surface self-potential tomography. Geophys Prospect 45:653–681

Paul MK (1965) Direct interpretation of self-potential anomalies caused by inclined sheets of infinite extensions. Geophysics 30:418–423

Pekşen E, Yas T, Kayman AY, Özkan C (2011) Application of particle swarm optimization on self-potential data. J Appl Geophys 75(2):305–318

Poli R, Kennedy J, Blackwell T (2007) Particle swarm optimization: an overview Swarm Intell, 1–25

Rao BSR, Murthy IVR, Reddy SJ (1970) Interpretation of self-potential anomalies of some simple geometrical bodies. Pure appl Geophys 78:60–77

Revil A, Ehouarne L, Thyreault E (2001) Tomography of self-potential anomalies of electrochemical nature. Geophys Res Lett 28(23):4363–4366

Revil A, Naudet V, Nouzaret J, Pessel M (2003) Principles of electrography applied to self-potential sources and hydrogeological applications. Water Resources Res. 39:1114. https://doi.org/10.1029/2001WR000916

Salmon S (2011). Particle Swarm Optimization in Scilab Available at http://forge.scilab.org/index.php/p/pso-toolbox/downloads/

Satyanarayana Murty BV, Haricharan P (1985) Nomogram for the complete interpretation of spontaneous potential profiles over sheet-like and cylindrical two-dimensional sources. Geophysics 50:1127–1135

Schiavone D, Quarto R (1984) Self-potential prospecting in the study of water movement. Geoexploration 22:47–58

Sharma SP (2012) VFSARES- a very fast simulated annealing FORTRAN program for interpretation of 1-D DC resistivity sounding data from various electrode array. Comput Geosci 42:177–188

Shi Y and Eberhart RC (1998). Parameter selection in particle swarm optimization Proc. 7th international conference on evolutionary programming VII (New York) pp 591–600

Sındırgı, P Özyalın Ş (2019) Estimating the location of a causative body from a self-potential anomaly using 2D and 3D normalized full gradient and Euler deconvolution. Turkish J Earth Sci 28:640–659. https://doi.org/10.3906/yer-1811-14

Sındırgı P, Pamukçu O, Özyalın Ş (2008) Application of normalized full gradient method to self potential (SP) data. Pure appl Geophys 165:409–427

Sözbilir H, Uzel B, Sümer Ö, Eski S, Softa M, Tepe Ç, Özkaymak Ç, Baba A (2018). Seismic Sources Of (14th January-20th March 2017) Çanakkale-Ayvacık Earthquake Swarm. Eskişehir Technical University Journal of Science and Technology B- Theoritical Sciences, Special Issue of 4th international earthquake engineering and seismology conference, vol 6, 1–17. https://doi.org/10.20290/aubtdb.498805

Storn R, Price KV (1995) Differential evolution-A simple and efficient adaptive scheme for global optimization over continuous spaces. Technical Report TR-95-012. International Computer Science Institute, Berkeley, CA

Storn R and Price K (1997). Differential evolution-a simple and efficient heuristic for global optimization over continuous spaces. J Glob Optim 11:341–359

Sundararajan N, Arun Kumar I, Mohan NL, Seshagiri Rao SV (1990) Use of the Hilbert transform to interpret self-potential anomalies due to two-dimensional inclined sheets. Pure appl Geophys 133:117–126

Yüngül S (1950) Interpretation of spontaneous-polarization anomalies caused by spherical ore bodies. Geophysics 15:237–246

Chapter 5
Estimation of the Buried Model Parameters from the Self-potential Data Applying Advanced Approaches: A Comparison Study

Mahmoud Elhussein and Khalid S. Essa

Abstract A comparison study using the least-squares minimization method, particle swam optimization method, and neural network method for interpreting self-potential data for typical shaped-models (spheres and cylinders). This interpretation process contains the delineation buried sources parameters, which are the amplitude factor, the depth to the structure, the source origin location, the angle of polarization, the shape factor. The stability of the suggested methods was tested on two synthetic data with and without noise and real data set from USA. The methods estimate the different structures parameters efficiently and accurately.

Keywords Self-potential data · Least-squares · Particle swarm · Neural network

5.1 Introduction

Self-potential method can be considered as one of the most effective geophysical techniques in solving different geophysical problems (Sundararajan et al. 1998; Drahor 2004; Mehanee 2014; Essa 2020; Elhussein 2020). Self-potential anomalies produced by natural potential difference which resulted due to the oxidation-reduction process of mineralized rocks which in contact with the ground water (Essa et al. 2008; Essa 2020).

To apply the self-potential technique in solving the different geophysical problems, the different subsurface geological bodies was approximated to simple geometrical bodies (like, sphere, cylinder and thin sheet) (Essa 2011; Mehanee 2014, 2015; Biswas 2017; Essa and Elhussein 2017; Sungkono and Warnana 2018; Essa 2020). To estimate the different parameters (like, amplitude coefficient, depth, polarization angle and body origin), different techniques were created and developed to overcome the ill-posedness and non-uniqueness problems (Tarantola 2005; Sharma and Biswas 2013; Essa 2019). From these techniques, gradient techniques (Abdelrahman et al. 2004, 2009b; Essa and Elhussein 2017), moving average techniques (Abdelrahman et al. 2006a; Mehanee et al. 2011; Essa 2019), characteristic curves and nomograms

M. Elhussein · K. S. Essa (✉)
Faculty of Science, Geophysics Department, Cairo University, Giza 12613, Egypt

A. Biswas (ed.), *Self-Potential Method: Theoretical Modeling and Applications in Geosciences*, Springer Geophysics, https://doi.org/10.1007/978-3-030-79333-3_5

(Yungul 1950; Fitterman 1979; Essa 2007; Abdelrahman et al. 2009a), nonlinear and liner least squares techniques (Abdelrahman et al. 2006b; Essa et al. 2008), Euler deconvolution method (Agarwal and Srivastava 2009); most of the previous methods require a priori information, other methods estimate the different parameters with high uncertainty as the accuracy of the estimated parameters mainly depends upon the accuracy of the regional-residual separation. Nowadays new recent techniques have been developed, like particle swarm optimization (Essa 2019, 2020), genetic-price technique (Di Maio et al. 2019), black-hole technique (Sungkono and Warnana 2018).

This chapter review different techniques applied to the different synthetic and real field self-potential data to estimate the different bodies parameters.

5.2 Methodology

5.2.1 Forward Modeling

The SP anomaly (V) caused by simple geometrical structures at any given point (p) (Bhattacharya and Roy 1981; Essa 2019) (Fig. 5.1) is given by:

$$V(x_j) = K \frac{(x_j - d)\cos\theta - z\sin\theta}{\left((x_j - d)^2 + z^2\right)^q}, \quad i = 0, 1, 2, 3, \ldots, N \tag{5.1}$$

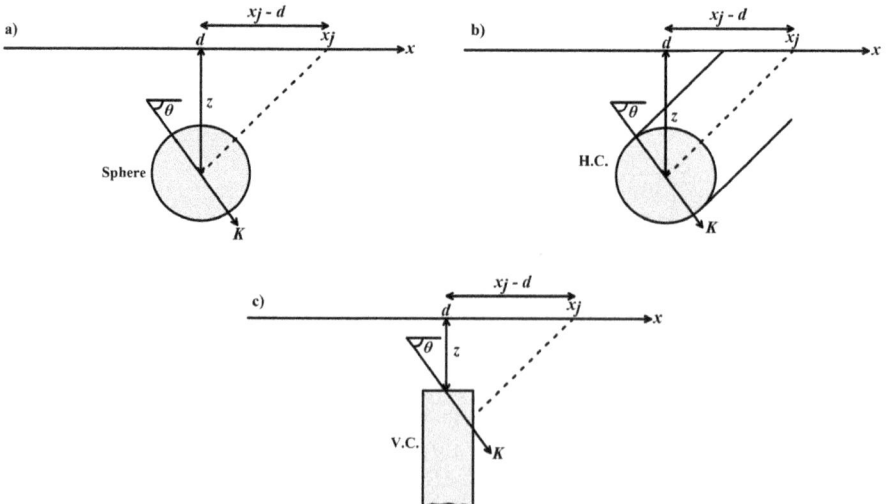

Fig. 5.1 A sketch showing the different geometrical shapes and their parameters: **a** sphere, **b** horizontal cylinder and **c** vertical cylinder

where K is the amplitude factor (mV x m^{2q-1}), z is the depth to the structure (m), d is the source origin (m), θ is the angle of polarization (degrees), q is the shape factor (dimensionless) which takes the following values: 1.5 for sphere, 1 for horizontal cylinder and 0.5 for vertical cylinder (Essa et al. 2008; Di Maio et al. 2016; Essa 2019).

5.2.2 Least Squares Inversion Technique

Essa et al. (2008) developed a least square inversion approach to estimate the different bodies parameters, by determining the depth applying the nonlinear equation:

$$\delta(z) = 0, \tag{5.2}$$

After estimating the depth, the angle of polarization is then calculated by the least square, also, the amplitude factor is then determined from the estimate depth and the polarization angle.

5.2.3 Particle Swarm Optimization

Essa (2019) developed a method based upon the PSO algorithm and the second moving average for estimating the different structures parameters. PSO is stochastic in its nature, the idea of PSO is based upon a group of birds or fishes looking for food, the group of birds represent the models, and the paths of particles represent the solutions (Essa 2019). The algorithm starts with random models, then the location and the velocity of the particles are updated using the following formulas, respectively.

$$x_i^{k+1} = x_j^k + V_j^{k+1} \tag{5.3}$$

$$V_j^{k+1} = c_3 V_j^k + c_1 rand\left(T_{best} - x_j^{k+1}\right) + c_2 rand\left(J_{best} - x_j^{k+1}\right), \tag{5.4}$$

where x_j^k is the location of jth particle at the iteration kth; V_j^k is the velocity of the jth model at the iteration kth; $rand$ is any random number between [0, 1]; c_1 and c_2 are cognitive and social factors and equal 2 (Parsopoulos and Vrahatis 2002; Singh and Biswas 2016; Essa 2019); c_3 is the inertial coefficient which governs the velocity of the model and usually takes a value less than one; T_{best} is the best location for individual model, and J_{best} is the global best location for any model in the group.

5.2.4 Neural Network Algorithm

Al-Garni (2009) proposed an approach based mainly on neural network (modular algorithm) to estimate the different structures parameters.

5.3 Synthetic Examples

5.3.1 Sphere Model

A noise free SP anomaly was generated using sphere model with the following parameters: $K = 1200$ mV x m^2, $z = 6$, $\theta = 45°$, $d = 55$ m, $q = 1.5$ and the profile length $= 100$ m (Fig. 5.2).

The different previous techniques were applied to estimate the different parameters. First the least square inversion technique was applied to the SP profile and the parameters were estimated accurately with no error (Table 5.1), then the PSO technique produce the parameters with 0% error (Table 5.1), Finally, the data were subjected to neural network and the parameters were estimated efficiently (Table 5.1).

To test the effect of noisy data on the different techniques, a 10% random noise was added to the previous SP model. For least square inversion technique, the estimated parameters are: $K = 1020$ mV x m^2, $z = 6.5$, $\theta = 47°$; while for PSO technique, the estimated parameters are: $K = 1140$ mV x m^2, $z = 5.8$, $\theta = 44.5°$, $d = 54.9$ m, $q = 1.45$; and in case of neural network, the estimated parameters are: $K = 1350$ mV x m^2, $z = 6.3$, $\theta = 45.7°$, $q = 1.57$ (Table 5.2). The error of the estimated parameters is shown in (Table 5.2).

5.3.2 Horizontal Cylinder Model

A noise free SP anomaly was generated using horizontal cylinder model with the following parameters: $K = 900$ mV x m, $z = 6.5$, $\theta = 40°$, $d = 60$ m, $q = 1$ and the profile length $= 100$ m (Fig. 5.3).

The different previous techniques were applied to estimate the different parameters. First the least square inversion technique was applied to the SP profile and the parameters were estimated accurately with no error (Table 5.1), then the PSO technique produce the parameters with 0% error (Table 5.2), Finally, the data were subjected to neural network and the parameters were estimated efficiently (Table 5.2).

To test the effect of noisy data on the different techniques, a 10% random noise was added to the previous SP model. For least square inversion technique, the estimated parameters are: $K = 1000$ mV x m, $z = 7.1$, $\theta = 41.5°$; while for PSO technique,

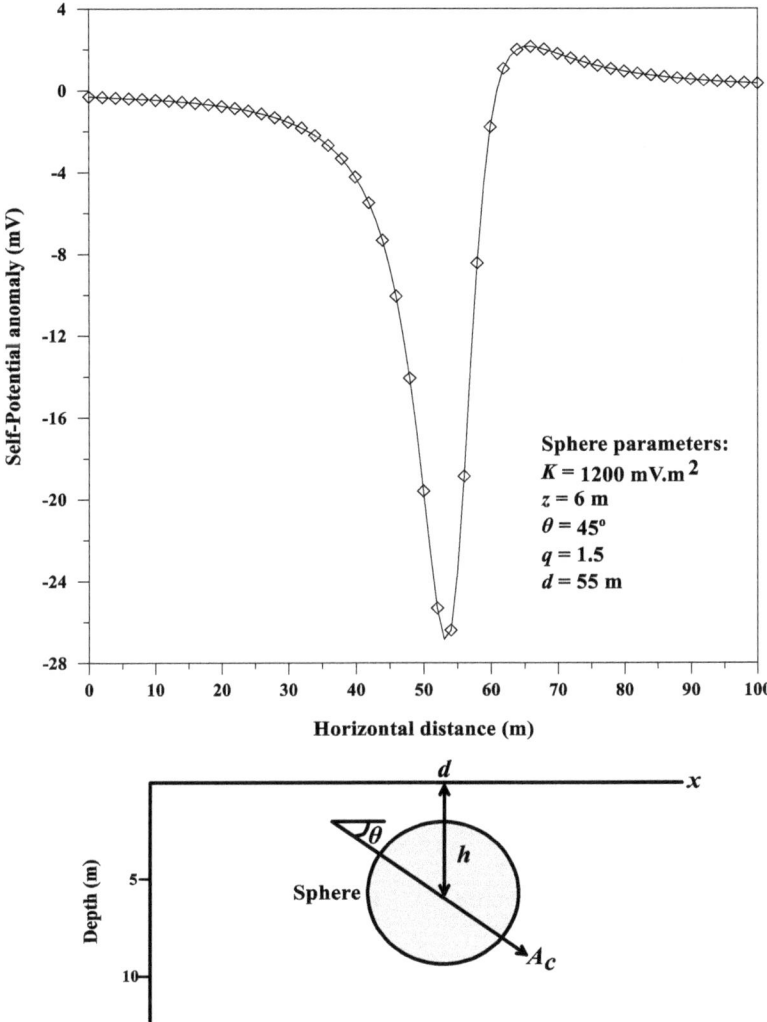

Fig. 5.2 Self-potential anomaly profile of sphere model ($K = 1200$ mV $\times$ m^2, $z = 6$ m, $\theta = 45°$, $q = 1.5$, and $d = 55$ m) and profile length 100 m

the estimated parameters are: K $= 960$ mV x m, z $= 6.6$, $\theta = 40.2°$, d $= 60.11$ m, q $= 1.04$; and in case of neural network, the estimated parameters are: K $= 1010$ mV x m, z $= 6.3$, $\theta = 39.7°$, q $= 0.9$ (Table 5.2). The error of the estimated parameters is shown in (Table 5.2).

Table 5.1 A correlation between results obtained from different methods applied to the self-potential anomaly of sphere model ($K = 1200$ mV $\times$ m^2, $z = 6$ m, $\theta = 45°$, $q = 1.5$, and $d = 55$ m)

Methods parameters	Essa et al. (2008) method		Al-Garni (2009) method		Essa (2019) method	
	Noise-freee					
	Results	Error (%)	Results	Error (%)	Results	Error (%)
K (mV x m^2)	1200	0	1200	0	1200	0
z (m)	6	0	6	0	6	0
θ (degree)	45	0	45	0	45	0
q (dimensionless)	–	–	1.5	0	1.5	0
d (m)	–	–	–	–	55	0
Results (after adding 10% random noise)						
	Results	Error (%)	Results	Error (%)	Results	Error (%)
K (mV x m^2)	1020	15	1350	12.5	1140	5
z (m)	6.5	8.33	6.3	5	5.8	3.33
θ (degree)	47	4.44	45.7	1.56	44.5	1.11
q (dimensionless)	–	–	1.57	4.67	1.45	3.33
d (m)	–	–	–	–	54.9	0.18

Table 5.2 A correlation between results obtained from different methods applied to the self-potential anomaly of H.C. model ($K = 900$ mV $\times$ m, $z = 6.5$ m, $\theta = 40°$, $q = 1$, and $d = 60$ m)

Methods parameters	Essa et al. (2008) method		Al-Garni (2009) method		Essa (2019) method	
	Noise-freee					
	Results	Error (%)	Results	Error (%)	Results	Error (%)
K (mV x m)	900	0	900	0	900	0
z (m)	6.5	0	6.5	0	6.5	0
θ (degree)	40	0	40	0	40	0
q (dimensionless)	–	–	1	0	1	0
d (m)	–	–	–	–	60	0
Results (after adding 10% random noise)						
	Results	Error (%)	Results	Error (%)	Results	Error (%)
K (mV x m)	1000	11	1010	12.2	960	4
z (m)	7.1	9.23	6.3	3.08	6.6	1.54
θ (degree)	41.5	3.75	39.7	0.75	40.2	0.5
q (dimensionless)	–	–	0.9	10	1.04	4
d (m)	–	–	–	–	60.11	0.18

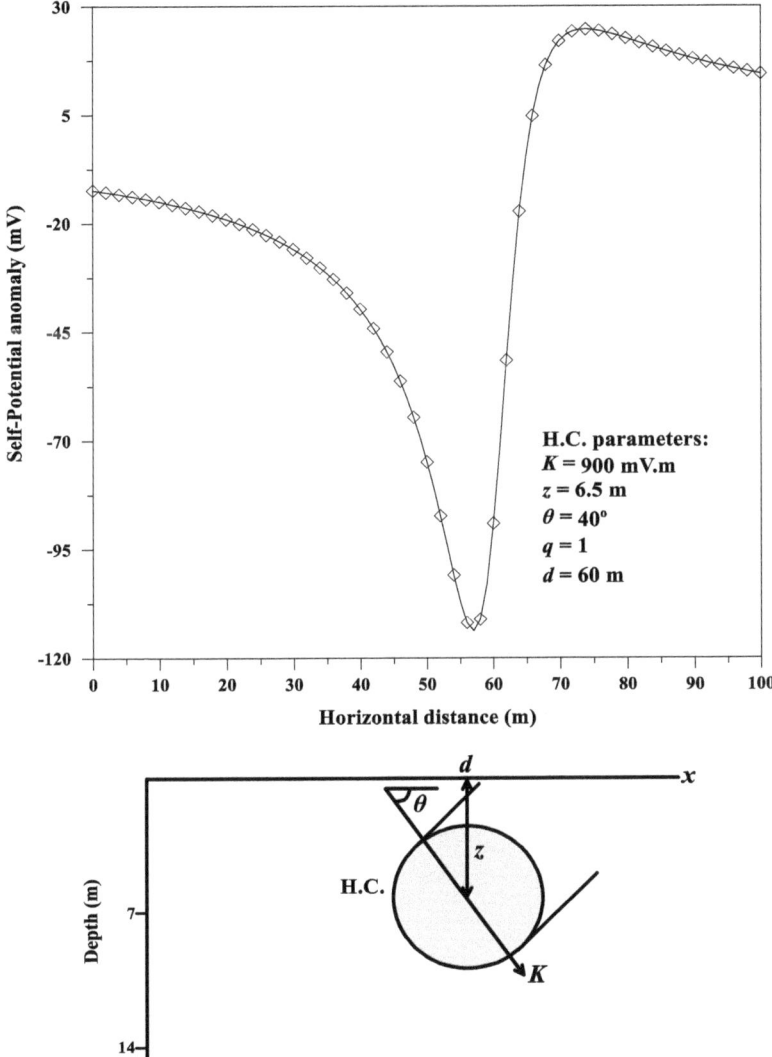

Fig. 5.3 Self-potential anomaly profile of H.C. model ($K = 900$ mV $\times$ m, $z = 6.5$ m, $\theta = 40°$, $q = 1$, and $d = 60$ m) and profile length 100 m

5.4 Field Example

5.4.1 Malachite Mine, USA Real Data

Malachite mine is composed of amphibolite belt which surrounded by gneiss and schist (Essa 2019). Self-potential profile was designed and measured by Heiland et al. (1945), the profile was taken above massive sulfide ore body which located in the Malachite mine. The profile length was 164 m, digitized at 1.25 m (Fig. 5.4). The SP profile was then subjected to the three different techniques to determine and compare between the parameters estimated from these different methods (Table 5.3). From Table 5.3 the parameters estimated using least square inversion method (Essa et al. 2008) are: $K = 275.39$ mV, $z = 12.87$, $\theta = 103.58°$; while the parameters estimated by using PSO technique (Essa 2019) are: $K = 236.53$ mV, $z = 13.74$, $\theta =$

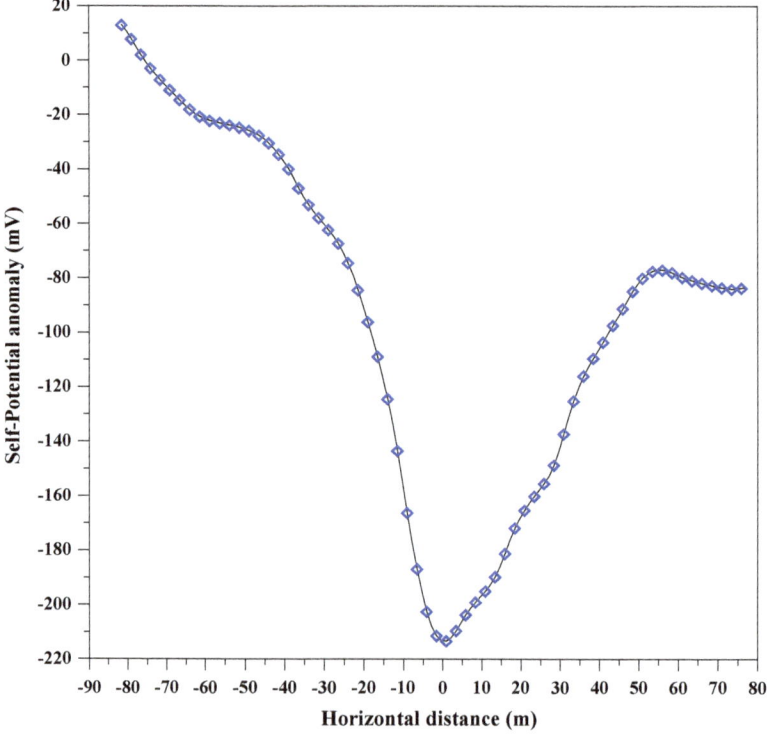

Fig. 5.4 Self-potential anomaly profile of Malachite mine, USA field example

Table 5.3 A correlation between results obtained from different methods applied to the self-potential anomaly of Malachite mine, USA field example

Methods parameters	Essa et al. (2008) method	Al-Garni (2009) method	Essa (2019) method
K (mV)	275.39	268.41	236.53
z (m)	12.87	13.2	13.74
θ (degree)	103.58	105	99.31
q (dimensionless)	–	0.63	0.45
d (m)	–	–	0.20

99.31° d $= 0.20$ m, q $= 0.45$; finally, the parameters estimated using neural network (Al-Garni 2009) are: $K = 268.41$ mV, $z = 13.2$, $\theta = 105°$, $q = 0.63$.

5.5 Conclusions

A comparative study was made in this chapter to see the differences between different methods in application to the self-potential data from different geological structures (Sphere, horizontal cylinder and vertical cylinder). the different methods are least-square (Essa 2008), neural network (Al-Garni 2009) and PSO (Essa 2019). These different methods were applied to two different synthetic data without and with 10% random noise and one real data from USA. The methods estimate the different structures parameters (K, z, d, θ and q) efficiently and accurately.

References

Abdelrahman EM, El-Araby TM, Essa KS (2009a) Shape and depth determination from second moving average residual self-potential anomalies. J Geophys Eng 6:43–52

Abdelrahman EM, Essa KS, Abo-Ezz ER, Soliman KS (2006a) Self-potential data interpretation using standard deviations of depths computed from moving average residual anomalies. Geophys Prospect 54:409–423

Abdelrahman EM, Essa KS, El-Araby TM, Abo-Ezz ER (2006b) A least-squares depth-horizontal position curves method to interpret residual SP anomaly profile. J Geophys Eng 3:252–259

Abdelrahman EM, Saber HS, Essa KS, Fouda MA (2004) A least-squares approach to depth determination from numerical horizontal self-potential gradients. Pure appl Geophys 161:399–411

Abdelrahman EM, Soliman KS, Abo-Ezz ER, Essa KS, El-Araby TM (2009b) Quantitative interpretation of self-potential anomalies of some simple geometric bodies. Pure appl Geophys 166:2021–2035

Agarwal B, Sirvastava S (2009) Analyses of self-potential anomalies by conventional and extended Euler deconvolution techniques. Comput Geosci 35:2231–2238

Al-Garni MA (2009) Interpretation of spontaneous potential anomalies from some simple geometrically shaped bodies using neural network inversion. Acta Geophys 58:143. https://doi.org/10.2478/s11600-009-0029-2

Biswas A (2017) A review on modeling, inversion and interpretation of self-potential in mineral exploration and tracing paleo-shear zones. Ore Geol Rev 91:21–56

Di Maio R, Piegari E, Rani P, Carbonari R, Vitagliano E, Milano L (2019) Quantitative interpretation of multiple self-potential anomaly sources by a global optimization approach. J Appl Geophys 162:152–163

Di Maio R, Rani P, Piegari E, Milano L (2016) Self-potential data inversion through a genetic-price algorithm. Comput Geosci 94:86–95

Drahor MG (2004) Application of the self-potential method to archaeological prospection: some case histories. Archaeol Prospect 11:77–105

Elhussein M (2020) A novel approach to self-potential data interpretation in support of mineral resource development. Nat Resour Res. https://doi.org/10.1007/s11053-020-09708-1

Essa KS (2007) Gravity data interpretation using the s-curves method. J Geophys Eng 4:204–213

Essa KS (2011) A new algorithm for gravity or self-potential data interpretation. J Geophys Eng 8:434–446

Essa KS (2019) A particle swarm optimization method for interpreting self potential anomalies. J Geophys Eng 16:463–477

Essa KS (2020) Self potential data interpretation utilizing the particle swarm method for the finite 2D inclined dike: mineralized zones delineation. Acta Geod Geophys 55:203–221

Essa KS, Elhussein M (2017) A new approach for the interpretation of self-potential data by 2-D inclined plate. J Appl Geophys 136:455–461

Essa KS, Mehanee S, Smith P (2008) A new inversion algorithm for estimating the best fitting parameters of some geometrically simple body from measured self-potential anomalies. Explor Geophys 39:155–163

Fitterman DV (1979) Calculations of self-potential anomalies near vertical contacts. Geophysics 44:195–205

Heiland CA, Tripp MR, Wantland D (1945) Geophysical surveys at the Malachite Mine. Jefferson County Colorado Amer Inst Min Metall Eng 164:142–154

Mehanee S (2014) An efficient regularized inversion approach for self-potential data interpretation of ore exploration using a mix of logarithmic and non-logarithmic model parameters. Ore Geol Rev 57:87–115

Mehanee S (2015) Tracing of paleo-shear zones by self-potential data inversion: case studies from the KTB, Rittsteig, and Grossensees graphite-bearing fault planes. Earth Planets Space 67:14

Mehanee S, Essa KS, Smith P (2011) A rapid technique for estimating the depth and width of a two-dimensional plate from self-potential data. J Geophys Eng 8:447–456

Parsopoulos KE, Vrahatis MN (2002) Recent approaches to global optimization problems through particle swarm optimization. Nat Comput 1:235–306

Sharma SP, Biswas A (2013) Interpretation of self-potential anomaly over 2D inclined structure using very fast simulated annealing global optimization: an insight about ambiguity. Geophysics 78:WB3–WB15

Singh A, Biswas A (2016) Application of global particle swarm optimization for inversion of residual gravity anomalies over geological bodies with idealized geometries. Nat Resour Res 25:297–314

Sundararajan N, Srinivasa Rao P, Sunitha V (1998) An analytical method to interpret self-potential anomalies caused by 2D inclined sheets. Geophysics 63:1551–1555

Sungkono Warnana DD (2018) Black hole algorithm for determining model parameter in self-potential data. J Appl Geophys 148:189–200

Tarantola A (2005) Inverse problem theory and methods for model parameter estimation. PA, Society for Industrial and Applied Mathematics (SIAM), Philadelphia

Yungul S (1950) Interpretation of spontaneous polarization anomalies caused by spheroidal orebodies. Geophysics 15:163–256

Chapter 6
Determining the Structure Factor and Parameters of a Buried Polarized Structure from Self Potential Anomalies

Coşkun Sari and Emre Timur

Abstract The self-potential or spontaneous polarization (SP) method is based on the surface measurement of natural potentials resulting from electrokinetic, electrochemical and thermoelectric reactions in the subsurface. This method in geophysics refers to an electrical surveying method used for looking at natural electrical anomalies in the ground. The self-potential method was the first electrical method primarily used for mineral exploration and is still used therein. Mathematical methods such as Bisection and Regula False have been used for several years for interpreting SP data. In this study, it is aimed to compare these two interpretation methods for evaluating SP data. Theoretical studies were carried out for four different models and the outcomes were presented. Also the methods were used to evaluate a field data from Turkey and the results were compared with previous studies. It is determined that both methods give similar results in accordance with the geological structure in the area.

Keywords Self-potential · RF method · Modeling · Geological structure

6.1 Introduction

The Self Potential (SP) Method in geophysical exploration is mainly based on the measurement of potential differences created by natural electrokinetic, electrochemical and thermoelectric sources. The mineralization potentials measured in the areas where mineralization exists are used to investigate the mineral sulfides, which are concealed as natural potential (SP) anomalies, affected by topographic effects in the study areas. However, a self potential anomaly can sometimes be easily separated from topography and regional influences and it is possible to model it as a single

C. Sari (✉) · E. Timur
Engineering Faculty, Department of Geophysics, Dokuz Eylül University, Buca, İzmir, Turkey
e-mail: coskun.sari@deu.edu.tr

E. Timur
e-mail: emre.timur@deu.edu.tr

© The Author(s), under exclusive license to Springer Nature Switzerland AG 2021 165
A. Biswas (ed.), *Self-Potential Method: Theoretical Modeling and Applications in Geosciences*, Springer Geophysics, https://doi.org/10.1007/978-3-030-79333-3_6

polarized structure. Interpretation of self potential data often contains many uncertainties. Overburden structures with different geometric shapes can create the similar self potential fields on earth's surface. Besides, when the electric dipole moment is a constant value and the bounded surface can be identified by a known geometrical shape, a singular solution is obtained from the self potential data. This point of view has been numerically proven by many researchers using various polarized structures classified in four types such as vertical cylinder, horizontal cylinder, sphere and curved thin plates in mining research studies (Yüngül 1950; Banerjee 1971; Fitterman 1979; Bhattacharya and Roy 1981; Abdelrahman and Sharafeldin 1997). In these interpretation methods, the structural factor of the theoretical structure that creates the self potential anomaly is necessary and required as preliminary information. In this study, firstly the structural factor of the buried structure is determined from the self potential data and the environment parameters such as the depth of the structure, the electric dipole moment and the polarization angle of the structure are determined. Estimation of the structural factor is a problem of calculating the solution of a nonlinear relation in $f(q) = 0$ form is applied on the theoretical data and applied to the field data, respectively.

6.2   Theory of the Method

Generalized equation of self-potential anomaly created by many polarized structures can be presented as

$$V(x_i, z, \theta, q) = K \frac{x_i \cdot \cos\theta + z \cdot \sin\theta}{(x_i^2 + z^2)^q}; \quad i = 1, 2, 3, \ldots, N \qquad (6.1)$$

Yüngül (1950), Bhattacharya and Roy (1981). In this equation, z is the depth of the structure, θ is the polarization angle, x_i is the distance or location of measurement, K is electrical dipol moment or polarization amplitude and q is structural shape factor.

The shape factor of a sphere is $q = 1.5$ and the shape factor of the semi-infinite vertical cylinder is $q = 0.5$ for three-dimensional case (3-D), and the shape factor of the horizontal cylinder is $q = 1.0$ for the two-dimensional case. When the shape of the structure approaches a horizontal plate, the structure factor approaches to $q = 0$ (Fig. 6.1).

For all models, $V(0)$ defines the anomaly value at $x_i = 0$ which is the starting point of the measurement profile and it can be expressed as a function of q in Eq. (6.1) as;

$$K = \frac{V(0) \cdot z^{2q-1}}{\sin\theta} \qquad (6.2)$$

If the Eq. (6.1) is equal to 0 and x_0 is the distance between $V(x) = 0$ and starting point of the profile (Fig. 6.1), then we can obtain

Fig. 6.1 Schematic diagram of sphere and horizontal cylinder models

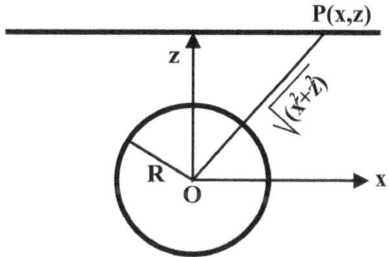

Horizontal cylinder - Sphere

$$\cot \theta = \frac{-z}{x_0} \tag{6.3}$$

After using Eqs. (6.2) and (6.3) in Eq. (6.1),

$$V(x_i, z, q) = \frac{V(0) \cdot z^{2q}(x_0 - x_i)}{x_0(x_i^2 + z^2)^q}, \tag{6.4}$$

Equation (6.4) defines the anomaly value of $x_i = A$ point with the following equation for all models.

$$V(A) = \frac{V(0) \cdot z^{2q}(x_0 - A)}{x_0(A^2 + z^2)^q}, \quad A = \pm 1, 2, 3, 4, 5 \tag{6.5}$$

By using this equation,
$M = \frac{x_0 \cdot V(A)}{V(0) \cdot (x_0 - A)}$ and the depth of the structure is defined as

$$z = \sqrt{\frac{A^2 \cdot M^{1/q}}{1 - M^{1/q}}} \tag{6.6}$$

In these equations, x_0 value is not equal to A, $V(0)$ and $V(A)$. Also it is not equal to 0 either. If the correlation (6.6) is replaced by the Eq. (6.4), the subsequent nonlinear correlation for the structural factor (q) is obtained.

$$V(x_i, q) = \frac{V(0) \cdot A^{2q} \cdot M \cdot (x_0 - x_i)}{x_0 \cdot (x_i^2 + M^{1/q} \cdot (A^2 - x_i^2))^q} \tag{6.7}$$

The requested and unknown structural factor parameter (q) defined in Eqs. (6.6) and (6.7) is achieved by minimalizing the following equation

$$\Phi(q) = \sum_{i=}^{N} \left[L(x_i) - V(0) \cdot A^{2q} \cdot M \cdot W(x_i, q)/x_0 \right]^2 \tag{6.8}$$

In this equation, $L(x_i)$ represents the measured value of the self potential anomaly at x_i point and it is possible to rearrange the Eq. (6.7). If derivative of $\Phi(q)$ by q is equal to 0, then

$$W(x_i, q) = \frac{(x_0 - x_i)}{(x_i^2 + M^{1/q}(A^2 - x_i))^q} \tag{6.9}$$

If derivative of $\Phi(q)$ by q is equal to 0, then a nonlinear equation is achieved.

$$f(q) = \sum_{i=1}^{N} \left\{ \left[L(x_i) - V(0) \cdot A^{2q} \cdot M \cdot W(x_i, q)/x_0 \right] \cdot W(x_i, q) \right.$$
$$\left. \times \left\{ S(x_i, q) \cdot M^{1/q} \cdot \ln M + 2q^2 \ln A \right\} \right\} = 0. \tag{6.10}$$

In this equation, it is possible to define that

$$S(x_i, q) = \ln \left\{ x_i^2 + M^{1/q}(A^2 - x_i^2) \right\}$$

The theory of the equations from (6.2) to (6.10) can be found in Abdelrahman and Sharafeldin (1997) in detail. It is possible to solve Eq. (6.10) for unknown q by using known solution methods (such as Newton-Raphson, Regula False, Bisection, Secant) for the solution of nonlinear equations. In these interpretation methods, it is aimed to obtain the solution for following equation where q_j is initial structural factor and q_f is corrected structural factor.

$$q_f = f(q_j) \tag{6.11}$$

The obtained q_f value is used as the q_j value in the next iteration of the analysis process. The calculation process is repeated until the condition $| q_f - q_j | \leq e$ (e, is a value very close to 0) is met or a defined iteration value is reached. When the (q) structural factor is determined, the depth (z) of the structure causing the self potential anomaly can be determined from the relation (6.6). When the parameter values (z) and (x_0) are known, the polarization angle (θ) can be calculated from the correlation (6.3). Since z, θ, q and $V(0)$ values are known, electric dipole moment (K) can be estimated from the Eq. (6.2).

6.2.1 Determination of X0 and V(0)

If the starting point of a self potential profile is known, x_0 and $V(0)$ values can be determined.

Fig. 6.2 An example of a self potential anomaly of a horizontal cylinder model. x_0 represents zero offset point, $V(0)$ represents the anomaly value at the origin, M and m are the locations of the maximum and minimum anomaly values respectively

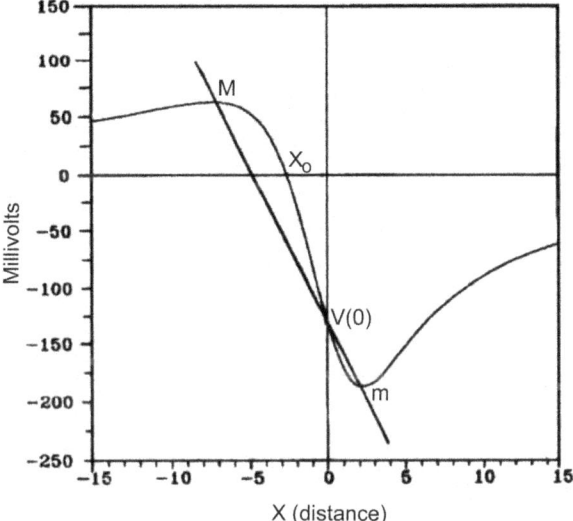

Otherwise, x_0 and $V(0)$ values can be obtained analytically through self potential anomaly (Stanley 1977).

As shown in Fig. 6.2, the point where the **M-m** line cuts the self potential anomaly is determined as $x_i = 0$ and the value of $V(0)$ is calculated at this point.

The base level of the anomaly is located between **M** and $V(0)$ above the point where the anomaly takes the smallest value (**m**). The location of x_0 can be defined with the following equation.

$$x_0 = (x_M - x_m)/2 \qquad (6.12)$$

6.3 Theoretical and Field Applications

For the application of the method, theoretical self potential anomaly values were calculated for a sphere model by taking the depth as $z = 4\ m$, polarization angles as $\theta = 30°$ and $60°$, electric dipole moment as $K = -10000\ mV$. x_0 and $V(0)$ values are estimated from the calculated anomaly values.

The depth of the structure, electrical dipole moment and polarization angle and primarily the structural factor of the underground medium are determined, using the anomaly values with the root finding methods of the nonlinear equations such as **Regula False** or **Bisection** (Appendix) by entering the detected anomaly values and x_0 and $V(0)$ values as input to the prepared computer program. The obtained solutions are presented in tables with the related comparative anomalies (Figs. 6.3 and 6.4).

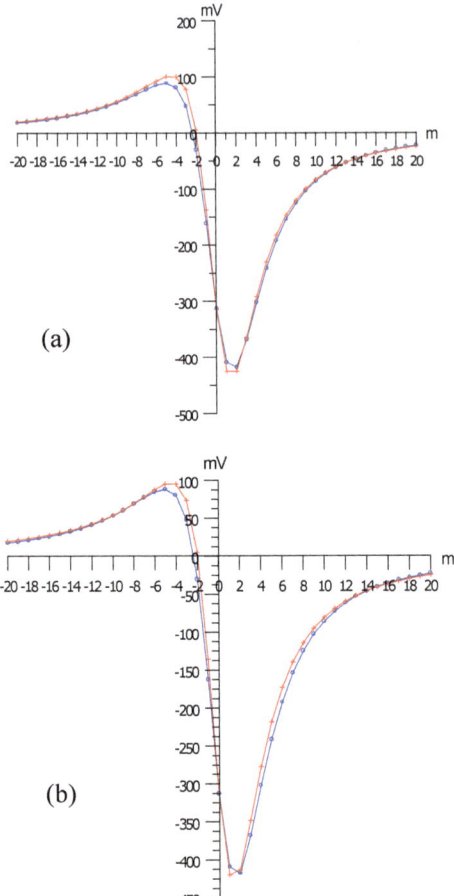

(a)

(b)

PARAMETER	Model	Regula False Solution	Bisection Solution
Structural Factor *(q)*	1.5	1.414	1.368
Depth *(z)*	4.0 m	3.43 m.	3.22 m.
Polarization Angle *(θ)*	30°	29.65°	31.17°
Electric Dipol Moment *(K)*	-10000	-10066.16	-10042.83

Fig. 6.3 a Observed and calculated and model parameters for Regula False solution using 30°
sphere model, **b** Observed and calculated and model parameters for Bisection solution. Blue curve
and red curve represent the observed and calculated models respectively

Similar procedure has been applied to the self potential anomaly values calculated
by considering the structure depth as $z = \boldsymbol{4\,m}$, polarization angles as $\boldsymbol{\theta} = 30°$ and
$\boldsymbol{60°}$, electric dipole moment as $\boldsymbol{K} = \boldsymbol{-10000\,mV}$ for a cylinder model. The obtained
outcomes are presented in tables with the related anomalies in comparison to each
other (Figs. 6.5 and 6.6).

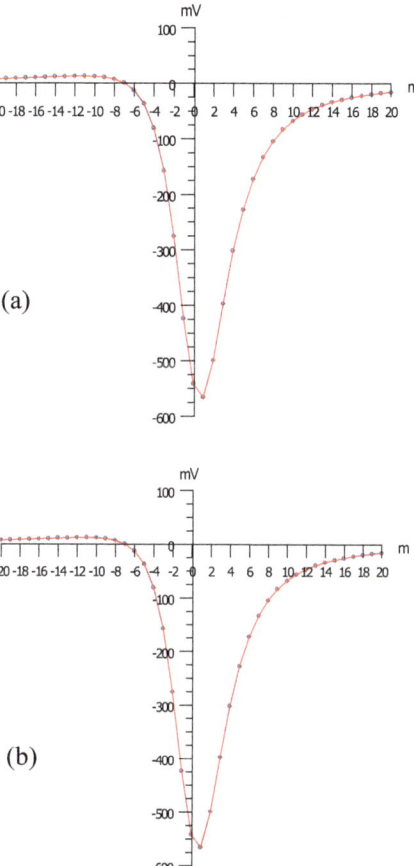

PARAMETER	Model	Regula False Solution	Bisection Solution
Structural Factor *(q)*	1.5	1.499	1.497
Depth *(z)*	4.0 m	4.00 m.	3.99 m.
Polarization Angle *(θ)*	60°	60.12°	60.16°
Electric Dipol Moment (K)	-10000	-9954.82	-9868.72

Fig. 6.4 **a** Observed and calculated and model parameters for Regula False solution using 60° sphere model, **b** Observed and calculated and model parameters for Bisection solution. Blue curve and red curve represent the observed and calculated models respectively

After the application of the method to the theoretical anomaly values, it has been applied to the self potential anomaly data of Süleymanköy interpreted by Yüngül (1950) in the Ergani Copper Mine region in Turkey. These anomalies were also evaluated with the theoretically explained method and the depth of the structure, the electrical dipole moment and polarization angle parameters of the overburden mass, especially the structural factor of the underground structure that constitutes the

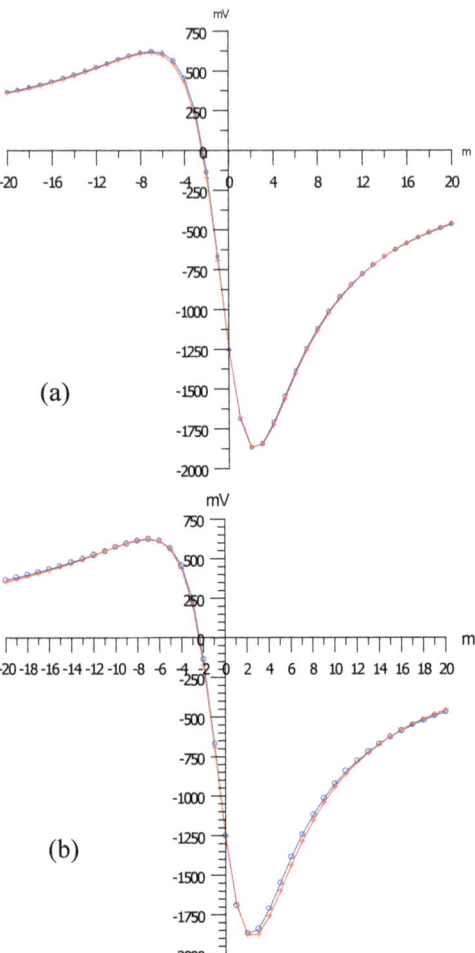

PARAMETER	Model	Regula False Solution	Bisection Solution
Structural Factor (q)	1.0	1.022	1.051
Depth (z)	4.0 m	4.22 m.	34.41 m.
Polarization Angle (θ)	30°	29.63°	28.55°
Electric Dipol Moment (K)	-10000	-11364.77	-13437.10

Fig. 6.5 a Observed and calculated and model parameters for Regula False solution using 30° cylinder model, **b** Observed and calculated and model parameters for Bisection solution. Blue curve and red curve represent the observed and calculated models respectively

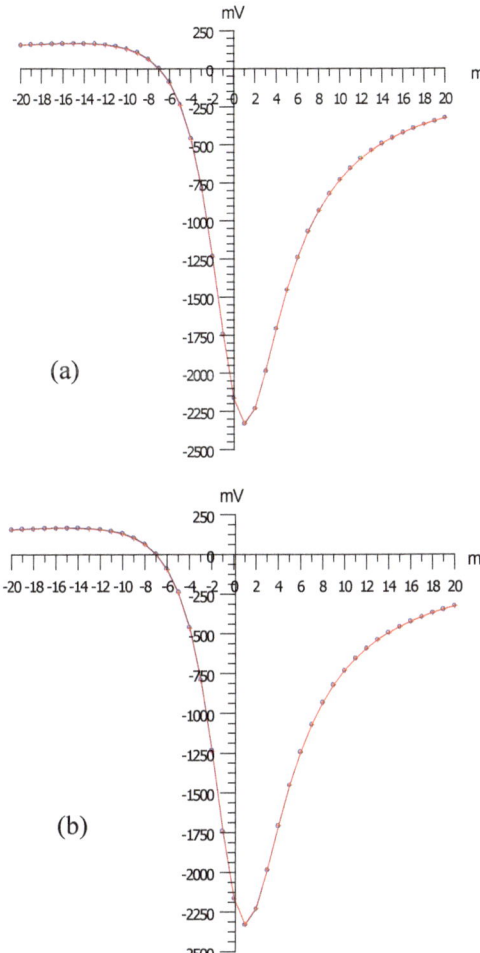

PARAMETER	Model	Regula False Solution	Bisection Solution
Structural Factor *(q)*	1.0	0.999	0.998
Depth *(z)*	4.0 m	4.01 m.	4.0 m.
Polarization Angle *(θ)*	60°	60.21°	60.23°
Electric Dipol Moment (K)	-10000	-9960.42	-9940.50

Fig. 6.6 a Observed and calculated and model parameters for Regula False solution using 60°
cylinder model, **b** Observed and calculated and model parameters for Bisection solution. Blue
curve and red curve represent the observed and calculated models respectively

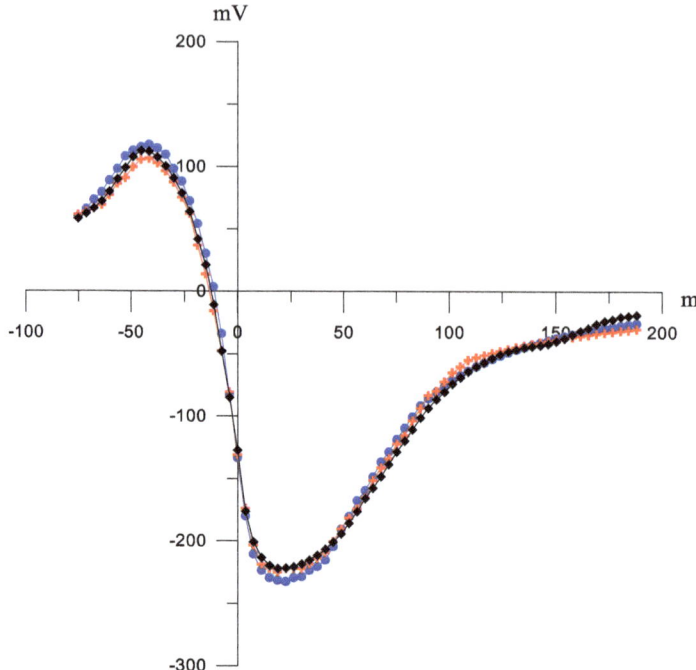

Fig. 6.7 Yüngül (1950) anomaly, Regula False and Bisection method results. Black line, blue line and red line indicate the SP anomaly, Regula False and Bisection results respectively

anomaly, were tried to be determined. The determined results were given by Yüngül (1950), Bhattacharya and Roy (1981) and Abdelrahman and Sharafeldin (1997) in comparison with the results obtained using nomogram methods (Fig. 6.7).

6.4   Conclusion

We do not have any information about the type of the structure when using auxiliary curves (nomograms) in order to define the structure parameters of a polarized structure embedded from SP anomalies. When using auxiliary curves, the structure is considered to be a cylinder or a sphere and evaluation is made in accordance with this condition. In the evaluations made using the nonlinear equation solution methods introduced in this study, the structure factor is determined first. As a result of the evaluation of self potential anomalies caused by model structures such as spheres and cylinders, the structure parameters related to the model structures were determined with great sensitivity and accuracy with both of the methods. Upon its success in theoretical studies, the methods were applied on field data and the outcomes were compared with the previous results. The problem of determining the depth of a horizontal cylinder or sphere embedded from the SP anomalies was changed to a proposed process of determining a solution of a non-linear equation. It is quite easy and practical to apply the described methods. Advantages of the method defined compared to previous methods using distance values and nomograms in only a few points are (1) using of all observation values, (2) the application of the methods are automated, and (3) less sensitive to errors in self potential anomalies.

Appendix: Determination of Roots of Non-linear Equations

Regula False Method

In this method, convergence to the real root is relatively slow, but since it is always convergence, it is more advantageous than simple iteration and Newton-Raphson methods. This method requires two initial values. Figure 6.8 indicates the graphical definition of the method. Let X_L and X_R be the two initial values. It can be assumed that X_L is located in the right side of the root. The line connecting the $[X_L, f(X_L)]$ and $[X_R, f(X_R)]$ points intersects with the X-axis at the X_M point. The $f(X_M)$ value is compared to $f(X_L)$ and $f(X_R)$ points. If X_L and X_R are located at different sides of the X_M point, then the signs of $f(X_L)$ and $f(X_R)$ are different. The corresponding points are also above and below the X-axis. After that, the signs of $f(X_M)$ and $f(X_L)$ are compared. If the signs are the same then the root (x) is not between X_L and X_M. If they have different signs, the root (x) is between X_L and X_M. Now, the X_M point becomes new X_R. The operation is repeated with the line connecting the new X_R and X_L points. In this way, the difference containing the root becomes smaller. Generally the procedure can be summarized with the following two steps.

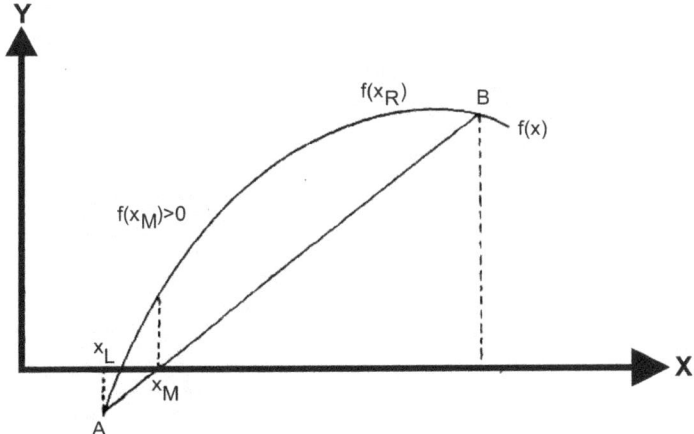

Fig. 6.8 Graphical representation of regula false method

1. If $f(X_L) \cdot f(X_M) < 0$ then X_R and X_M change their locations.
2. If $f(X_R) \cdot f(X_M) < 0$ then X_L and X_M change their locations.

The X_M can be defined with the following equations.

$$X_M = X_L - \frac{X_R - X_L}{f(X_R) - f(X_L)} \cdot f(X_L)$$

$$X_M = \frac{X_L f(X_R) - X_R f(X_L)}{f(X_R) - f(X_L)}$$

Bisection Method

Bisection method is very similar with the Regula False method, but it is much more simple. Let X_L and X_R be the two initial values. In this method there is no need to connect two points with a line The X_M value is calculated with the arithmetic mean of the initial values automatically using the following equation (Çağal 1998).

$$X_M = X_L + \frac{X_R - X_L}{2} = \frac{X_L + X_R}{2}$$

References

Abdelrahman EM, Sharafeldin SM (1997) A least squares approach to depth determination from self-potential anomalies caused by horizontal cylinders and spheres. Geophysics 62(1):44–48

Banerjee B (1971) Quantitative interpretation of self potential anomalies of some specific geometric bodies. Pure Appl Geophy Pageoph 90(1):138–152

Bhattacharya BB, Roy N (1981) A note on the use of a nomogram for self-potential anomalies. Geophys. Prosp. 29:102–107

Çağal B (1998) Sayısal Analiz. Birsen Publishing, 501 pp., İstanbul

Demidovich BP, Maron IA (1973) Computational mathematics, Mir Publ

De Witte L (1948) A new method of interpretation of self-potential field data. Geophysics 13(4):600–608

Fitterman DV (1979) Calculations of self-potential anomalies near vertical contacts. Geophysics 44(2):195–205

Heiland CA (1940) Geophysical exploration. Prentice-Hall, Inc., New York

Heiland CA (1940) Geophysical exploration, Hafner Publ. Co

Meiser P (1962) A method of quantitative interpretation of self-potential measurements. Geophys Prosp 10:203–218

Murty S, Haricharen P (1985) Nomogram for the complete interpretation of spontaneous potential profiles over sheet-like and cylindrical two-dimensional sources. Geophysics 50:1127–1135

Stanley JM (1977) Simplified magnetic interpretation of the geologic contact and thin dike. Geophysics 42(6):1236–1240

Yüngül S (1950) Interpretation of spontaneous polarization anomalies caused by spherical ore bodies. Geophysics 15:237–246

Chapter 7
Ensemble Kalman Inversion for Determining Model Parameter of Self-potential Data in the Mineral Exploration

Sungkono, Erna Apriliani, Saifuddin, Fajriani, and Wahyu Srigutomo

Abstract Self-potential (SP) method has been increasingly popular in geophysical exploration of mineral resources using an assumption that the causative bodies have idealized geometry (horizontal and vertical cylinders, sphere, and 2-D inclined sheet). In this study, ensemble Kalman inversion (EKI) is proposed to analyze the data and hence determine the associated model parameters. As indicated by its name, the algorithm is constructed based on the iterative ensemble Kalman filter. EKI is applied perform inversion of noisy synthetic and field SP data. The field data were the Neem-ka-Thana and Surda SP anomalies obtained from India, Malachite mine anomaly from USA, and KTB anomaly from Germany containing information of single and multiple anomalous bodies. The EKI exhibits high effectively and accuracy in determination of model parameters and model uncertainty.

Keywords Ensemble kalman inversion · Self-potential data · Model uncertainty · Ore deposit

Sungkono (✉) · Saifuddin
Department of Physics, Institut Teknologi Sepuluh Nopember (ITS), Jl. Arief Rachman Hakim 60111, Surabaya, Indonesia
e-mail: hening_1@physics.its.ac.id

E. Apriliani
Department of Mathematics, Institut Teknologi Sepuluh Nopember (ITS), Jl. Arief Rachman Hakim 60111, Surabaya, Indonesia

Fajriani
Department of Physics, Universitas Samudra, Jl. Prof. Syarief Thayeb, Meurandeh, Kota Langsa, Aceh, Indonesia

W. Srigutomo
Physics of Earth and Complex System, Department of Physics, Faculty of Mathematics and Natural Sciences, Institut Teknologi Bandung (ITB), Bandung, Indonesia

© The Author(s), under exclusive license to Springer Nature Switzerland AG 2021   179
A. Biswas (ed.), *Self-Potential Method: Theoretical Modeling and Applications in Geosciences*, Springer Geophysics, https://doi.org/10.1007/978-3-030-79333-3_7

7.1 Introduction

The self-potential method is categorized as a passive geophysical method since its sources are generated from difference of natural potential in the subsurface. The potential is caused by several phenomena including thermoelectric, electrochemical, thermoelectric, and electro kinetic fields within the Earth's interior (Biswas 2017; Revil and Jardani 2013). The SP method has been widely used for subsurface characterization such as groundwater and mining explorations, geothermal investigation, archeological study, paleo-shear zone identification, and cavity detection (Al-Saigh et al. 1994; Arora et al. 2007; Fernández-Martínez et al. 2010; Giang et al. 2018; Mauri et al. 2012; Mehanee 2015; Moore et al. 2011; Sungkono 2020a; Sungkono and Warnana 2018).

To delineate a subsurface profile from SP data, it is required several data analyses. The analyses methods consist of characteristic point method (Fedi and Abbas 2013), nomograms (Bhattacharya and Roy 1981), window curves method (Hafez 2005), Gauss-Newton (GN) inversion (Abdelrahman et al. 2004; Candra et al. 2014; Mehanee 2014), signal analysis including Hilbert transform, continuous wavelet transform, horizontal gradient (Abdelrahman et al. 2004; Di Maio et al. 2017; Mauri et al. 2011; Sundararajan and Srinivas 1996), and inversion using global optimization (GO) methods such as neural networks (El-Kaliouby and Al-Garni 2009), particle swarm optimization (Monteiro Santos 2010), differential evolution variants (Balkaya 2013; Sungkono 2020b), genetic algorithm (GA) and simulated annealing (SA) (Göktürkler and Balkaya 2012), very fast SA (VFSA) (Biswas and Sharma 2017; Sharma and Biswas 2013), genetic price algorithm (GPA) (Di Maio et al. 2019, 2016), black hole algorithm (BHA) (Sungkono and Warnana 2018), whale optimization algorithm (WOA) (Abdelazeem et al. 2019), and flower pollination algorithm (FPA) (Sungkono 2020a). Additionally, combining method between horizontal derivative and PSO are also applied to interpret the SP data (Elhussein 2020; Essa 2020), where the horizontal derivative is used to eliminate regional effect, while the PSO is employed to invert the SP data.

Inversion of SP data is ill-posed problems and has non-unique solutions. Accordingly, different inversion approaches have been developed to handle these problems as described above. As description in the free lunch theorem in optimization (in the case is inversion problem), a method cannot handle all of problems (Wolpert and Macready 1997). Thus, in this paper an inversion method based on ensemble Kalman filter is proposed, further called ensemble Kalman inversion (EKI), for determining model parameter and its uncertainty assuming the sources of anomaly are caused by simple geometric structures (horizontal cylinder, vertical cylinder, sphere, and 2-D inclined sheet). The EKI has several advantages including does not require first derivative of function, ease implementation and application, and free tuning parameter (Chada et al. 2019, 2018).

7.2 Methodology

7.2.1 Ensemble Kalman Inversion

The ensemble Kalman Filter (EnKF) consists of Monte Carlo approach, which uses an ensemble size Ne to determine the statistical features including the covariance and mean of the model parameter (Evensen 2009, 2003; van Leeuwen and Evensen 1996). The covariance matrix is used to handle a unique solution in an inversion (Cho and Olivera 2014; Oliver and Chen 2009) obtained from the ensemble size. The EnKF with some iterations (called iterative EnKF) (Gu and Oliver 2007; Li and Reynolds 2009) can be applied for determining model parameters. However, this approach requires a Jacobian matrix in the Gauss-Newton to minimize the least-square approach. Furthermore, EnKF inversion (EKI) or iterative EnKF, with derivative-free is clearly proposed and described for solving inversion problems by previous authors (Chada et al. 2019, 2018; Iglesias 2016; Iglesias et al. 2013). However, EKI is linked to regularized least square-problems (Iglesias et al. 2013). It means that the EKI is possible for solving nonlinear inverse problems with free gradient of function calculation.

The EKI method for finding the solution of inverse problems of X given observations data d_{obs} of the form

$$d_{obs} = f(X) + \eta \tag{7.1}$$

where $f(X)$ indicates a forward modeling, while η is a noise of observed data. A probabilistic distribution $P(X|d_{obs})$ is the solution in the inverse problem. $P(X|d_{obs})$ denotes the model parameter X is given on observation data d_{obs}. The method is based on the Bayes' theorem as following

$$P(X|d_{obs}) \propto P(X)P(d_{obs}|X) \tag{7.2}$$

where $P(X)$ and $P(d_{obs}|X)$ indicate a prior distribution of the model parameter and a likelihood, respectively, while $P(X|d_{obs})$ denotes the posterior parameter distribution. Direct sampling of posterior parameter generally uses a Monte Carlo (Markov Chain or Sequential) sampling which requires millions of evaluations of the forward models (Sungkono and Santosa 2015; Zhang et al. 2020). Therefore, an efficient method for estimating the sampling model parameter X has applied using global optimization method (GOM) with a threshold (Sungkono 2020a; Sungkono and Warnana 2018).

Assuming that both the observation noises and prior are Gaussian distribution, the posterior in the Eq. (7.2) can be expressed as follows:

$$P(X|d_{obs}) \propto \exp(-S(X)) \tag{7.3}$$

where $S(X)$ is an objective function. In the inversion process using stochastic or deterministic approaches, it is generally to maximize posterior or minimize the objective function. The objective function for the EKI is presented in Eq. (7.4).

$$S(X_k) = \frac{1}{2}\left[r_k\left(C^d\right)^{-1}r_k^T\right] + \frac{1}{2}\left[\Delta X_k\left(C_k^X\right)^{-1}\Delta X_k^T\right] \tag{7.4}$$

where ΔX_k reflects change of states for kth iteration and r_k denotes the difference between observed d_{obs} and calculated $f(X_k)$ data for kth iteration $r_k = d_{obs} - f(X_k)$. Additionally, C^X and C^d, respectively, indicate the covariances of the prior model parameters and the noises disturbed the observed data. The covariance C^X is obtained from a limited size ensemble of model parameter. Using Gauss-Newton approach, the Eq. (7.2) can easily determine change of model parameter reflecting minimum objective function (Chada et al. 2018; Iglesias et al. 2013; Zhang et al. 2020).

$$\Delta X_k = K_{k-1}\{r_{k-1} + e_{k-1}\} \tag{7.5}$$

where the of e_{k-1} is a measurement error with zero mean and has covariance C_d, $e_{k-1} = \sqrt{C^d}N(0, 1)$, while K_{k-1} denotes a Kalman gain for (k−1)th iterations.

The Eq. (7.5) shows that the state changing controlled by Kalman gain, where the parameter can be represented by (Chada et al. 2018; Iglesias et al. 2013; Zhang et al. 2020):

$$K_{k-1} = C_{k-1}^{Xd}\left[C_{k-1}^{XX} + C^d I\right]^{-1} \tag{7.6}$$

where $C^{Xd} = \mathrm{cov}(X, d)$ indicates a cross-covariance between model parameters X and calculated model, while $C^{XX} = \mathrm{cov}(X, X)$ represents an auto-covariance of model parameters X. In addition, I denotes identity matrix. The Eq. (7.6) indicates that the Kalman gain performance is controlled by the auto-covariance of the model parameters derived from a limited size ensemble. It means that the EKI performance is highly controlled by size of the ensemble number. A small ensemble size can produce a significant error for calculating a covariance (Wang et al. 2010). In addition, the Kalman gain can be singular when the ensemble of the parameter models is identic and C^d is too small.

Furthermore, an updating model parameter for kth iteration X_k can be written as follows:

$$X_k = X_{k-1} + \beta \Delta X_k \tag{7.7}$$

where β denotes the step length parameter, which can be determined using standard line search. The value of β can be used a random number of −1 or 1 (Zhang et al. 2020) or the value can be set between 0 and 1 (Liu et al. 2020). In this paper, the β value uses a random number between 0 and 1 for all ensemble numbers. It means that different value of step length parameter for each ensemble member, where the condition can improve the efficiency of the algorithm (Wang et al. 2010).

In the above, EKI generally consists of three steps (Chada et al. 2019, 2018, 2019; Iglesias 2016), such as: (1) generate an ensemble size of model parameters; (2) calculate forward modeling for each ensemble member of model, (3) update each ensemble member of model parameter using Eq. (7.7). In this paper, to improve the EKI performance, both a greedy selection and a boundary handling are applied, that is to select each ensemble member for existing in the future iterations and keep the model parameter in the desired ranges, respectively. the Greedy selection is used because the updated model parameter in the EKI does not guarantee to have better the objective function as compared to the previous model parameter before updated (Wang et al. 2010).

Furthermore, the EKI process requires some parameters including an ensemble number Ne, a covariance of noise disturbed observed data C^d, and search bounds of model parameters $[X_{min}, X_{max}]$. The EKI algorithm firstly generates the ensemble number of the model parameters X, which uses random approaches in the search spaces $[X_{min}, X_{max}]$, where X_{min} and X_{max} denote lower and upper bounds, respectively. The model parameters $X = [X^{[1]}, X^{[2]}, \ldots, X^{[Ne]}]$ contain the ensemble number Ne, where each model parameter $X^{[i]}$ accommodates the number of the estimated model parameter, generally called d dimension, $X^{[i]} = (X^{[i],1}, X^{[i],2}, \ldots, X^{[i],d})$. The second step, a forward modeling for each ensemble of the model parameter is estimated to calculate objective function. The third step, auto- and cross-covariance matrices (C^{XX} and C^{Xd}) are determined for estimation of Kalman gain (Eq. 7.6). The fourth step, the ensemble member of parameter model is updated for kth iteration (using Eq. 7.9) and checked the boundary handling of the model parameter. The boundary handling uses in the paper as follows:

$$
X_k = \begin{cases} 2X_{min} - X_k & if \quad X_k < X_{min} \\ 2X_{max} - X_k & elseif \quad X_k > X_{max} \\ X_k & else \end{cases} \tag{7.8}
$$

Furthermore, the objective function for each ensemble is calculated and applied using a greedy selection approach for the model parameter for selecting to further iteration. The selection can be expressed as following

$$
X_k = \begin{cases} X_k & if \ obj(X_k) \leq obj(X_{k-1}) \\ X_{k-1} & else \end{cases} \tag{7.9}
$$

where $obj(X_k)$ indicates the objective function of model parameter for iteration kth.

The last step checks whether iteration has to stop, if the iteration number reaches the maximum iteration, the iteration is stop, while the others is repeat the loop on go to the third step.

7.2.2 Forward Modeling

Forward modeling $F(X)$ determines calculated data for giving model parameter X. The Self-Potential (SP) anomaly is assumed the idealized body (e.g. sphere, horizontal cylinder and vertical cylinder), at a point x_i can be determined as (Mehanee 2014; Monteiro Santos 2010)

$$v(x_i) = K \frac{(x_i - D)\cos(\theta) + h\sin(\theta)}{\left((x_i - D)^2 + h^2\right)^q} \tag{7.10}$$

where K and θ, respectively, denote a polarization magnitude and an angle, while the anomalous source's center for depth and position are represented by h and D, respectively. In addition, q denotes a shape factor, which is 1.5, 1.0 and 0.5 for a sphere, horizontal cylinder, and vertical cylinder, respectively. Furthermore, model parameters X for the forward modeling contain K, θ, h, D and q. Moreover, SP anomaly sourced by 2D inclined sheet (Fig. 7.1 c) can be expressed as follow (Biswas and Sharma 2014a):

$$v(x_i) = K \log \left[\frac{\{(x_i - D) - a\cos\theta\}^2 + (h - a\sin\theta)^2}{\{(x_i - D) + a\cos\theta\}^2 + (h + a\sin\theta)^2} \right] \tag{7.11}$$

where θ describes the angle of inclination, while a indicates the half-width of the sheet. The other is same with notation in the Eq. (7.11). Furthermore, the forward modeling $F(X)$ for calculated SP data sourced by multiple anomalies is estimated using the summation from each responses resulted by individual anomalies (Biswas and Sharma 2014c).

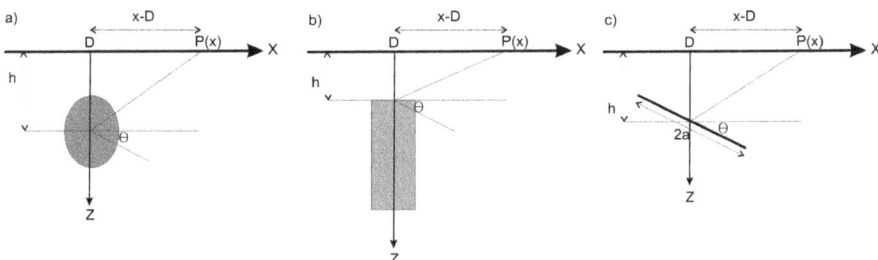

Fig. 7.1 Visualization of Model parameters from SP anomaly for sphere and horizontal cylinder (**a**), vertical cylinder (**b**), and 2D inclined sheet (**c**) in subsurface

7.2.3 Inversion Using EKI

As description above, EKI requires two parameters including ensemble numbers and covariance noises disturbed the observed data C^d. The ensemble number in the EKI can be set as 50, 100, 150, etc., the higher ensemble number needs the higher time consuming for forward calculation. Thus, it requires long iterations for the ensemble number to converge. Furthermore, C^d generally does not exist, consequently in this paper C^d designed as $C^d = \sigma d_{obs}$, where $\sigma = 0.02$ for a high amplitude of SP data, while $\sigma = 0.2$ for a low amplitude of SP data. The step is done so that the Eq. (7.6) is not singular.

Furthermore, Eq. (7.7) shows that the inversion process using EKI is updated using model parameters X_k. In this case, the model parameters X_k consists of K, θ, h, D, and q for all anomalous sources assuming idealized bodies, while if the SP anomaly is sourced by inclined sheet, the model parameter contains K, x_1, z_1, x_2, and z_2 for all anomalous sources.

Obtained ensemble number of model parameters from EKI can be used to estimate uncertainties of model parameters (Cho and Olivera 2014; Gu and Oliver 2007), but sometimes the model parameters from the last iteration is over fitting. Thus, the uncertainty in the paper is assumed as limit of acceptability approach (Sungkono 2020a; Vrugt and Beven 2018) or objective function's trade-off (Fernández-Martínez et al. 2013; Fernández-Muñiz et al. 2019; Laby et al. 2016). Consequently, before conducting a greedy selection (a step in EKI algorithm), an objective function for each ensemble member of model parameters X is calculated using Eq. (7.12) (Monteiro Santos 2010):

$$Obj(X_k) = \frac{2\|d_{Obs} - F(X_k)\|}{\|d_{Obs} - F(X_k)\| + \|d_{Obs} + F(X_k)\|} \tag{7.12}$$

The geophysical data inversion does not have a unique solution due noise containing in observed data, a physical assumption in a forward modeling, and also an inherent theoretical relationship between observed and model parameters. Consequently, several model parameters with different combination that have calculated data match the observed data with some tolerances, called posterior distribution model (PDM), which is can be expressed ($|d_{dobs} - F_i(X)| \leq \sigma d_{obs}$) (Vrugt and Beven 2018). Where σ is standard deviation of noise in observed data.

As a description in the section about EKI, the EKI indicates that exploration and exploitation capabilities have existed. The exploration capability depends on ensemble number and the covariance noises disturbed the data, where both parameters are directly proportional to exploration capability of the EKI, while exploitation property indicated by step length parameter in Eq. (7.7), where the parameter is inversely proportional to exploitation property in EKI. In addition, a greedy selection in Eq. (7.9) also improves exploitation capability. The exploration capability is to avoid trap in local minima, while exploitation capability is to speed up the

convergence. EKI is expected to have balance in both properties so that EKI is able
to quickly find a solution correlated to global minima

7.3 Synthetic Model

EKI performance is tested before application to field SP data. Based on the described
above, the EKI depends on the ensemble numbers Ne and assumption of covariance
noises C^d disturbs in observed data. In the EKI process, Ne sets 50, 500 maximum
iteration is used, and $C^d = 0.02d_{obsi}$. To evaluate the EKI algorithm, the method is
applied to synthetic data sourced an anomaly (horizontal cylinder). After that, the
EKI is also tested to several noisy data. The last but not least, the performances of the
EKI algorithm is evaluated for multiples anomalies (sphere and semi-infinite vertical
cylinder). Because, EKI is based on Monte Carlo approach, each process of EKI has
a different result. Consequently, EKI has proceeded five times and the best fitting is
used for the analysis.

7.3.1 EKI in Single Anomaly

To evaluate EKI performances for SP data inversion, synthetic data with and without
10% of a Gaussian noise are reconstructed by a true model (2D inclined sheet) in
Table 7.1. In the inversion, search spaces of model parameters also presented in the
Table 7.2. Figure 7.2a shows the median of an objective function revealed by EKI.
The figure indicates that the median of the objective function relatively decreases
with increasing iteration until convergence. The condition is resulted by a greedy
selection to improve exploitation performance of EKI, in the standard EKI this step

Table 7.1 EKI algorithm applied to 2-D inclined sheet model with and without 10% Gaussian noises

Parameters	Ranges	True models	Noise-free	10% noise
K (mVm)	−1000–1000	50	50.00 ± 0.06	53.53 ± 14.13
D(m)	0–100	55	55.00 ± 0.01	55.05 ± 0.53
a (m)	1–40	10	10.00 ± 0.01	10.62 ± 0.96
θ (°)	0–180	150	150.00 ± 0.02	151.58 ± 1.34
h (m)	0–80	12	12.00 ± 0.01	11.12 ± 2.22
Minimum of the objective function			7.87E-07	7.47E-02
Median of the objective function			8.21E-06	7.48E-02
Interquartile of the objective function			2.81E-05	2.00E-04

Table 7.2 EKI algorithm tested to the two synthetic models (sphere and vertical cylinder) with and without 10% Gaussian noise

Parameters	Souces	K (mVm)	D (m)	h (m)	θ (°)	q
Ranges	1	0–2000	−150–150	0–100	0–180	0.1–1.8
	2	−700–700	50–150	0–100	0–180	0.1–1.8
True parameters	1	1000	−100	7	30	1.5
	2	−400	30	30	60	1
EKI for free noise	1	1000.69 ± 9.64	−100.00 ± 0.01	7.00 ± 0.01	30.01 ± 0.04	1.5 ± 0.00
	2	−401.45 ± 5.82	30.00 ± 0.02	30.01 ± 0.07	59.99 ± 0.07	1.00 ± 0.00
EKI for 10% noise added	1	1488.08 ± 98.89	−99.83 ± 0.03	7.47 ± 0.07	29.60 ± 0.17	1.56 ± 0.01
	2	−553.37 ± 20.38	31.34 ± 0.05	31.52 ± 0.18	59.21 ± 0.15	1.04 ± 0.00

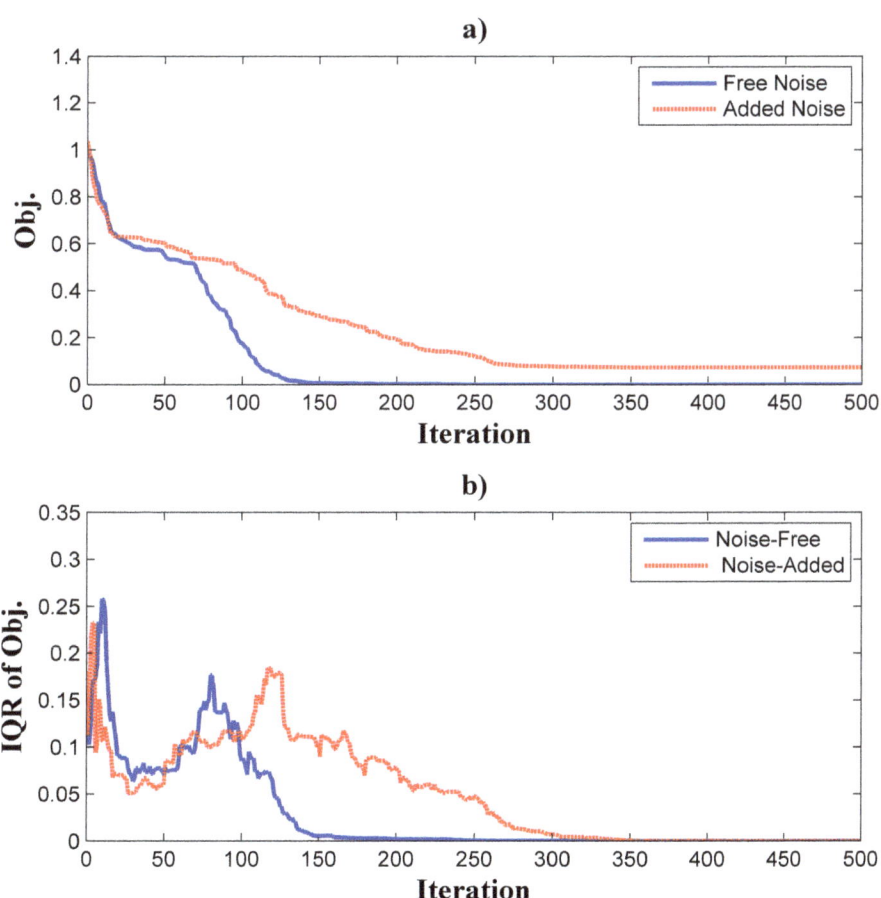

Fig. 7.2 The statistics **a** Median; (**b** interquartile) of objective function resulted by EKI in SP data inversion with and without noise

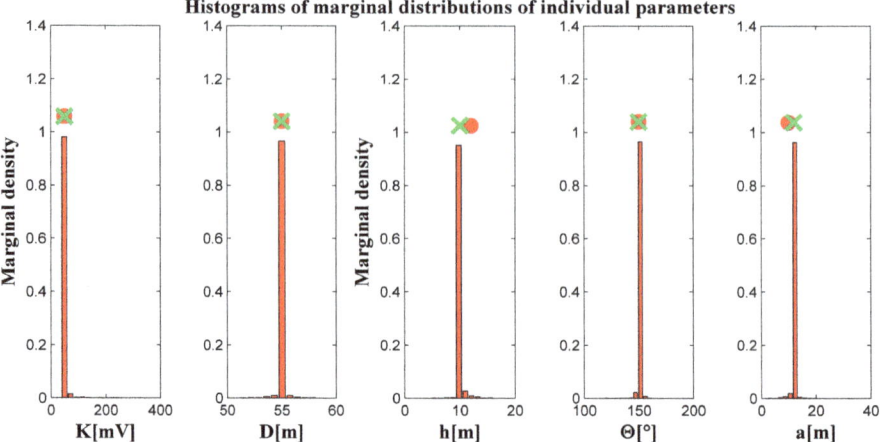

Fig. 7.3 Histogram of DPM for each parameter of noise-free SP data revealed by EKI. The highest probability of PDM correlates with median of PDM (crosses) and true model parameter (dots). Consequently, EKI is capable to determine PDM

does not exists. The figure also indicates the noise-added data can shift the objective function (Fernández-Martínez et al. 2014a, b; Sungkono 2020b). Furthermore, Fig. 7.2b indicates interquartile (iqr) of the objective function for each iteration, which indicates that the EKI has exploration capability until 150th and 300th iteration for the noise-free and the noise-added data, respectively. Figure 7.2b does not always decrease with increasing iteration. It means that the EKI has good explorative capability. Consequently, the solution of model parameter cannot trap in the local minimum.

Additionally, a PDM is constructed by EKI with a tolerance as described in Fig. 7.3. Crosses indicate the median of PDM, while dots denote true model parameters. The Figure demonstrates that the median of PDM is very close to true model parameter. Figure 7.4a, c show data with and without noises are fitted to calculate data from the median of PDM resulted by EKI process, respectively. Both figures demonstrate that the median of calculated data has very good fitting. Table 7.1 shows the statistically (median±iqr) of model parameter from PDM determined by EKI, which is visualized in Fig. 7.4b, d. The results demonstrates three things including EKI can provide PDM using a tolerance of an objective function, the EKI is capable and accurately to solve inversion of SP data, and the noises-added in the data can increase the uncertainty of the model parameter.

7.3.2 EKI in Multiple Anomalies

An SP anomaly is usually caused by more than one of sources. In the section, EKI is applied for solving two anomalous sources with (10% of Gaussian noise) and without

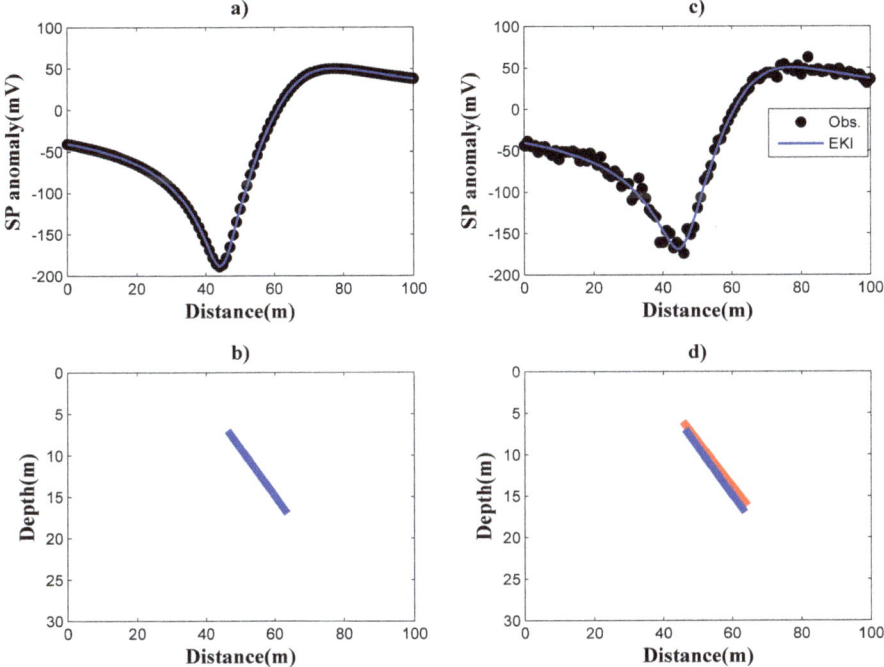

Fig. 7.4 Inversion of synthetic SP data with and without noises results uses EKI. **a, c** Fitting of synthetic and calculated data (median of PDM) for noise-free and noise-added, respectively; **b, d** The sketches are showing the model parameter (blue lines) determined by EKI for noise-free and noise-added of SP data, respectively. The model parameter (blue lines) inverted by EKI is very close to true model (red lines)

noise-added as in the Fig. 7.5a, c, respectively. Model parameters of both data are optimizing using EKI with search ranges as in Table 7.4. Table 7.4. also demonstrates that the EKI is successfully applied for two anomalous SP data inversion, where the true model parameter is closed to the medians of PDM with and without noise-added. The results can be viewed in the Fig. 7.5b, d for without and with noises added, respectively. Again, Table 7.4 also shows the noises-added can increase the uncertainty of model parameter in the inversion result.

7.4   Field Examples

In order to demonstrate EKI performance in the real data, three fields of SP data in different areas (e.g. Neem-Ka-Thana anomaly in India, Malachite mine in USA, and Surda anomaly, India). In the inversion process, EKI applies 500 and 1000 of maximum iteration for single and multiple anomalous sources, respectively, C^d set as

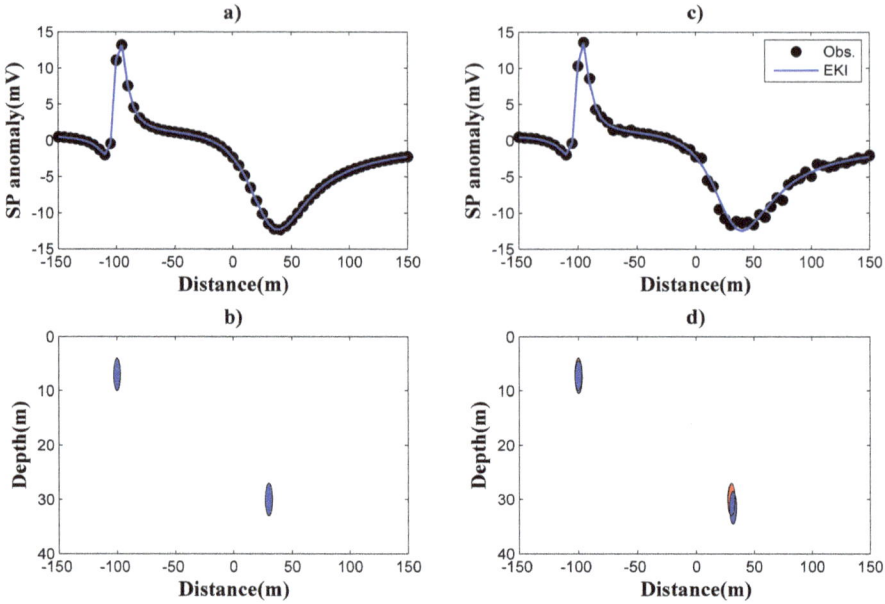

Fig. 7.5 The description is like on the Fig. 7.4, but for two sources of SP anomalies

$C^d = 0.02d_{obsi}$ and 50 is used in ensemble number. The EKI process are conducted five times to know consistency and stability of the algorithm.

7.4.1   Neem-Ka-Thana, India

The SP anomaly in the Neem-ka-Thana was sourced by a deposit of copper belt in the Ahirwala, India (Reddi et al. 1982). The deposit is commonly restricted within shear planes and faults, which has concentrated between 0.6 to 1.2%. The measured SP anomaly in Neem-Ka-Thana can be shown in Fig. 7.6a, c. The anomalies (Fig. 7.6a, c) were analyzed by various authors (Agarwal and Srivastava 2009; Balkaya 2013; Biswas and Sharma 2015; Göktürkler and Balkaya 2012; Sungkono 2020b). The interpretation of the data can be classified into two parts including a single body (main anomaly located around 170 m in distance) (Balkaya 2013; Biswas and Sharma 2015; Göktürkler and Balkaya 2012; Sungkono 2020b) and multiple anomalies (two peaks negative anomalies before main negative anomaly) (Biswas and Sharma 2015; Sungkono 2020b). A global optimization method (GOM) is generally applied to determine model parameters of the SP data (Balkaya 2013; Biswas and Sharma 2015; Göktürkler and Balkaya 2012; Sungkono 2020b).

The SP data are inverted using EKI considering that the anomaly contains an anomaly (body around the high negative anomaly) with search spaces ranges can be

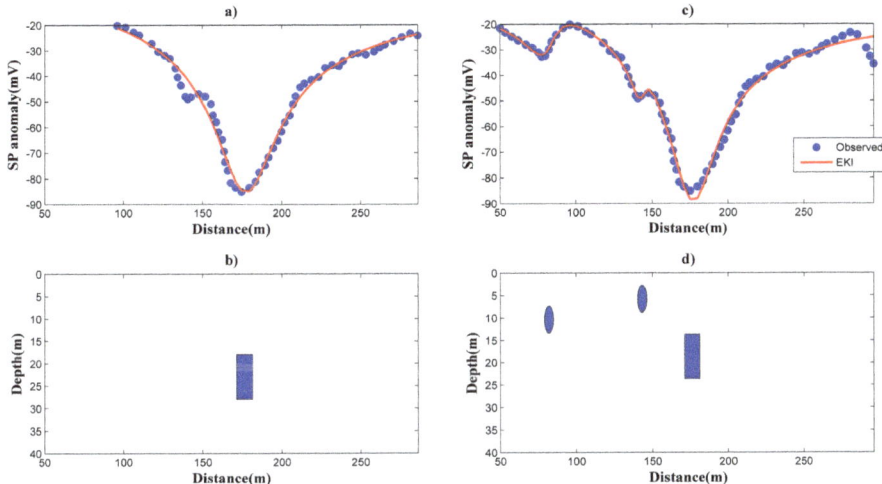

Fig. 7.6 Inversion result of Neem-ka-Thana anomaly uses EKI with single and multiple sources assumption; **a, c** comparison of observed and calculated data from median of PDM uses assumption of single and multiples anomalous, respectively; **c, d** subsurface model revealed by EKI for single and multiples anomalous assumption

shown in the Table 7.3. The inversion result s are shown in Table 7.3 and Fig. 7.6b, while observed and calculated SP data are compared in Fig. 7.6a. Table 7.3 indicates that the distances and the shape of the anomaly body for EKI are good agreement with those of other GOMs. The Table also shows that source of the body using EKI has a depth of 17.97±0.33 m, where the result is comparable with those of other GOMs (10.8 to 18.81 m) (Balkaya 2013; Biswas and Sharma 2015; Göktürkler and Balkaya 2012; Sungkono 2020b). Moreover, the estimated depth from EKI is relatively closed with drilling information, where directional drilling found ore deposits located between 10 and 15 m in the depth from the top of the sources (Srivastava and Agarwal 2009).

Again, EKI is applied to determine model parameter of Neem-ka-Thana anomaly considering that the SP data is sourced by three ores bodies. The search range for multiple sources anomalies are shown in Table 7.4. The SP data has again simultaneously inverted considering multiple for all the bodies. The inversion result using EKI for the anomaly can be shown in Table 7.4 and described in Fig. 7.6d. The Table also indicates that the positions (distances and depths) of three ore bodies revealed by EKI are closed to those of other GOMs (VSFA, μ JADE, vectorized random mutation factor in DE or MVDE). Furthermore, the depth of third anomaly resulted by MVDE and EKI is more appropriate with that of drilling information compared to others, which the depth of ore deposits is between 10 and 15 m from the top (Srivastava and Agarwal 2009).

Table 7.3 Comparison of parameters results estimated of Neem-ka-Thana anomaly between EKI and other methods considering an anomaly body

Model parameters	Search Space	VFSA (Biswas and Sharma 2015)	DE (Balkaya 2013)	Göktürkler and Balkaya (2012)			Sungkono (2020)		EKI
				GA	PSO	SA	MDE	μJADE	
k (mVm)	−100–0	−32.2 ± 0.6	−48.50	−53.99	−49.53	−44.62	−48.38 ± 5.07	−49.93 ± 24.79	−48.55 ± 0.80
D (m)	150–200	177.8 ± 0.2	176.8	176.84	176.77	176.92	178.32 ± 2.68	176.66 ± 0.33	176.32 ± 0.14
h(m)	0–50	10.8 ± 0.6	17.3	18.6	17.6	16.34	18.81 ± 1.31	17.91 ± 4.35	17.97 ± 0.33
θ (°)	0–180	89.6 ± 0.1	88.05	87.83	88	88.25	88.95 ± 1.65	88.06 ± 0.76	87.94 ± 0.08
q	0–2	0.5	0.4	0.42	0.4	0.38	0.41 ± 0.02	0.41 ± 0.08	0.4 ± 0.00

Table 7.4 Comparison of parameters results are determined by different approaches for SP anomaly of Neem-ka-Thana, India, considering three ores bodies

Methods	Sources	K (mVm)	D (m)	h(m)	θ (°)	q
Search Spaces	1	−100:0	50:100	0:50	0:180	0:2
	2	−200:0	100:150	0:50	0:180	0:2
	3	−300:0	150:200	0:50	0:180	0:2
VFSA (Biswas 2017)	1	−28.5 ± 1.4	69.20 ± 3.0	18.80 ± 2.00	88.40 ± 5.80	0.50
	2	−140.0 ± 0.4	138.10 ± 3.6	10.20 ± 3.50	87.90 ± 19.3	1.00
	3	−81.4 ± 2.5	174.60 ± 1.0	16.40 ± 2.00	82.40 ± 0.80	0.50
MVDE (Sungkono 2020)	1	−77.81 ± 16.13	82.70 ± 4,24	8.90 ± 1.75	147.88 ± 27.02	0.89 ± 0.02
	2	−98.55 ± 110.38	137.92 ± 24.92	23.38 ± 41.39	129.11 ± 52.29	1.5 ± 0.48
	3	−33.41 ± 2.25	176.40 ± 0.43	14.21 ± 0.74	88.70 ± 0.04	0.32 ± 0.01
μJADE (Sungkono 2020)	1	−100 ± 0.00	81.03 ± 0.56	16.23 ± 2.09	126.16 ± 2.51	0.77 ± 0.02
	2	−199.4 ± 1.83	145.62 ± 1.09	7.56 ± 0.18	179.99 ± 0.03	1.092 ± 0.03
	3	−164.40 ± 2.90	172.21 ± 0.47	28.14 ± 1.30	81.55 ± 1.06	0.59 ± 0.00
EKI	1	−61.84 ± 11.83	82.23 ± 0.23	10.35 ± 0.27	141.25 ± 2.31	0.80 ± 0.03
	2	−103.89 ± 12.04	143.27 ± 0.21	5.76 ± 0.17	143.12 ± 2.14	1.10 ± 0.03
	3	−40.41 ± 0.47	175.99 ± 0.13	13.66 ± 0.19	88.13 ± 0.14	0.34 ± 0.00

7.4.2 Malachite Mine, Jefferson County, Colorado, USA

The second example of field data study SP's anomaly was measured in the Jefferson County, Colorado, USA, where the SP data (Fig. 7.7a, c) are associated with the Malachite mine in the Jefferson County, Colorado, USA. The anomaly indicates that main anomaly (peak of negative) is around 0 m in distances and the others may be around both edges in distance (negative peaks). Consequently, The anomaly was analyzed considering as single body (Abdelrahman et al. 2004; Balkaya 2013; Biswas and Sharma 2015; Fedi and Abbas 2013; Mehanee 2014; Tlas and Asfahani 2013) and multiple bodies (Biswas and Sharma 2015).

Using assumption that the SP data in the malachite mine considers a body, EKI is applied with search ranges as Table 7.5. Table shows the model parameter revealed by EKI comparing with other methods including a horizontal gradient approach (HGA) (Abdelrahman et al. 2004), DEXP (Fedi and Abbas 2013), DE (Balkaya 2013), inversion using Fair function approach (FFA) (Tlas and Asfahani 2013), Gauss-Newton (GN) (Mehanee 2014), and VFSA (Biswas and Sharma 2015), PSO (Essa 2019). The depth of the ore body has revealed 15.5 m by HGA, 13.6 m by EPM, 19.2 m by DE, 15.6 m by FFA, 12 m by GN, 15.2 m by VFSA, 21.42 by PSO, while

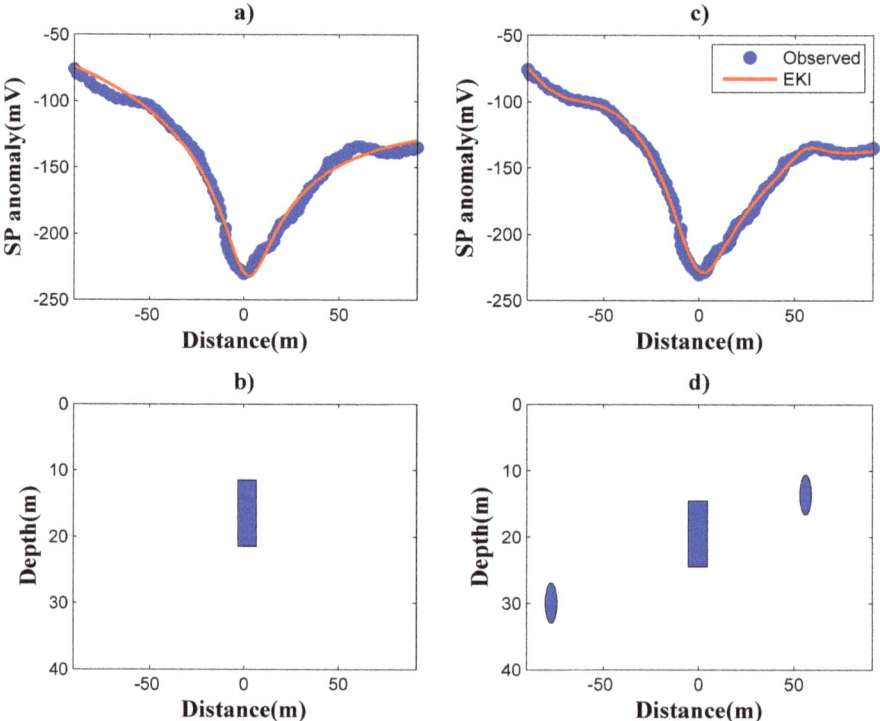

Fig. 7.7 The description is like on the Fig. 7.6, but for Malachite mine of SP data inversion

Table 7.5 Inversions results for SP anomaly of Malachite mine anomaly with their uncertainties estimated by EKI, which is compared with other methods using a sources anomaly assumption

Model parameters	Search range	HGA (Abdelrahman et al. 2004)	DEXP (Fedi and Abbas 2013)	FFA (Tlas and Asfahani 2013)	DE (Balkaya 2013)	GN (Mehanee 2014)	VFSA (Biswas and Sharma 2015)	EKI
K	−500–0	−227	–	−220.6	−436	209	−224.4 ± 1.2	−52.81 ± 0.48
D	−10–10	–	–	–	2.32*	–	−0.96 ± 0.2	1.96 ± 0.41
h	0–20	15.5	13.6	15.6	19.2	12	15.2 ± 0.1	11.45 ± 0.27
θ	0–180	83.4	–	80.2	83.6	85	80.7 ± 0.1	88.10 ± 0.09
q	0–2	–	0.5	0.501	0.62	0.5	0.5	0.20 ± 0.00

the depth is 11.45±0.27 m by EKI. The depth determined by EKI indicates fair agreement with that of the drilling information (13.7 m) (Dobrin and Savit 1988). Moreover, Table 7.5 also indicates that the distance of anomaly body determined by EKI is good agreement with those of DE, PSO and VFSA, while the shape factor of EKI is sufficiency different with others, although the shape factor indicates that the ore body is a vertical cylinder.

Furthermore, interpretation of Malachite mine anomaly uses EKI multiple anomalous bodies considering the peak negative anomalies considered as the center of anomalies. The search ranges uses in the inversion is based on Biswas and Sharma (2015), which is presented in the Table 7.6. Figure 7.7c indicates good fittings between the observed and median of PDM from EKI considering multiple anomalous and Fig. 7.6d demonstrates the subsurface structure for multiple bodies. Table 7.6 reflects the model parameters of anomalous revealed by EKI are comparable with those of VFSA. In addition, the shape bodies of VFSA also have same meaning with EKI result, while the depth of second anomaly from EKI is closer with drilling information (13.7 m) than VFSA result. It indicates that the source of the ore bodies have been successfully estimated using EKI.

7.4.3 Surda Anomaly, Portugal

Surda anomaly was measured SP data in Surda area, which is correlated with Rakha mines, Singhbhum copper belt, Jharkhand, India. The anomaly was interpreted using global optimization (GO) including NN (El-Kaliouby and Al-Garni 2009), GPA (Di Maio et al. 2016), PSO (Monteiro Santos 2010), VFSA (Sharma and Biswas 2013), and BHA (Sungkono and Warnana 2018). In this paper, EKI inverts the SP anomaly using the parameter range as in the Table 7.7. Furthermore, the results are compared to those of other GOs.

Figure 7.8a shows the observed SP data comparing with calculated data from median of PDM, resulted by EKI, indicates that the comparison of data are closed. The EKI inversion results and the other approaches are tabulated in Table 7.7. The result shows that the model parameters inverted by EKI is good agreement with those of other GOs. Thus, the model parameter of anomaly has been accurately determined using EKI.

7.5 Conclusion

Ensemble Kalman Inversion (EKI) has been developed and applied for SP data inversion assuming models of a simple geometry such as horizontal and vertical cylinders, sphere, and inclined sheet. The method was tested on noise-free and noise-added synthetic data for a single and multiple anomalous sources. Additionally, the EKI has also implemented on several fields SP data for ore bodies identifications

Table 7.6 Inversions results for SP anomaly of Malachite mine anomaly with their uncertainties estimated by EKI, which is compared with other methods using three sources anomalies assumption

Model parameters	Body 1			Body 2			Body 3		
	Search range	VFSA	EKI	Search range	VFSA	EKI	Search range	VFSA	EKI
K	−1000–0	−598.3 ± 102.6	−371.74 ± 107.67	−500–0	−253.4 ± 30.5	−61.49 ± 1.45	−1000–0	−664.8 ± 103.8	−548.14 ± 104.01
D	−90−−70	−78.7 ± 1.3	−77.37 ± 0.51	−10−10	−1.0 ± 0.4	−0.39 ± 0.24	40−70	59.5 ± 1.9	55.98 ± 1.31
h	0−30	12.7 ± 2.7	29.95 ± 0.28	0−20	16.9 ± 1.0	14.51 ± 0.38	0−20	11.8 ± 1.7	13.59 ± 0.78
θ	0−90	20.8 ± 7.5	89.69 ± 1.34	45−135	78.6 ± 0.7	84.21 ± 0.38	−90−0	−59.4 ± 13.1	−88.34 ± 6.07
q	0−2	1.2 ± 0.1	0.79 ± 0.05	0−2	0.52 ± 0.0	0.26 ± 0.01	0−2	1.2 ± 0.1	1.18 ± 0.04

Table 7.7 Comparison parameter model from Surda anomaly, are resulted by different approaches using simple polarized as source

Parameter models	Ranges of inversion	NN (El-Kaliouby and Al-Garni 2009)	PSO (Monteiro Santos 2010)	VFSA (Sharma and Biswas 2013)	GPA (Di Maio et al. 2016)	BHA (Sungkono and Warnana 2018)	EKI
K (mV)	90–180	130.8	98.36	121.3 ± 3.9	128.67	102.67 ± 0.99	102.75 ± 0.44
D (m)	−20–40	5.8	−3.87	1.1 ± 0.4	1.72	−2.89 ± 0.3	−2.88 ± 0.06
h (m)	10–40	27.8	31.4	30.4 ± 0.1	31.94	31.16 ± 0.14	31.20 ± 0.04
a (m)	10–30	19.5	28.8	22.4 ± 0.6	21.92	27.43 ± 0.26	27.33 ± 0.13
θ (deg)	20–50	50.9	45.98	50.8 ± 0.2	52.47	45.55 ± 0.26	45.95 ± 0.07

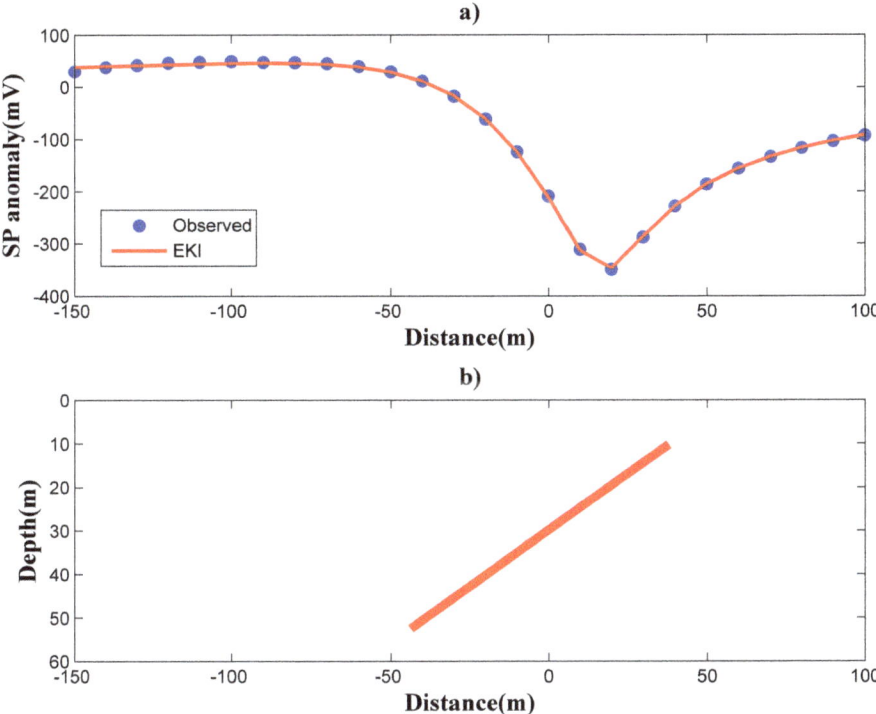

Fig. 7.8 Inversion results of Surda anomaly uses EKI; a) Fitting SP data between observed (dots) and calculated (median of PDM) data (lines) using EKI

including Neem-ka-Thana, Malachite mine, and Surda anomalies. The results of the algorithm demonstrate that is good agreement with those of other methods result and geological or drilling information.

Acknowledgements Computer codes used are available upon request to the corresponding author.

References

Abdelazeem M, Gobashy M, Khalil MH, Abdrabou M (2019) A complete model parameter optimization from self-potential data using Whale algorithm. J Appl Geophys 170: https://doi.org/10.1016/j.jappgeo.2019.103825

Abdelrahman EM, Saber HS, Essa KS, Fouda MA (2004) A least-squares approach to depth determination from numerical horizontal self-potential gradients. Pure Appl Geophys 161:399–411. https://doi.org/10.1007/s00024-003-2446-5

Agarwal BNP, Srivastava S (2009) Analyses of self-potential anomalies by conventional and extended Euler deconvolution techniques. Comput Geosci Progress Trans Spat Datasets web Environ 35:2231–2238. https://doi.org/10.1016/j.cageo.2009.03.005

Al-Saigh NH, Mohammed ZS, Dahham MS (1994) Detection of water leakage from dams by self-potential method. Eng Geol 37:115–121. https://doi.org/10.1016/0013-7952(94)90046-9

Arora T, Linde N, Revil A, Castermant (2007) Non-intrusive characterization of the redox potential of landfill leachate plumes from self-potential data. J Contam Hydrol 92:274–292

Balkaya Ç (2013) An implementation of differential evolution algorithm for inversion of geoelectrical data. J Appl Geophys 98:160–175. https://doi.org/10.1016/j.jappgeo.2013.08.019

Bhattacharya BB, Roy N (1981) A Note on the use of a nomogram for self-potential anomalies. Geophys Prospect 29:102–107. https://doi.org/10.1111/j.1365-2478.1981.tb01013.x

Biswas A (2017) A review on modeling, inversion and interpretation of self-potential in mineral exploration and tracing paleo-shear zones. Ore Geol Rev 91:21–56. https://doi.org/10.1016/j.ore georev.2017.10.024

Biswas A, Sharma SP (2017) Interpretation of self-potential anomaly over 2-D inclined thick sheet structures and analysis of uncertainty using very fast simulated annealing global optimization. Acta Geod Geophys 52:439–455. https://doi.org/10.1007/s40328-016-0176-2

Biswas A, Sharma SP (2015) Interpretation of self-potential anomaly over idealized bodies and analysis of ambiguity using very fast simulated annealing global optimization technique. Surf Geophys 13:179–195. https://doi.org/10.3997/1873-0604.2015005

Biswas A, Sharma SP (2014a) Optimization of self-potential interpretation of 2-D inclined sheet-type structures based on very fast simulated annealing and analysis of ambiguity. J Appl Geophys 105:235–247. https://doi.org/10.1016/j.jappgeo.2014.03.023

Biswas A, Sharma SP (2014b) Resolution of multiple sheet-type structures in self-potential measurement. J Earth Syst Sci 123:809–825. https://doi.org/10.1007/s12040-014-0432-1

Candra AD, Srigutomo W, Sungkono, Santosa BJ (2014) A complete quantitative analysis of self-potential anomaly using singular value decomposition algorithm. Presented at the 2014 IEEE international conference on smart instrumentation, measurement and applications (ICSIMA), pp 1–4. https://doi.org/10.1109/ICSIMA.2014.7047419

Chada NK, Iglesias MA, Roininen L, Stuart AM (2018) Parameterizations for ensemble Kalman inversion. Inverse Probl. 34: https://doi.org/10.1088/1361-6420/aab6d9

Chada NK, Schillings C, Weissmann S (2019) On the incorporation of box-constraints for ensemble Kalman inversion. Found. Data Sci. 1:433. https://doi.org/10.3934/fods.2019018

Cho H, Olivera F (2014) Application of multimodal optimization for uncertainty estimation of computationally expensive hydrologic models. J. Water Resour Plan Manag 140:313–321. https://doi.org/10.1061/(ASCE)WR.1943-5452.0000330

Di Maio R, Piegari E, Rani P (2017) Source depth estimation of self-potential anomalies by spectral methods. J Appl Geophys 136:315–325. https://doi.org/10.1016/j.jappgeo.2016.11.011

Di Maio R, Piegari E, Rani P, Carbonari R, Vitagliano E, Milano L (2019) Quantitative interpretation of multiple self-potential anomaly sources by a global optimization approach. J Appl Geophys 162:152–163. https://doi.org/10.1016/j.jappgeo.2019.02.004

Di Maio R, Rani P, Piegari E, Milano L (2016) Self-potential data inversion through a Genetic-Price algorithm. Comput Geosci 94:86–95. https://doi.org/10.1016/j.cageo.2016.06.005

Dobrin MB, Savit CH (1988) Introduction to geophysical prospecting, 4th edn. McGraw-Hill, New York

Elhussein M (2020) A novel approach to self-potential data interpretation in support of mineral resource development. Resour. Res, Nat. https://doi.org/10.1007/s11053-020-09708-1

El-Kaliouby HM, Al-Garni MA (2009) Inversion of self-potential anomalies caused by 2D inclined sheets using neural networks. J Geophys Eng 6:29. https://doi.org/10.1088/1742-2132/6/1/003

Essa KS (2020) Self potential data interpretation utilizing the particle swarm method for the finite 2D inclined dike: mineralized zones delineation. Acta Geod Geophys 55:203–221. https://doi.org/10.1007/s40328-020-00289-2

Essa KS (2019) A particle swarm optimization method for interpreting self-potential anomalies. J Geophys Eng 16:463–477. https://doi.org/10.1093/jge/gxz024

Evensen G (2009) Data assimilation: the ensemble Kalman filter, 2nd ed. 2009. ed. Springer

Evensen G (2003) The Ensemble Kalman Filter: theoretical formulation and practical implemen-
tation. Ocean Dyn 53:343–367. https://doi.org/10.1007/s10236-003-0036-9
Fedi M, Abbas MA (2013) A fast interpretation of self-potential data using the depth from extreme
points method. Geophysics 78:E107–E116. https://doi.org/10.1190/geo2012-0074.1
Fernández-Martínez JL, Fernández-Muñiz Z, Pallero JLG, Pedruelo-González LM (2013) From
Bayes to Tarantola: New insights to understand uncertainty in inverse problems. J Appl Geophys
98:62–72. https://doi.org/10.1016/j.jappgeo.2013.07.005
Fernández-Martínez JL, García-Gonzalo E, Naudet V (2010) Particle swarm optimization applied
to solving and appraising the streaming-potential inverse problem. Geophysics 75:WA3–WA15.
https://doi.org/10.1190/1.3460842
Fernández-Martínez JL, Pallero JLG, Fernández-Muñiz Z, Pedruelo-González LM (2014a) The
effect of noise and Tikhonov's regularization in inverse problems. Part I: the linear case. J Appl
Geophys 108:176–185. https://doi.org/10.1016/j.jappgeo.2014.05.006
Fernández-Martínez JL, Pallero JLG, Fernández-Muñiz Z, Pedruelo-González LM (2014b) The
effect of noise and Tikhonov's regularization in inverse problems. Part II: the nonlinear case. J
Appl Geophys 108:186–193. https://doi.org/10.1016/j.jappgeo.2014.05.005
Fernández-Muñiz Z, Hassan K, Fernández-Martínez JL (2019) Data kit inversion and uncertainty
analysis. J Appl Geophys. https://doi.org/10.1016/j.jappgeo.2018.12.022
Giang NV, Kochanek K, Vu NT, Duan NB (2018) Landfill leachate assessment by hydrological and
geophysical data: case study NamSon, Hanoi Vietnam. J Mater Cycles Waste Manag 20:1648–
1662. https://doi.org/10.1007/s10163-018-0732-7
Göktürkler G, Balkaya Ç (2012) Inversion of self-potential anomalies caused by simple-geometry
bodies using global optimization algorithms. J Geophys Eng 9:498–507. https://doi.org/10.1088/
1742-2132/9/5/498
Gu Y, Oliver DS (2007) An Iterative Ensemble Kalman Filter for multiphase fluid flow data
assimilation. SPE J 12:438–446. https://doi.org/10.2118/108438-PA
Hafez MA (2005) Interpretation of the self-potential anomaly over a 2D inclined plate using a
moving average window curves method. J Geophys Eng 2
Iglesias MA (2016) A regularizing iterative ensemble Kalman method for PDE-constrained inverse
problems. Inverse Probl 32: https://doi.org/10.1088/0266-5611/32/2/025002
Iglesias MA, Law KJH, Stuart AM (2013) Ensemble Kalman methods for inverse problems. Inverse
Probl. 29: https://doi.org/10.1088/0266-5611/29/4/045001
Laby DA, Sungkono, Santosa BJ, Bahri AS (2016) RR-PSO: fast and robust algorithm to invert
Rayleigh waves dispersion. Contemp Eng Sci 9:735–741. https://doi.org/10.12988/ces.2016.6685
Li G, Reynolds AC (2009) Iterative Ensemble Kalman Filters for data assimilation. SPE J 14:496–
505. https://doi.org/10.2118/109808-PA
Liu K, Huang G, Jiang Z, Xu X, Xiong Y, Huang Q, Šimůnek J (2020) A gaussian process-based
iterative Ensemble Kalman Filter for parameter estimation of unsaturated flow. J Hydrol 589:
https://doi.org/10.1016/j.jhydrol.2020.125210
Mauri G, Williams-Jones G, Saracco G (2011) MWTmat—application of multiscale wavelet tomography on potential fields. comput. geosci., geospatial cyber infrastructure for polar research
geospatial cyber infrastructure for. Polar Res 37:1825–1835. https://doi.org/10.1016/j.cageo.
2011.04.005
Mauri G, Williams-Jones G, Saracco G, Zurek JM (2012) A geochemical and geophysical investigation of the hydrothermal complex of Masaya volcano. Nicaragua. J. Volcanol. Geotherm. Res.
227–228:15–31. https://doi.org/10.1016/j.jvolgeores.2012.02.003
Mehanee SA (2015) Tracing of paleo-shear zones by self-potential data inversion: case studies from
the KTB, Rittsteig, and Grossensees graphite-bearing fault planes. Earth Planets Space 67:14.
https://doi.org/10.1186/s40623-014-0174-y
Mehanee SA (2014) An efficient regularized inversion approach for self-potential data interpretation
of ore exploration using a mix of logarithmic and non-logarithmic model parameters. Ore Geol
Rev 57:87–115. https://doi.org/10.1016/j.oregeorev.2013.09.002

Monteiro Santos FA (2010) Inversion of self-potential of idealized bodies' anomalies using particle swarm optimization. Comput Geosci 36:1185–1190. https://doi.org/10.1016/j.cageo.2010.01.011

Moore JR, Boleve A, Sanders JW, Glaser SD (2011) Self-potential investigation of moraine dam seepage. J Appl Geophys 74:277–286. https://doi.org/10.1016/j.jappgeo.2011.06.014

Oliver DS, Chen Y (2009) Improved initial sampling for the ensemble Kalman filter. Comput Geosci 13:13–27. https://doi.org/10.1007/s10596-008-9101-2

Reddi AGB, Madhusudan IC, Sarkar B, Sharma JK (1982) An album of geophysical responses from base metal belts of Rajasthan and Gujarat (Calcutta: Geological Survey of India). Miscellaneous Publication

Revil A, Jardani A (2013) The self-potential method: theory and applications in environmental geosciences. Cambridge University Press, Cambridge

Sharma SP, Biswas A (2013) Interpretation of self-potential anomaly over a 2D inclined structure using very fast simulated-annealing global optimization—an insight about ambiguity. Geophysics 78:WB3–WB15. https://doi.org/10.1190/geo2012-0233.1

Srivastava S, Agarwal BNP (2009) Interpretation of self-potential anomalies by Enhanced Local Wave number technique. J Appl Geophys 68:259–268

Sundararajan N, Srinivas Y (1996) A modified Hilbert transform and its application to self potential interpretation. J Appl Geophys 36:137–143. https://doi.org/10.1016/S0926-9851(96)00048-1

Sungkono (2020a) Robust interpretation of single and multiple self-potential anomalies via flower pollination algorithm. Arab J Geosci 13. https://doi.org/10.1007/s12517-020-5079-4

Sungkono (2020b) An efficient global optimization method for self-potential data inversion using micro-differential evolution. J Earth Syst Sci 129:178. https://doi.org/10.1007/s12040-020-01430-z

Sungkono, Santosa BJ (2015) Differential evolution adaptive metropolis sampling method to provide model uncertainty and model selection criteria to determine optimal model for rayleigh wave dispersion. Arab J Geosci 8:7003–7023. https://doi.org/10.1007/s12517-014-1726-y

Sungkono, Warnana DD (2018) Black hole algorithm for determining model parameter in self-potential data. J Appl Geophys 148:189–200. https://doi.org/10.1016/j.jappgeo.2017.11.015

Tlas M, Asfahani J (2013) An approach for interpretation of self-potential anomalies due to simple geometrical structures using fair function minimization. Pure Appl Geophys 170:895–905. https://doi.org/10.1007/s00024-012-0594-1

van Leeuwen PJ, Evensen G (1996) Data Assimilation and Inverse Methods in Terms of a Probabilistic Formulation. Mon Weather Rev 124:2898–2913. https://doi.org/10.1175/1520-0493(1996)124%3c2898:DAAIMI%3e2.0.CO;2

Vrugt JA, Beven KJ (2018) Embracing equifinality with efficiency: limits of Acceptability sampling using the DREAM(LOA) algorithm. J Hydrol 559:954–971. https://doi.org/10.1016/j.jhydrol.2018.02.026

Wang Y, Li G, Reynolds AC (2010) Estimation of Depths of Fluid Contacts by History Matching Using Iterative Ensemble-Kalman Smoothers. SPE J 15:509–525. https://doi.org/10.2118/119056-PA

Wolpert DH, Macready WG (1997) No Free Lunc Theorems for optimization. IEEE Trans Evol Comput 1:67–82

Zhang J, Vrugt JA, Shi X, Lin G, Wu L, Zeng L (2020) Improving simulation Efficiency of MCMC for inverse modeling of hydrologic systems with a Kalman-Inspired proposal distribution. Water Resour Res 56:e2019WR025474. https://doi.org/10.1029/2019WR025474

Chapter 8
Advanced Analysis of Self-potential Anomalies: Review of Case Studies from Mining, Archaeology and Environment

Lev V. Eppelbaum

Abstract Self-potential (SP) method is one of the most non-expensive and unsophisticated geophysical methods. However, its application is limited due to absence of reliable interpreting methodology, first for the complex geological-environmental conditions. The essential disturbances appearing in the SP method and some ways for their removal (elimination) before the quantitative analysis are discussed. A brief review of the available interpretation methods is presented. For the magnetic method of geophysical prospecting, have been developed special quantitative procedures applicable under complex physical-geological environments (oblique polarization, uneven terrain relief and unknown level of the normal field). Earlier detected common peculiarities between the magnetic and SP fields have been extended. These common aspects make it possible to apply the advanced procedures developed in magnetic prospecting to SP method. Besides the reliable determination of the depth of anomalous targets, these methodologies enable to calculation of corrections for the non-horizontal SP observations and direction of the polarization vector. For classification of SP-anomalies is proposed to apply a new parameter—'self-potential moment'. The quantitative procedures (improved modifications of the characteristic point, tangent techniques and areal method) have been successfully tested on SP models and employed in numerous real situations in mining, archaeological, environmental and technogenic geophysics. The obtained results indicate the practical importance of the developed interpretation methodologies.

Keywords Disturbances · Quantitative analysis · Complex physical-geological environments · Self-potential moment · Ore deposits · Archaeological sites · Underground caves · Underground pipes

L. V. Eppelbaum (✉)
Faculty of Exact Sciences, Department of Geophysics, Tel Aviv University, Ramat Aviv, 6997801
Tel Aviv, Israel
e-mail: levap@tauex.tau.ac.il

Azerbaijan State Oil and Industry University, 20 Azadlig Ave., Baku AZ1010, Azerbaijan

© The Author(s), under exclusive license to Springer Nature Switzerland AG 2021   203
A. Biswas (ed.), *Self-Potential Method: Theoretical Modeling and Applications in Geosciences*, Springer Geophysics, https://doi.org/10.1007/978-3-030-79333-3_8

8.1 Introduction

The Self-Potential (SP) method is based on the study of natural electric fields (in some sources this method is named as 'spontaneous polarization'). The term "natural" here means that this field does not create by any external artificial sources. Permanent electric fields arise in the course of redox, filtration, and diffusion-adsorption processes in the upper part of geological section. The registration of these fields is the goal of the SP method, and the geophysical interpretation of the parameters generating this field is the main purpose of SP data examination. An oxidizing object (e.g., ore body, archaeological or other target) is a galvanic cell, the occurrence of which requires: (1) the contact of electric conductors with different types of conductivity (electronic and ionic), and (2) the difference in the redox conditions at different contact points of these conductors. An appearance of these conditions is usually impossible without the underground water contact (Sato and Mooney 1960).

In the geological section, the conditions for the formation of a galvanic cell arise on targets with electronic conductivity, if these bodies occur in the water-saturated rocks with ionic conductivity. The change in the redox conditions at the contact of the electronic conductor (anomalous target) and the surrounding medium is associated with a decrease in the oxygen content with a depth.

Fox's (1830) SP observations at copper vein deposits in Cornwall (England) laid the foundation of the application of all electric methods in geophysics as a whole. SP is an effective, prompt and comparatively simple geophysical method. Equipment for SP method is one of the most non-expensive in the applied geophysics (Table 8.1).

Conventional equipment employed in the SP method consists of microVoltmeter, pair of non-polarizable electrodes, cable and $CuSO_4$ solution (the latter is necessary for the better contact of employed electrodes with the environment).

Without hesitation, ground penetration radar (GPR) and electric resistivity tomography (ERT) are more powerful geophysical tools, which can theoretically produce a lot more detailed geophysical-archaeological information. However, they are much more expensive and, most importantly, water content in subsurface strongly complicates application of these methods. At the same time, presence of water is only positive factor for the SP method, since it enables to increase SP anomaly intensity (Semenov 1980; Parasnis 1986).

In this investigation is considered SP method employment in mining geophysics (e.g., Semenov 1980; Corry 1985; Babu and Rao 1988; Lile 1996; Golkdie 2002; Bhattacharya et al. 2007; Dmitriev 2012; Fedi and Abbas 2013; Biswas and Sharma 2016; Eppelbaum 2019a, b; Eppelbaum 2021), archaeological geophysics (e.g., Wynn and Sherwood 1984; Mauriello et al. 1998; Eppelbaum et al. 2003a, b; Drahor 2004; Di Maio et al. 2010; Shevnin et al. 2014; De Giorgi and Leucci 2017, 2019;

Table 8.1 Averaged prices for geophysical potential field equipment

Method	Gravity	Magnetic	Resistivity	Self-Potential
Price of equipment, US $	60,000–110,000	20,000–25,000	35,000–55,000	150–200

Eppelbaum 2020), environmental geophysics (e.g., Corwin 1990; Quarto and Schi-avone 1996; Jardani et al. 2006a; Eppelbaum 2007; Srigutomo et al. 2010; Chen et al. 2018; Gusev et al. 2018; Oliveti and Cardarelli 2019) and technogenic geophysics (e.g., Castermant et al. 2008; Fomenko 2010; Onojasun and Takum 2015; Cui et al. 2017). Application of quantitative analysis in the SP method for solving other geological-geophysical problems is beyond the scope of this study.

8.2 Self-potential Observations: Common Disturbances

8.2.1 Different Kinds of Noise in SP Observations

The main kinds of noise appearing in the SP method are shown in a block-scheme (Fig. 8.1). Some of these noise effects are considered below in detail.

8.2.1.1 Electrode Noise in the SP Method

Although the fact that SP electrode is called as "non-polarizable", after some time it accomplishes some polarization effects from the surrounding media. However, taking into account that we measure the value $\Delta U = (U_1 - U_2)$, the most is important is to keep not absolute non-polarizability, but an equivalent polarization on both of the employed electrodes. For inspecting this equivalent, the following procedure can be employed in field conditions (of course, direct measurements in a physical laboratory are more precise). We can write a trivial equation for the first electrode: $U_1 + e_1$ (U_1 is the first "medium" signal, and e_1 is the noise accumulated in the first electrode). For the second electrode, correspondingly we have $U_2 + e_2$ (U_2 is the second "medium" signal, and e_2 is the noise of accumulated in the second electrode). According to (Semenov 1980), we measure

$$\Delta U_1 = (U_1 + e_1) - (U_2 + e_2). \tag{8.1}$$

Let us will change electrodes by their places. In this case we will obtain

$$\Delta U_2 = (U_1 + e_2) - (U_2 + e_1). \tag{8.2}$$

After this, calculating a difference between ΔU_1 and ΔU_2, we receive

$$\partial U = \Delta U_1 - \Delta U_2 = [U_1 + e_1 - U_2 - e_2] - [U_1 + e_2 - U_2 - e_1] = 2(e_1 - e_2) \tag{8.3}$$

or

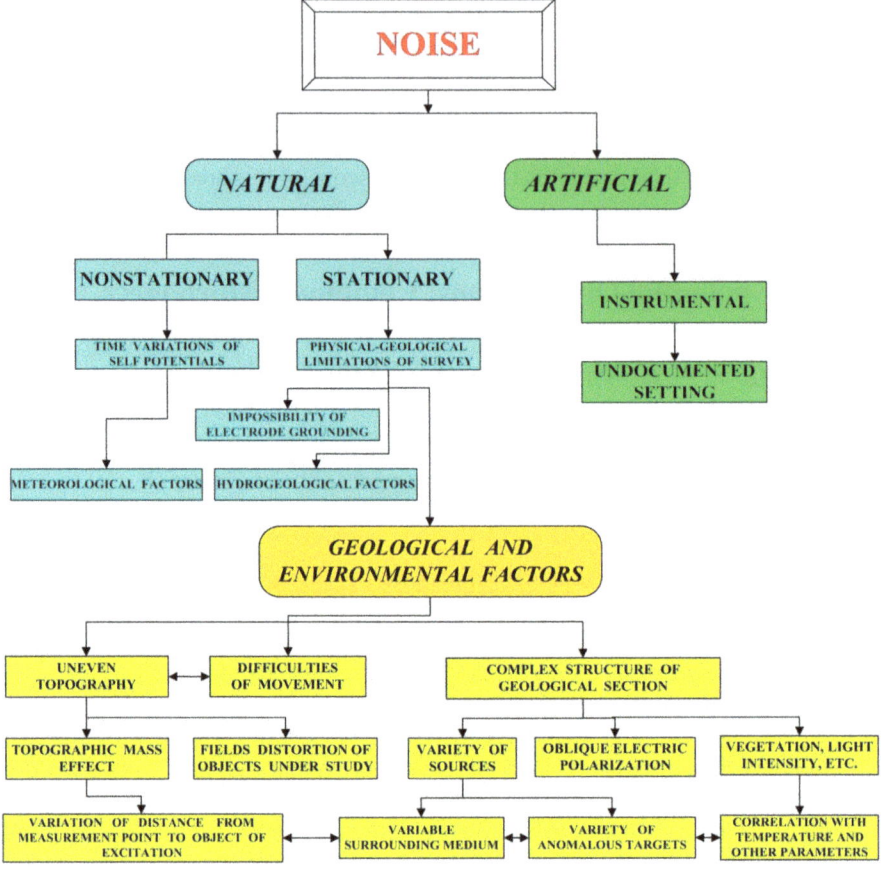

Fig. 8.1 General scheme of disturbances in the SP method

$$(e_1 - e_2) = \frac{\partial U}{2}. \qquad (8.4)$$

If the value $(e_1 - e_2)$ is significant (e.g., ≥ 3 mV), the noised electrodes must be replaced by new ones.

A similar methodology for the electrode noise detection was suggested by Perrier and Pant (2005).

8.2.1.2   Temporal Variations in SP Method

Parasnis (1986) has been carried out SP measurements along the same profiles in the Akulla region (Sweden) seven times during 1960–1967 years. These measurements indicate a good repeatability despite of the fact that they were conducted under different climatic conditions. Similar investigations performed by other researchers (e.g., Semenov 1980) generally confirm a good repeatability of the different SP time observations.

For the estimation of accuracy ε of SP field measurements the following trivial formula often employed in various geophysical methods may be applied

$$\varepsilon = \frac{\sqrt{\sum_{i=1}^{N} \left(\Delta U_{SP}^{init} - \Delta U_{SP}^{rep}\right)^2}}{N},$$

where N is the total number of SP observations, 'init' means the ordinary measurements, and 'rep' means the repeated measurements. The number of repeated measurements should be at least 8–10% of the total number pf observations. If the value of ε exceeds some a priori assumed value (this value usually depend on the concrete spread of the SP amplitudes), results of SP survey can be rejected as non-reliable ones.

8.2.1.3 Terrain Relief Correction

In the SP method, terrain relief influence is two-fold. On the one hand, over the positive topographic forms can be created negative SP anomalies caused by electromotive force (this phenomena strongly depends on the peculiarities of underground water circulations). Comparison of the SP graphs with the topographic data usually allows to perform identifying anomalies of this type by the characteristic mirror images (Khesin et al. 1996).

From other side, as follows from the very detailed SP measurements of Ernstson and Schrerer (1986), at the inclined topographic surface, the SP field increases directly with increases in the relief highs (Fig. 8.2). In the last case, for the elimination of the terrain relief influence, a correlation method developed in magnetic prospecting

Fig. 8.2 SP observations at inclined relief (after Ernstson and Schrerer (1986), with small modifications) (Middle Keuper of the Steigerwald highlands, 60 km east of Würzburg, Germany)

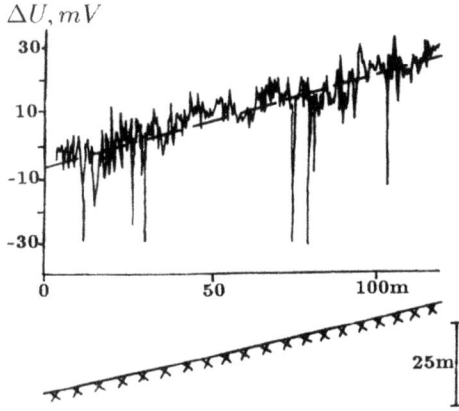

(Khesin et al. 1996) and VLF studies (Eppelbaum and Mishne 2011) can be applied. The essence of this method is as follows. The method employs for removing the terrain relief effect from the observed field ΔU_{obser} a linear least-squares relation ΔU_{appr} (application of other relationships is also possible):

$$\Delta U_{appr} = c + bh,$$

where h is the height of relief, b is the angle coefficient, and c is the free member.

Value ΔU_{appr} approximates the observed field as a function of elevation h (SP anomalous zones usually do not include to the correlation field) and then we obtain the corrected (residual) field ΔU_{corr}, where the relief influence is essentially eliminated:

$$\Delta U_{corr} = \Delta U_{obser} - \Delta U_{appr}.$$

It should be noted that this correction only eliminates the effect of the inclined topographic masses with certain electric properties. Special methods allowing to calculate the difference in altitudes of SP observations to the anomalous target on an inclined profile, are considered in Sect. 4.2 'SP observations on an inclined profile'.

Wang and Geng (2015) studied the problem of terrain correction in the SP method in detail on several field examples. They decided that the mechanism of SP anomalies formed by terrains is rather complex, and therefore it is difficult to obtain the corresponding analytical formulas. The authors applied three types of relief fitting: (1) linear, (2) quadratic and (3) exponential. After comprehensive analysis, Wang and Geng (2015) concluded that the linear fitting is more optimal since it does not create fictitious anomalies. Thus, the aforementioned investigation confirms the application of the aforementioned linear least-squares relation.

8.2.1.4 Calculation of SP Anomaly Distortion Due to Observations on Uneven Surface

SP anomalies (as and anomalies of other potential fields) distort due to observations on uneven surfaces (and correspondingly, from different distances to anomalous objects). This disturbing effect usually is calculated at the end of the interpretation process (see Sect. 8.4.2).

8.2.1.5 Net Justification in Areal Observations (Elimination of Temporal Variations)

Net justification of SP data (elimination of temporal variations caused by different natural factors) is conventionally performed by the use of procedure identical to justification of observations in gravity and magnetic prospecting (e.g., Telford et al.

1990). Some other strategies for removing this effect are presented, for instance, in Revil and Jardani (2013).

8.2.1.6 Influence of Meteorological Factors

Many scientists note that the rains increase the intensity of SP anomalies increases (e.g., Semenov 1980; Parasnis 1986; Revil and Jardani 2013). Therefore, occasionally an artificial irrigation of site intended for SP research is recommended.

8.2.1.7 Presence of Magmatic Associations

Obviously development of magmatic associations (or other kinds of hard geological rocks) in a site destined for field investigations does not allow for the grounding of SP electrodes. The same reason may limit the water circulation at subsurface, which can weaken, or even to completely cancel a generation of the SP anomalies.

8.2.1.8 Some Environmental Factors

The SP anomaly level may affect some environmental factors. One of these factors is the shadowing of a part of the investigated area. For instance, Revil and Jardani (2013) have documented the fact that the difference between the SP electrodes placed in cold and warm media may exceed 10 mV. Another factor is a presence of some hygrophilous plants (e.g., hazel and almonds) whose roots can pick over a lot of moisture from the upper part of the geological section (thereby hindering the generation of SP anomalies).

According to Semenov (1980), polarization of the electrodes ε and difference of their temperatures have the following relationship:

$$e(mV) \approx 0.75\Delta T (^{\circ}C),$$

where e is the electrodes' polarization and ΔT is the difference of the electrodes' temperature.

Ernstson and Schrerer (1986) have been monitored SP and temperature anomalies during 15 months (in 1980–1981) (Fig. 8.3). The correlation between SP and soil temperature is interpreted as result of the influence of thermal diffusivity and convection processes in subsurface (Ernstson and Schrerer 1986). Between these parameters a correlation of $r = 0.64$ was established. It is not a high relationship, but in any case should be taken into account.

Perrier and Morat (2000) also detected a correlation between the amplitudes of the SP variations and soil temperatures. According to these authors, the state of the soil in the first 30 cm seems to play an important role. Perrier and Morat (2000)

Fig. 8.3 A correlation between temperature and SP observations: **a** temperature, **b** SP intensity, **c** correlation between these parameters (**a** and **b**–after Ernstson and Schrerer (1986)). SP observations were carried out in the Middle Keuper of the Steigerwald highlands 60 km east of Würzburg, Germany

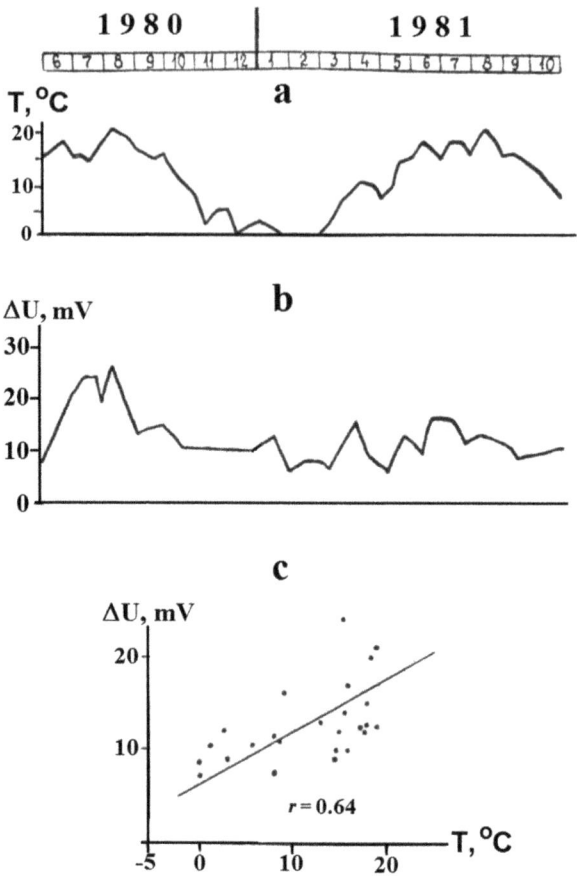

concluded that the joint monitoring of electric potential and temperature appears to be a powerful tool to monitor the underlying soil processes.

Jardani et al. (2008) proposed the following model for explanation of SP pattern generating by geothermal flow:

$$\text{grad} \cdot (\sigma \, \text{grad} \Pi) = \text{grad} \cdot \mathbf{J}_s = \zeta, \tag{8.5}$$

where σ is the conductivity (in S/m), $\mathbf{J}_s$ is the current density vector (in A/m^2), and ζ is the volumetric current density (in A/m^3).

An essential relationship between the SP data and temperature values recorded in boreholes at intermediate and large depths was established by Jardani and Revil (2009) in the Cerro Prieto geothermal field (Baja California, USA). Existence of this phenomenon under normal thermophysical conditions is not studied yet in detail.

8.3 Review of Quantitative Interpretation Methods

An extensive literature is devoted to the interpretation of self-potential anomalies. Unfortunately, it is not possible to cite even all the most important publications in this review, and the author apologizes to those researchers whose excellent publications were not included in this overlook.

The calculation of theoretical anomalies due to SP has long been based primarily on Petrovsky's (1928) solution which was derived for a vertically polarized sphere (Zaborovsky 1963). Later on, some substantial solutions for sheet-like bodies and inclined plates were obtained (Semenov 1980). The electric polarization vector was generally considered to be directed along the sheet-like body dipping (along the longer axis of the conductive body).

Other methods of SP anomaly quantitative interpretation include anomalous body with a simple geometrical shape which approximates the anomaly source. Its parameters (i.e. the depth, the angle between the horizon and the direction of the polarization vector) are usually determined: (1) graphically using characteristic points of the anomaly plot, or (2) by trial-and-error method consisting of visual comparison of the observed anomaly with a set of master curves (Semenov 1980).

Zaborovsky (1963), Semenov (1980) and Murty and Haricharan (1984) applied to SP anomaly, generated by plate and calculated along the profile across its strike, the following formula:

$$U(x) = \frac{j\rho}{2\pi} \ln \frac{r_1^2}{r_2^2}, \tag{8.6}$$

where j is the current per unit length, ρ is the host medium resistivity, r_1 and r_2 are the distances from the plate left and right ends to the observation point. The interpretation procedures based on use of Eq. (8.5), are undoubtedly useful for simple geological (environmental) models.

Fitterman (1979) gave the method of SP anomaly calculation for field sources of an arbitrary shape based on numerical integration using Green's function. This potentially promising approach is highly computer intensive even at the modern computers and docs not provide sufficient accuracy.

There are a number of recent interpretation techniques based on minimizing the difference between observed and theoretical anomalies. The minimization is achieved by sequential optimization of the interpretation parameters through computer-aided iterations. These techniques are inapplicable for complex geological sections.

Quantitative procedures suggested by Rao and Babu (1983) for SP anomalies from 2D sheet-like bodies does nor calculate such factors as oblique polarization and uneven terrain relief.

Eskola and Hongisto (1987) have proposed a methodology of SP interpretation constructed by use of the macroscopic physical model for the mineral SP effect. This model is based on the SP electrochemical principles described by Logn and Bolviken (1974). A further evolution of this methodology may be perspective one.

A few groups of authors have performed spatial-frequency analysis of SP anomalies produced by polarized bodies of various geometrical shapes: Rao et al. (1982), Rao and Mohan (1984), Banerjee and Pal (1990), Skianis et al. (1991, 1995).

A series of publications (Abdelrahman and Sharafeldin 1997; Abdelrahman et al. 1997, 1998; El-Araby 2004; Essa et al. 2008) provided a large number of methodological approaches based mainly on application of the gradient analysis and calculation of derivatives. However, these approaches by their general usefulness, have not caused a quantitative jump in this field.

Gibert and Pessel (2001) have applied the continuous wavelet transform for the localization of SP anomalies. The wavelet analysis provided both an estimate of the location and of the nature of the target responsible for a given self-potential signal. The wavelet-based techniques as well as analytic signals were employed by Sailhac and Marquis (2001) to interpret SP anomalies caused by subsurface fluid flow in the Mt. Etna.

Patella (1997) suggested an application a tomographic presentation of SP images. It consisted of scanning the section through SP profiles, by the unit strength elementary charge, which is given a regular grid of coordinates within the section. At the each point of the section the charge occurrence probability function is calculated. The complete set of calculated grid values is employed to draw colored sections. This method was evaluated by Di Maio et al. (2016).

Mendonça (2008) has employed the Green's functions to simplify the evaluation of SP anomalies from buried conductors. This approach was used by this author to simulate geoelectric targets in mineral exploration and to obtain current source terms by inverting an SP data set.

Srivastava and Agarwal (2009) developed the 'Enhanced Local Wave number technique' wherein the nature of the causative source has been determined by computing structural indices based on its horizontal location and depth. This approach was tested on several mineral deposits with complex ore body distribution.

An attempt to apply the neural network approach to compute the shape factor and depth of the causative target from SP anomaly was undertaken in Al-Garni (2017). Without denying the general promising of this approach, it should be noted some inconclusiveness of the examples given in this work.

Gobashy et al. (2019) proposed a method based on utilizing the optimization algorithm, which as an effective heuristic solution to the inverse problem of SP field due to a 2D inclined bed. The realization of this algorithm in complex physical-geological conditions is under question. Rao et al. (2020) proposed a global optimization methodology expressed in development of inversion algorithm for 2D inclined plates. Absence of geological sections in the mentioned work complicates examination of this methodology.

Hristenko and Stepanov (2012) have demonstrated a system of 2D modeling of SP field from several different anomalous bodies with introduced Gaussian noise under conditions of uneven relief. Some applied transformations allowed to exclude effect of near-surface geological inhomogeneties from the total SP field and to enhance anomalies from the buried targets.

The procedure based on the interpretation of self-potential anomalies due to simple geometrical structures using Fair function minimization (Tlas and Asfahani 2013), is of certain interest. Giannakis et al. (2019) suggested a hybrid optimization scheme for SP measurements due to multiple sheet-like bodies. This procedure demands a wide verification on concrete field examples. A special interest presents the recent research of Oliveti and Cardarelli (2019) who developed the least square subspace preconditioned method to compute the known Tikhonov solution to reliable detecting the depth and the shape of shallow electrical current density sources.

Fedi and Abbas (2013) proposed the 'depth from extreme points method' and employed it on several models and field examples. The method yields estimates of the source horizontal location, depth (top or center), and geometry.

Kilty (1984) published a paper that acknowledged the analogy between the current density of SP and magnetic induction. This author suggested interpreting SP anomalies based on the conventional methods developed for magnetic prospecting. However, the trivial methodologies are not acceptable for complex physical-geological conditions when we should calculate influence of inclined relief and oblique polarization as well as superposition of SP anomalies of various orders. A similar approach, but with the improved interpretation methodology was proposed by Khesin et al. (1996). Eppelbaum and Khesin (2012) proposed a new elaboration of the interpretation process. The present research shows a final generalization of this approach.

8.4   Some Common Aspects of Magnetic and SP Fields

The magnetic field is a potential one (when value of target's magnetization is not very high) and satisfies Poisson's equation:

$$\mathbf{U_a} = -\mathrm{grad}\,V, \tag{8.7}$$

where $\mathbf{U_a}$ is the anomalous magnetic field and V represents the magnetic potential.

SP polarization is generated by the spontaneous manifestation of electric double layers on contacts of various geological (or environmental and artificial) objects. The electric fields $\mathbf{E}$ of the electric double layer l caused by natural electric polarization are defined as the gradient of a scalar potential Π_i:

$$\mathbf{E_{SP}} = -\mathrm{grad}\,\Pi_i. \tag{8.8}$$

The potential Π_i satisfies Laplace's equation everywhere outside the layer l (Zhdanov and Keller 1994).

Analytical expressions for some interpreting models for magnetic and SP fields are presented in Table 8.2.

Table 8.2 Magnetic and SP fields: comparison of analytical expressions for some interpreting models

Field	Analytical expression	
Magnetic	*Thin bed (TB)* (9) $Z_v = 2J2b\frac{z}{x^2+z^2}$	*Point source (rod)* (10) $Z_v = \frac{mz}{(x^2+z^2)^{3/2}}$
Self-Potential	*Horizontal circular cylinder (HCC)* (11) $\Delta U =$ $2\frac{\rho_1}{\rho_1+\rho_2}U_0 r_0 \frac{z}{x^2+z^2}$	*Sphere* (12) $\Delta U =$ $\frac{2\rho_1}{2\rho_2+\rho_1}U_0 R^2 \frac{z}{(x^2+z^2)^{3/2}}$

Here Z_v is the vertical magnetic field component at vertical magnetization; J is the magnetization; b is the horizontal semi-thickness of TB; m is the magnetic mass (point pole magnetic charge); ρ_1 is the host medium resistivity; ρ_2 is the anomalous object (HCC or sphere) resistivity; U_0 is the potential jump at the source body/host medium interface; r_0 is the polarized cylinder radius; R is the sphere radius; x is the current coordinate; z is the depth of the upper edge of TB (center of HCC or sphere) occurrence.

Formulas describing potential character of magnetic (Eq. 8.7) and SP (Eq. 8.8) fields are identical ones. The proportionality of analytical expressions (9) and (11), (10) and (12) for magnetic and SP fields is obvious. It allows to employing in SP data analysis advanced interpretation methods developed in magnetic prospecting (we assume that the SP polarization vector is analogue of the vector of magnetization). It is supposed that the majority of interpretation methodologies developed for magnetic and gravity fields are applicable for the SP method.

For instance, results of SP physical modeling on several buried spheres demonstrate practical applicability of the downward continuation (Fig. 8.4). SP anomalies are practically non-detectable at large distances from the anomalous bodies (combined effect from four spheres give only regional anomalous background), but separate SP anomalies can clearly be recognized at more close distances to the hidden spheres (Fig. 8.4).

In several works was demonstrated application of some advanced procedures from the most developed potential fields (gravity and magnetics), to the self-potential data. Akgün (2001) applied the Hilbert transform (usually employed in gravity and magnetic prospecting) for analysis of SP data. Sindirgi et al. (2008) successfully tested on the SP anomalies method of the total normalized gradient developed in gravity prospecting. This investigation has been continued in (Sindirgi and Özyalin 2019) with the application of the Euler deconvolution. Agarwal and Srivastava (2009) successfully verified the extended Euler deconvolution techniques in some mineral deposits and observations in deep borehole. Biswas (2018) suggested to studying SP anomalies a 2D analytic signal developed in magnetic prospecting.

Of special interest is the recently published research of Sungkono (2020), where the author proposed to employ posterior distribution model of the SP anomaly inversion. The advantages of this approach are that the SP data may contain single and multiples of SP sources and this method does not require prior assumptions over the shape of the anomaly source.

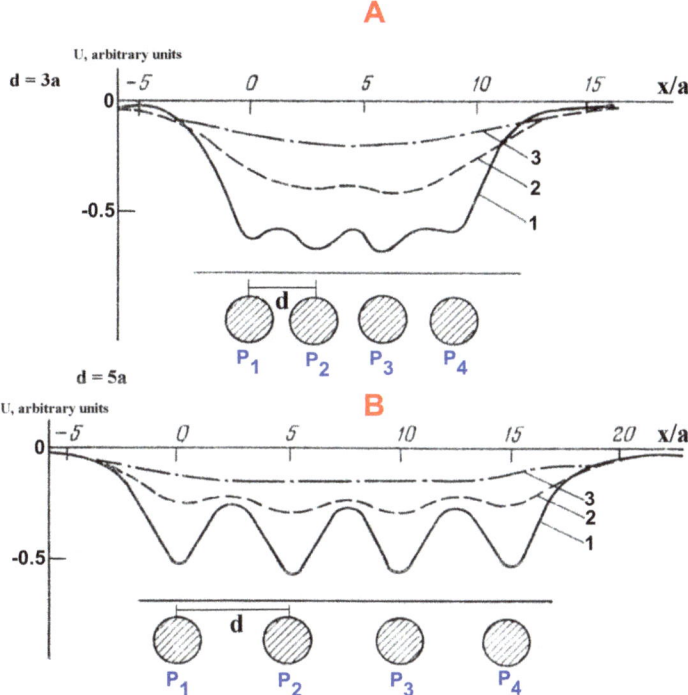

Fig. 8.4 The model SP curves over a group of the polarized spheres. The horizontal distance d between spheres: A–$3a$, B–$5a$ (after Tarasov (1961), with modifications). The spheres are geometrically and physically identical: $P_1 = P_2 = P_3 = P_4$. The levels of the SP observations: (1) $z_0 = 2a$, (2) $z_0 = 3a$, (3) $z_0 = 5a$, z_0 is the depth from surface to the center of the sphere, x/a is the relative distance

8.4.1 Quantitative Analysis of SP Anomalies by the Use of Advanced Methodologies Developed in Magnetic Prospecting

The improved methods for SP anomaly analysis include characteristic point, tangent and areal methods (these methods are described in detail in the publications suggested to magnetic anomaly interpretation: e.g., Khesin et al. 1996; Eppelbaum et al. 2000, 2001; Eppelbaum and Mishne 2011; Eppelbaum and Khesin 2012; Eppelbaum 2015). Formulas for interpretation SP anomalies by the use of characteristic point method are shown in Table 8.3. Several figures below display some peculiarities of the characteristic point and tangent methods application.

Preliminary results for SP anomalies to estimating HCC radius and length of horizontal upper edge may be obtained from 3D magnetic field modeling (taking into account a common similarity of these fields).

Table 8.3 Formulas for quantitative interpretation of magnetic anomalies from bodies approximated by thin bed (TB) and horizontal circular cylinder (HCC) using the improved characteristic point method (after Eppelbaum and Mishne (2011), with modifications)

Parameters necessary for examination	Parameters derived from anomalies from models		Formulas for calculation of parameters necessary for quantitative analysis	
	Thin bed	HCC	Thin bed	HCC
Generalized angle θ	$d_1, d_2\ d_{1r}$ $d_1, d_5\ d_{1r}, d_5$ $d_1 = x_{min} - x_{max}$ $d_2 = (x_{0.5\Delta U_A})_r - (x_{0.5\Delta U_A})_l$ $d_5 = x_r - x_l$ $\Delta U_A = \Delta U_{max} - \Delta U_{min}$		$\tan(\theta) = d_2/d_1$ $\sin(\theta/3) = d_5/\sqrt{3}d_1$	$\cot(\theta/3) = \sqrt{3}\dfrac{(d_{1l}+d_{1r})}{(d_{1l}-d_{1r})}$ $\dfrac{d_5}{d_{1r}} = \dfrac{\sqrt{2}\cos(\theta/2)-1}{\sqrt{3}\cos(60^o+\theta/3)}$
Depth h_0, h_c	$d_1, d_2, \theta\ d_{1r}, \theta$ d_5, θ		$h_0 = \sqrt{d_1 d_2/k_{1,2}},\ h_c = d_{1r}/k_{1r},\ \text{where } k_{1r}$ where $k_{1,2} = 2\sqrt{3}\dfrac{\cos(60^o + \theta/3)}{\cos\theta}$ $= \dfrac{2}{\sqrt{\sin\theta}\cos\theta}$ $h_c = \dfrac{d_{1r}}{d_{1r} - d_{1r}(\Delta h)}\Delta h$ $h = d_5/k_5,\ \text{where}$ $k_5 = 2\sqrt{3}\dfrac{\sin(\theta/3)}{\sin\theta}\ k_5 = 2\sqrt{2}\dfrac{\cos(\theta/2)-1}{\cos\theta}$	

(continued)

Table 8.3 (continued)

Parameters necessary for examination	Parameters derived from anomalies from models		Formulas for calculation of parameters necessary for quantitative analysis	
	Thin bed	HCC	Thin bed	HCC
Horizontal displacement x_0, x_c	$h, \theta, x_{max}, x_{min,r}$, $(x_{0.5\Delta U_A})_r$, $(x_{0.5\Delta U_A})_l$		$x_0 = 0.5(x_{max} + x_{min}) - h\cot\theta$ $x_0 = h\tan\left(\frac{\theta}{2}\right)$	$x_c = 0.5(x_{max} + x_{min,r}) - h_c\frac{\sin(60° + \theta/3)}{\cos\theta} + h_c\tan\theta$ $x_c = 0.5(x_r + x_l) + h_c\tan\theta - \sqrt{2}h_c\frac{\sin(\theta/2)}{\cos\theta}$
Normal background ΔU_{backgr}	$\Delta U_{min}, \Delta U_A, \theta$		$\Delta U_{backr} = \Delta U_{min} + \Delta U_A\frac{k_0}{1+k_0}$, where $k_0 = \frac{1-\cos\theta}{1+\cos\theta}, k_0 = \frac{\cos^3(60°+\theta/3)}{\cos^3(\theta/3)}$	
Self-potential moment	$\Delta U_a, h_0, h_c, Q$		$M_{\Delta U} = \frac{1}{2}\Delta U_a h_0$	$M_{\Delta U} = \frac{\Delta U_a h_c^2}{\left(3\sqrt{3}/2\right)\cos(30° - \theta/3)}$

Indices "0" and "c" designate the thin bed (TB) and horizontal circular cylinder (HCC) models, respectively. Values h_0 and h_c are the depths to upper edge of TB and center of the HCC, respectively. Parameter Δh designates measurements of self-potential field at different depths of the electrodes' grounding

The improved versions of tangent and areal methods are presented in detail in Eppelbaum et al. (2000, 2001b), Eppelbaum and Mishne (2011) and Eppelbaum (2015, 2019a).

8.4.2 SP Observations on an Inclined Profile

When potential geophysical anomalies are observed on an inclined profile, the obtained parameters characterize some fictitious body (Eppelbaum 2019a). The transition from the parameters of fictitious target to those of real target is realized using the following expressions:

$$\left\{ \begin{array}{l} h_r = h_f + x_{0f} \tan \omega_0 \\ x_r = -h_f \tan \omega_0 + x_{0f} \end{array} \right\}, \tag{8.13}$$

where h is the depth of the body upper edge occurrence (or HCC (sphere) center), x_0 is the shifting of the anomaly maximum from the projection of the center of the anomalous body to the earth's surface (produced by an oblique polarization), and ω_0 is the angle of the terrain relief inclination ($\omega_0 > 0$ when the inclination is toward the positive direction of the x-axis), the subscripts "r" and "f" stand for parameters of real and fictitious bodies, respectively.

The direction of the electric self-polarization vector ϕ_p is calculated from the expression

$$\varphi_p = 90° - \theta, \tag{8.14}$$

and on an inclined relief

$$\varphi_{p,r} = 90° - \theta + \omega_0. \tag{8.15}$$

The performed calculations of the vector ϕ_p direction on concrete field examples indicates that for the interpreting models closed to the model of inclined (vertical) thin bed, the direction of this vector approximately coincides with the anomalous body dipping. It enables to obtain a supplementary interpretation parameter.

Besides the geometric parameters of the anomalous target, the self-potential moment can also be determined (see Table 8.3). For the models of thin bed and HCC, the self-potential moment can be calculated by the use of Eq. (8.16a) and (8.16b), respectively (see also Tables 8.3 and 8.4)

$$M_{\Delta U} = \frac{1}{2} \Delta U_a h_0, \tag{8.16a}$$

Table 8.4 Nomenclature of variables applied for quantitative analysis of SP anomalies due to model of thin bed and horizontal circular cylinder (see Table 8.3)

Variable	Description
θ	Generalized angle reflecting the degree of SP anomaly asymmetry as a function relation of an anomalous body depth of occurrence, geometric form, value of polarization
x_0	Horizontal displacement of projection of the middle of the upper edge of thin bed to the earth's surface due to oblique polarization
x_c	Horizontal displacement of projection of the center of the HCC to the earth's surface due to oblique polarization
h_0	Depth to the upper edge of thin bed
h_c	Depth to the center of HCC
ΔU_{max}	Maximum value of SP anomaly
ΔU_{min}	Minimum value of SP anomaly
ΔU_A	Total amplitude of SP anomaly
d_1	Difference of extremum abscissae for thin bed
d_{1r}	Difference of extremum abscissae for HCC
d_2	Difference of semi amplitude point abscissae
d_5	Difference of inflection point abscissae
x_r	Right inflection abscissae point
x_l	Left inflection abscissae point
ΔU_{backr}	Normal background level of SP anomaly
$M_{\Delta U}$	Self-potential moment for the models of thin bed or HCC

$$M_{\Delta U} = \frac{\Delta U_a h_c^2}{\left(3\sqrt{3}/2\right) \cos\left(30^\circ - \theta/3\right)}, \qquad (8.16b)$$

where ΔU_a is the amplitude of SP anomaly (in mV), h_0 is the depth of the upper edge of thin bed (in meters), h_c^2 is the squared depth to the center of the HCC (in m^2), and θ is the some generalized angle (see Tables 8.3 and 8.4). The self-potential moment, by analogy with the magnetic field analysis, can be used for classification of various SP anomalies (and, correspondingly, hidden targets).

Initial methodologies for quantitative analysis of magnetic anomalies under complex physical-geological conditions for the model of the thick bed were presented in Khesin et al. (1996) and Eppelbaum et al. (2000, 2001) and its significant evaluation (including the intermediate models between the thick and thin beds)–in Eppelbaum (2015).

For observation on inclined profile, the real self-potential moment can be calculated as follows:

$$M_{\Delta U,r} = M_{\Delta U,f} \cos \omega 0. \qquad (8.17)$$

Here the subscripts "r" and "f" stand for a parameter of real and fictitious self-potential moments, respectively.

Undoubtedly, the calculation of all aforementioned parameters from SP data should be joined to a unified computerized interpreting system with a minimal participation of an interpreter.

For the testing of some SP anomalies, software for the 3D computation of the magnetic field may be applied. In this case, magnetic vector orientation can be utilized as analogue of self-potential vector.

8.5   Quantitative Analysis of SP Anomalies

Thus, the developed interpretation system in the SP method is applicable for complex physical-geological conditions: oblique polarization, inclined relief and unknown level of the SP normal field.

8.5.1   Testing on Theoretical Models

First of all the aforementioned interpretation methods were successfully tested on the SP anomalies from models presented in Semenov (1980), Göktürkler and Balkaya (2012) and Hristenko and Stepanov (2012).

8.5.2   Mining Geophysics

The self-potential method often enough has been employing in ore deposits of different kind (e.g., Stern 1945; Yüngül 1954; Sengupta et al. 1969; Logn and Bolviken 1974; Cowan et al. 1975; Semenov 1980; Nayak 1981; Corry 1985; Eskola and Hongisto 1987; Babu and Rao 1988; Lile 1996; Eppelbaum and Khesin 2002; Goldie 2002; Bhattacharya et al. 2007; Mendonça 2008; Srivastava and Agarwal 2009; Dmitriev 2012; Fedi and Abbas 2013; Biswas and Sharma 2016; Alizadeh et al. 2017; Erofeev et al. 2017; Safipour et al. 2017; Eppelbaum 2019a, b; Eppelbaum 2021; Zhu et al. 2021).

8.5.2.1   Chyragdere Sulfur Deposit (Central Azerbaijan)

It is interesting to compare SP studies carried out in the Ghyragdere sulfur deposit (central Azerbaijan) during several years: 1930, 1937 and 1938 (Fig. 8.5). This figure shows that the mining works in the underground shaft (1930–1938) strongly distort the SP field observed at the earth's surface (distance from the observation points to

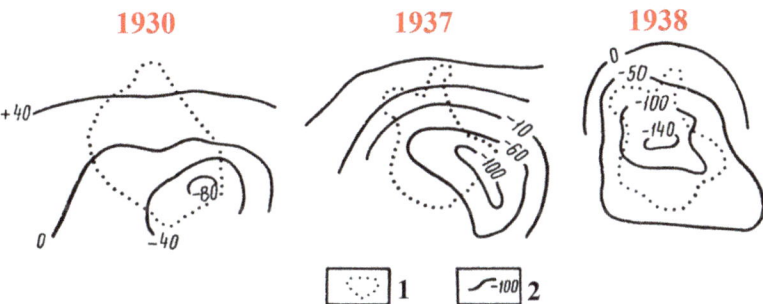

Fig. 8.5 Displacement of self-potential isolines during exploitation of the new shaft of Chyragdere sulfur deposit (Lesser Caucasus) (after Eppelbaum and Khesin 2012, with modifications). (1) stock contour, (2) isolines of self-potential field (in milliVolts)

ore deposit consisted several tens of meters). This testifies to the tight correlation between the mining processes and SP anomalies. It would be fascinating to compare the volumes and contours of the mined ore with the SP isolines, separately for the abovementioned years, but these documents have been lost over the past years.

8.5.2.2   Sariyer Sulphide-Pyrite Deposit (near Istanbul, Turkey)

Yüngül (1954) documented the results of the survey in the Sariyer area (Istanbul). The performed interpretation indicates that the obtained position of *HCC* center is in the line with geometrical and physical parameters of the sulphide-pyrite ore body (Fig. 8.6). Here and in some other figures, displayed parameters d_3 and d_4 relate to the improved tangent method (this method is described in detail, for instance, in Eppelbaum et al. (2001)). Calculating the self-potential moment by the use of Eqs. (8.16b) and (8.17), we obtain $M_{\Delta U} = 31800\,\text{mV} \cdot \text{m}^2$. The calculated direction of self-potential vector by use of Eq. (8.15) is estimated as vertical one (Fig. 8.6).

8.5.2.3   Polymetallic Deposit (Rudnyi Altai, Russia)

Figure 8.7 displays results of SP anomaly quantitative interpretation using characteristic points and tangent methods (areal method based on the calculation of the area occupied by SP anomaly has also been applied). The interpretation results, as can easily see from Fig. 8.7, have a good agreement with the ore body location. The self-potential moment (here model of a thin bed was selected and Eqs. (8.16a) and (8.17) were applied) $M_{\Delta U} = \frac{1}{2}60\,\text{mV} \cdot 6.5\,\text{m} \cdot 0.93 = 181\,\text{mV} \cdot \text{m}$. The calculated direction of self-potential vector (Eq. (8.15)) practically coincides with the polymetallic body dipping (Fig. 8.7).

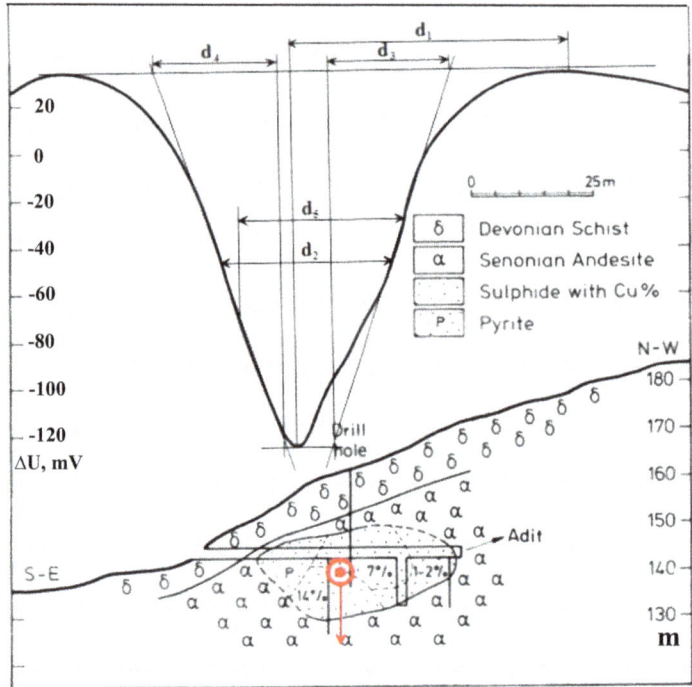

Fig. 8.6 Quantitative interpretation of *SP* anomaly by the characteristic point and tangent methods in the Sariyer area, Turkey. The "Θ" symbol marks the obtained position of the ore body center (approximated by a HCC). Red arrow shows the orientation of self-polarization vector. Observed SP curve and geological section are taken from Yüngül (1954)

8.5.2.4   Katsdag Polymetallic Deposit (Azerbaijan)

Three SP anomalies were successfully interpreted in the Katsdag copper-polymetallic deposit (southern slope of the Greater Caucasus, Azerbaijan) under conditions of rugged terrain relief (Fig. 8.8). Anomalies 1 and 2 are intensive ones, but anomaly 3 is comparatively small. It is important to underline here an essential difference between the quantitative results of SP anomalies analysis calculated without and with estimation of the rugged relief influence. The SP moment calculated for anomaly 1 (after applying Eqs. (8.16a) and (8.17)) is $M_{\Delta U} = \frac{1}{2} 180\,\text{mV} \cdot 20\,\text{m} \cdot 0.984 = 3450\,\text{mV} \cdot \text{m}$.

8.5.2.5   Filizchai Polymetallic Deposit (Azerbaijan)

An intensive SP anomaly (almost 500 mV) was observed in the portion of the Filizchai copper-polymetallic field (southern slope of the Greater Caucasus, Azerbaijan) under the conditions of highly complex terrain relief (Fig. 8.9). The results of the interpretation (improved methods of characteristic points and tangents were applied)

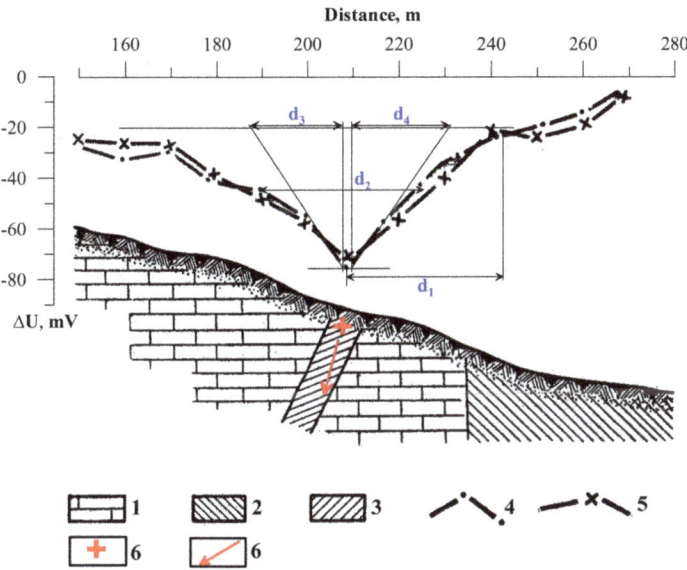

Fig. 8.7 Quantitative interpretation of *SP* anomaly over polymetallic body (Rudnyi Altai, Russia). Observed SP curve and geological section are taken from Zaborovsky (1963). (1) limestone, (2) shales, (3) copper-polymetallic ore body, SP observations carried out in (4) 1952 and (5) 1953, (6) determined position of the center of the upper edge of the ore body, (7) direction of the self-potential vector

also indicate significant difference of position of the upper edge of anomalous body calculated without rugged terrain relief influence (blue circle) and after calculation of this influence (red circle). The calculated SP moment (after Eq. (8.16a)) is $M_{\Delta U} = \frac{1}{2}440\,\mathrm{mV} \cdot 90\,\mathrm{m} = 19800\,\mathrm{mV} \cdot \mathrm{m}$. The employment of Eq. (8.17) gives us some decreased value: 14600 mV·m. It is a sufficiently high value of SP moment (for a thin bed model). However, such large SP anomalies of polymetallic ore origin are rarely observed. The direction of the self-potential vector was calculated by the use of Eq. (8.15). Position of this vector agrees well with the dipping of this pyrite-polymetallic body (Fig. 8.9).

8.5.2.6 Uchambo Ore Field (Georgia)

Figure 8.10 depicts the position of the HCC center (characteristic point, tangent and areal methods were applied), which evidently fixes the edge of a flat-lying orebody in the Uchambo polymetallic deposit (southern Georgia). The SP moment calculated using Eqs. (8.16b) and (8.17) is $M_{\Delta U} = 13480\,\mathrm{mV} \cdot \mathrm{m}^2$. The complex form of ore-body obviously did not allow to determine an exact position of the self-potential vector.

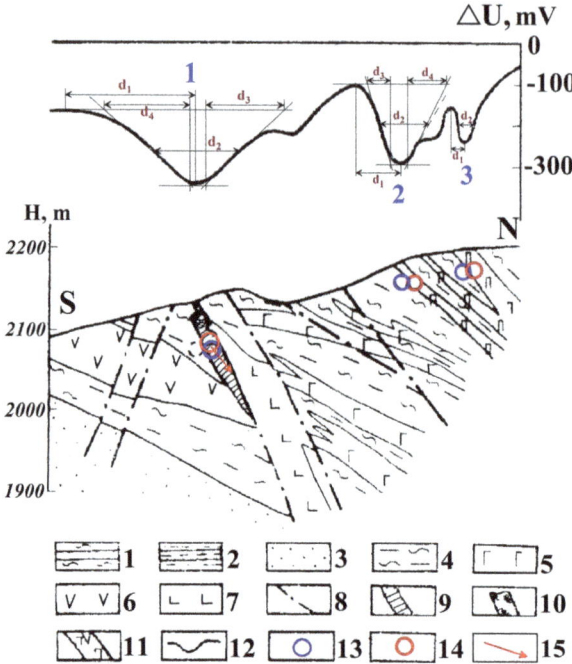

Fig. 8.8 Results of quantitative interpretation of *SP* anomalies in the area of Katsdagh copper-polymetallic deposits on the southern slope of the Greater Caucasus (Azerbaijan). (1) interbedding of sands and clay schists, (2) clay schists with the flysh packages, (3) clay sandstone; (4) sand-clay schists; (5) diabases, gabbro-diabases and diabasic porphyrites; (6) andesites and andesite-porphyrites; (7) dacitic porphyrites; (8) faults; (9) massive ore of pyrite-polymetallic composition; (10) oxidized ore; (11) zones of brecciation, crush and boudinage with lean pyrite-polymetallic ore; (12) *SP* curves; location of anomalous source: (13) without calculation of inclined relief influence, (14) after introducing correction for terrain relief, (15) position of the self-potential vector

8.5.2.7 Potentsialnoe Polymetallic Deposit (Rudnyi Altai, Russia)

Here three different interpreting models were utilized (thin bed, HCC and thick bed) (Fig. 8.11). All three applied models are suitable ones. The calculated SP moment (for the HCC model) is (after employing Eqs. (8.16b) and (8.17)) $M_{\Delta U} = 21890\,\text{mV}\cdot\text{m}^2$. The calculated position of the polarization vector (HCC) coincides with the dipping of the polymetallic body (Fig. 8.11).

8.5.2.8 Graphite Body (Southern Bavarian Woods, Germany)

The graphite subvertical body occurring in the gneisses produces sufficiently large SP anomaly—more than 600 mV (Fig. 8.12). Application of the aforementioned interpretation methods enabled to determine exactly position of the center of the upper edge of the anomalous body and to calculate position of the self-potential vector.

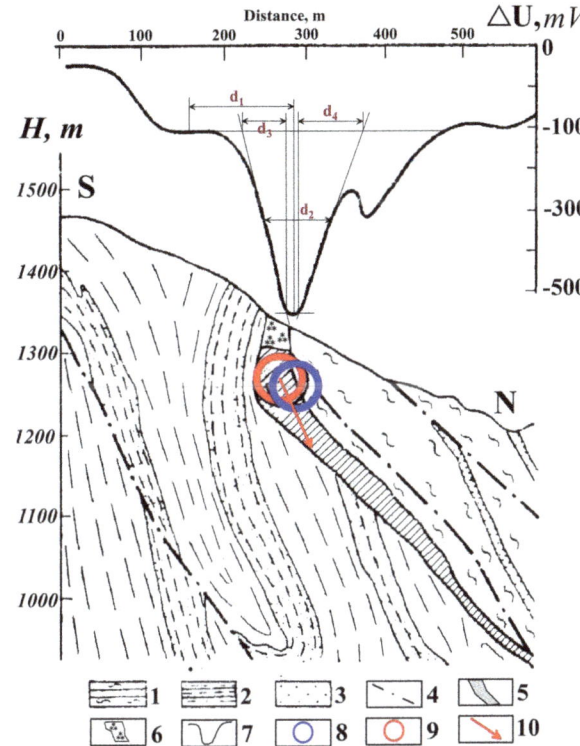

Fig. 8.9 Results of quantitative interpretation of SP anomaly in the area of Filizchay copper-polymetallic deposit in the southern slope of the Greater Caucasus (Azerbaijan) (revised after Eppelbaum and Khesin (2012)). (1) interbedding of sands and clay schists, (2) clay schists with the flysh packages, (3) clay sandstone, (4) faults; (5) massive ore of pyrite-polymetallic composition, (6) oxidized ore, (7) SP curves, location of the anomalous source: (8) without calculation of inclined relief influence, (9) after introducing correction for relief, (10) direction of self-potential vector

Both these determined parameters nicely agree with the body geometrical parameters (Fig. 8.12). The calculated self-potential moment (after assumed correction for the inclined relief) consists of 13465 mV·m.

8.5.2.9   Canyon Makhtesh Ramon (Negev Desert, Southern Israel)

The Makhtesh Ramon erosional–tectonic depression (canyon), 40 km long and approximately 8 km wide, is situated in the Negev Desert (southern Israel), 65 km southwest of the Dead Sea. On the basis of integrated geological-geophysical investigations, in this area were detected several tens of microdiamonds (largest sample is 1.35 mm) and a large amount of mineral-satellites of diamond (Eppelbaum et al. 2006). Many geological-geophysical indicators showed that at least a part of the indigenous sources (kimberlites or lamproites) of the aforementioned minerals can occur here in subsurface. The compiled SP map (Fig. 8.13a) displayed the presence of some anomalous zones. Quantitative analysis of the SP anomaly was carried out along profile A–B (Fig. 8.13b) crossing one of the mentioned zones. Results of the performed interpretation indicated that the anomalous inclined body (having a geometrical form close to the thin bed model) occurs at the depth of 40 m which agrees with the preliminary available geophysical data.

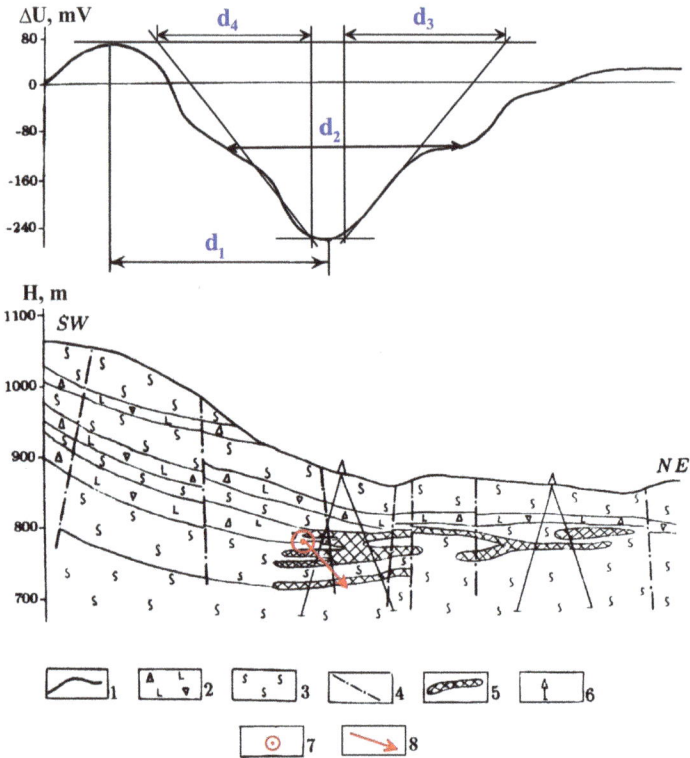

Fig. 8.10 Interpretation of SP anomaly by the method of characteristic points in the area of the Uchambo ore field of the Adjar group of copper-polymetallic deposits (Georgia, Lesser Caucasus). (1) SP observed values; (2) heteroclastic tuff breccia and their tuffs; (3) cover trachyandesite-basalts with pyroclastic interbeds; (4) disjunctive dislocations; (5) zones of increased mineralization; (6) drilled wells; (7) location of HCC center according to the interpretation results, (8) orientation of self-potential vector ((1–6) from Bukhnikashvili et al. (1974))

8.5.2.10 SP as a Component of Multimodel Approach

The multimodel approach to geophysical data analysis may be illustrated on example of quantitative analysis of different geophysical data. Quantitative interpretation is traditionally oriented to a single model for the hidden objects identification. In the case of the existence of several hypotheses relating to the parameters of the body causing the disturbance (i.e., the buried object) usually only one model was selected roughly presenting the object in the domain $\mathfrak{R}_x$ of k-dimensional space of the physical-geological factors. At the same time, many geological features are strongly disturbed by the various geological processes (erosion, tectonic-geodynamic activity, metamorphism, etc.) and can be reflected in different ways in various geophysical fields.

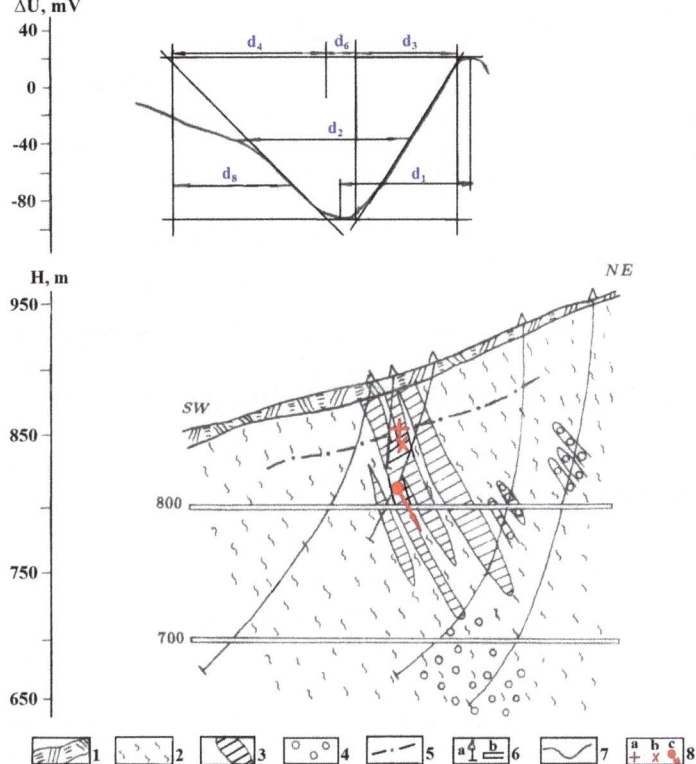

Fig. 8.11 Interpretation of SP anomaly by the developed techniques in the area of the Potentsialnoe ore deposit (Rudny Altai, Russia) (initial data from Semenov (1975)). (1) soil-vegetative layer; (2) alternation of lavas and tuffs of acid composition and chlorite-sericitic schists; (3) sulfide ores; (4) sulfide impregnation, pyritization; (5) level of ground waters; (6) drilling wells (**a**) and adits (**b**); (7) plot of SP; (8) interpretation results: **a**—upper edge of the thin bed, **b**—mid-point of the inclined thick bed's upper edge, **c**—center of a horizontal circular cylinder (arrow indicates the direction of the polarization vector obtained by interpretation)

Additional noise affecting quantitative interpretation includes the rugged terrain relief, anisotropy (polarization) of geological objects and heterogeneous host medium. As a consequence, the response function Ψ_i—geophysical field—may ambiguously represent the studied targets. Therefore, the domain $\Re_x$ may be divided into several subdomains $\Re_1, \Re_2,..., \Re_m$ and in each of them a single model will dominate (Eppelbaum 1987). In such a way we could develop m physical-geological models of the same target, each corrected for the separate subdomains $\Re_1, \Re_2,..., \Re_m$.

The multimodel approach can also be applied at varying levels of the geophysical field observations. Hence, different explanatory models may be used in the process of quantitative interpretation. Integrating several response functions Γ_i, yields a more accurate and reliable physical-geological model of the buried target.

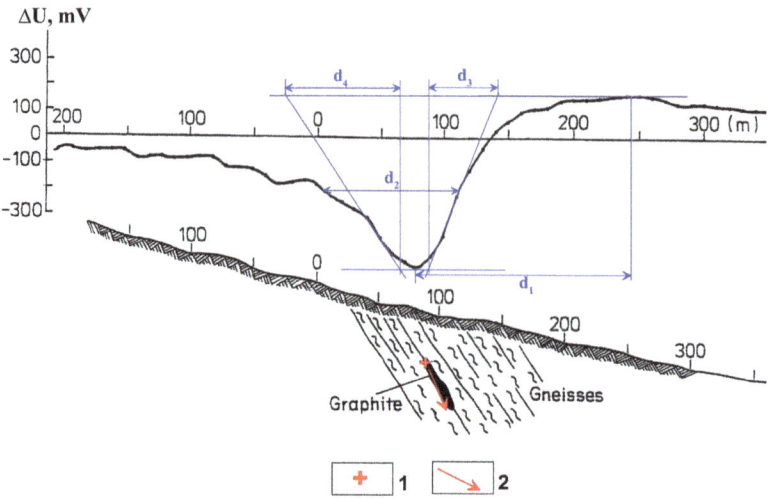

Fig. 8.12 Quantitative interpretation of SP anomaly over graphite body in the southern Bavarian woods, Germany (geological section and observed SP curve are taken from Meiser 1962). (1) position of the middle of the upper edge of thin bed, (2) orientation of the self-potential vector

Rapid methods of quantitative interpretation make it possible to determine the following parameters: position of the mass center of the anomaly-forming body by the Δg curve (Fig. 8.14a), position of the upper edge by the ΔZ curve (Fig. 8.14b) and position of the HCC center in the upper portion of the ore-body at the ground water level by the SP curve (Fig. 8.14c). Thus, the obtained specific models reflect the contrasting character of the physical properties of the target and the host medium. They allow obtaining exhaustive description of the geometric parameters of the buried target. Combining these three models (we have three response functions Γ_1, Γ_2 and Γ_3 from the subdomains $\Re_1$, $\Re_2$ and $\Re_3$), yields a combined model of the anomalous body (Fig. 8.14d), which is in a good agreement with the initial (prescribed) model.

8.5.3 Archaeological Sites

SP measurements are not frequently applied for searching and localization of archaeological targets (e.g., Wynn and Sherwood 1984; Mauriello et al. 1998; Eppelbaum et al. 2003a, b; Drahor 2004; Drahor et al. 2006; Di Maio et al. 2010; Shevnin et al. 2014; Tsokas et al. 2014; De Giorgi and Leucci 2017, 2019; Eppelbaum 2020). Obviously, absence of reliable methodologies for quantitative analysis of SP anomalies, weak SP anomalies and different kinds of noise impedes a wide employment of the self-potential method in archaeological prospection.

The territory of Israel contains more than 35,000 discovered archaeological sites of different age and origin. For SP observations several typical archaeological sites

Fig. 8.13 **a** SP map
observed in the western
Makhtesh Ramon (northern
Negev desert).
b Interpretation of the SP
anomaly along profile A–B,
western Makhtesh Ramon
(see Fig. 8.13a). The red
cross indicates the position
of the center of the upper
edge of the anomalous body,
and the red arrow indicates
the orientation of the
self-potential vector

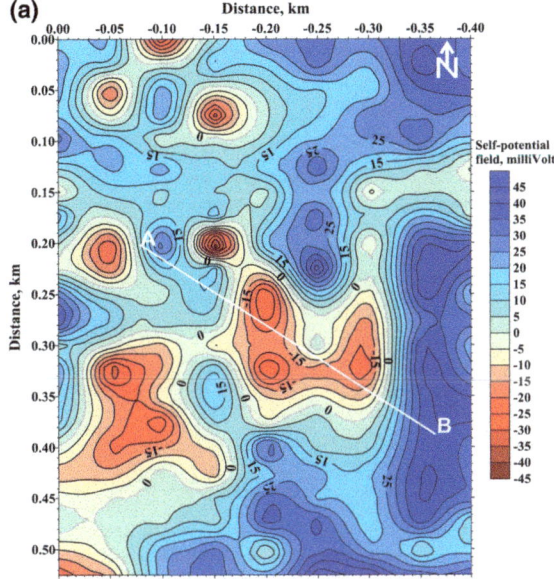

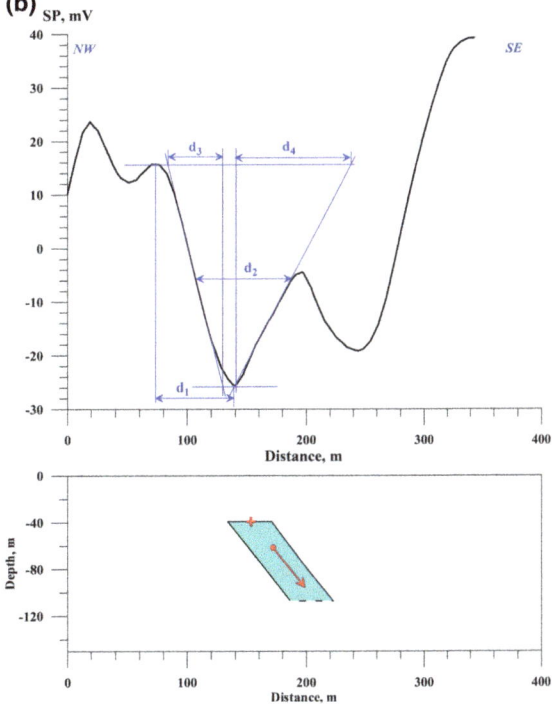

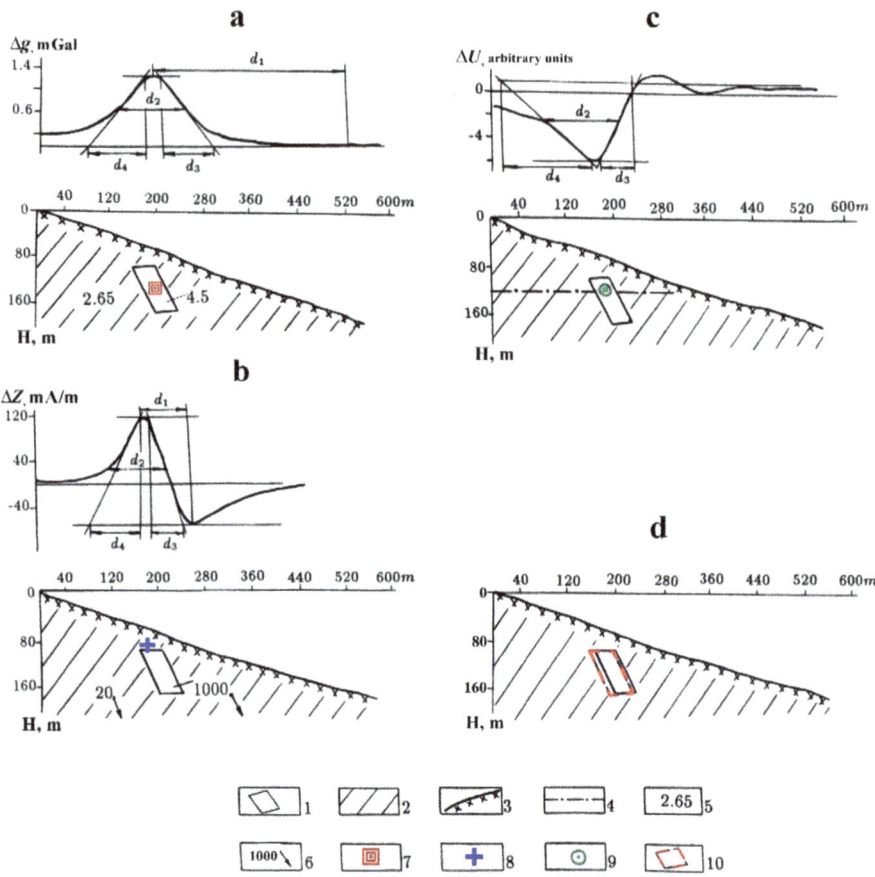

Fig. 8.14 Combined interpretation of the model gravity (Δg), magnetic (ΔZ) and self-potential (ΔU_{SP}) due to generalized ore body model of the Filizchai type under different approximations of the anomalous body: **a–c** results of the model fields interpretation, **d** anomalous object according to the results of integrated interpretation. (1) anomalous body; (2) host medium; (3) topography; (4) position of the ground water level; physical properties: (5) density (g/cm^3); (6) magnetization (mA/m); (7) mass center (for a horizontal circular cylinder) by Δg plot; (8) mid-point of the upper edge of an inclined thin bed by ΔZ plot; (9) position of the center of HCC inscribed into the upper portion of the anomalous body at the ground water level by ΔU_{SP} plot; (10) contour of the anomalous body obtained from the results of integrated quantitative interpretation

located in different regions of the country were selected (Eppelbaum et al. 2001a, b, 2004). All SP measurements were performed using microVoltmeter with the high input impedance and distinctive non-polarizable electrodes (Cu in CuSO$_4$ solution). The interpretation results obtained earlier at these sites were revised and generalized (the joined methodologies were employed) (Eppelbaum 2020).

8.5.3.1 Roman Site of Banias, Northern Continuation (Northern Israel)

The remains of the city of Banias are located in northern Israel, at the foot of Mt. Hermon. Banias was the principal city of the Golan and Batanaea regions in the Roman period and occupied an area of more than 250 acres (Kempinsky and Reich 1992). Here different ancient remains of Roman and other historical periods were found. The area of the present study is located several km north of the well-investigated Banias site (Meyers 1996). In the nearest vicinity of the area of geophysical examination (SP and magnetic surveys), the remains of an ancient Roman cemetery and aqueduct (Hartal 1997) were discovered. Mineralogical and geochemical analyses of the excavated Roman chambers indicated that these objects were composed from the special type of hot worked limestone.

SP observations were carried out by the grid of 1×1 m. A compiled SP map (Fig. 8.15) nicely indicates two anomalies. Interpretation profiles I–I and II–II were selected crossing centers of these anomalies (Fig. 8.16). The upper edges of the recognized anomalous targets occur at the depth of 1.1–1.3 meters. Presence of these anomalous sources was confirmed by a comprehensive magnetic data analysis. Angle φ_p for anomaly I consists of $75°$, but it is calculated from the opposite side (due to inversion of parameters d_3 and d_4 on the SP curve). SP moment for the anomaly I (thin bed) is $M_{\Delta U} = \frac{1}{2} 16.5\, \text{mV} \cdot 1.2\, \text{m} = 9.9\, \text{mV} \cdot \text{m}$. Interestingly that this SP

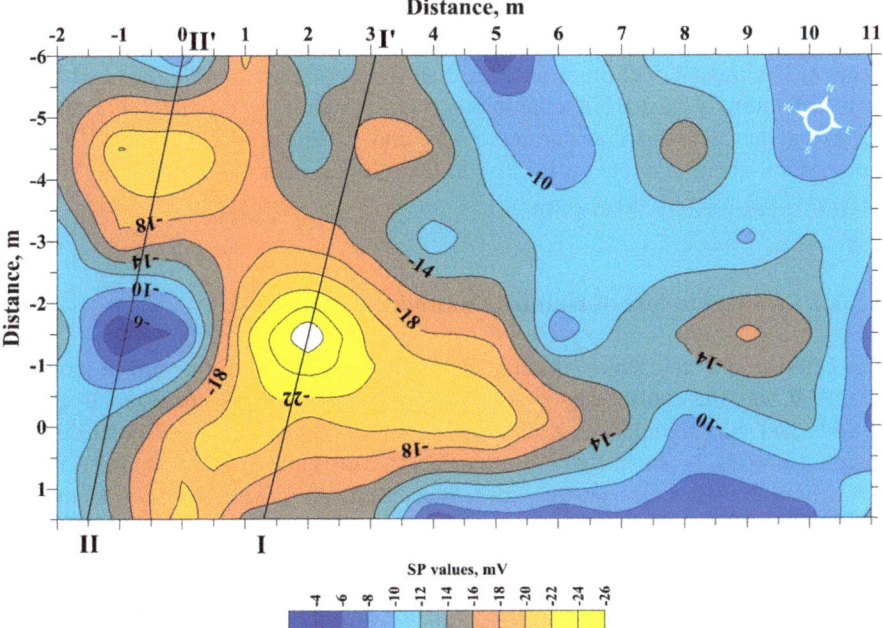

Fig. 8.15 Self-potential map observed in the Banias site (northern Israel) and location of interpreting profiles I–I' and II–II'

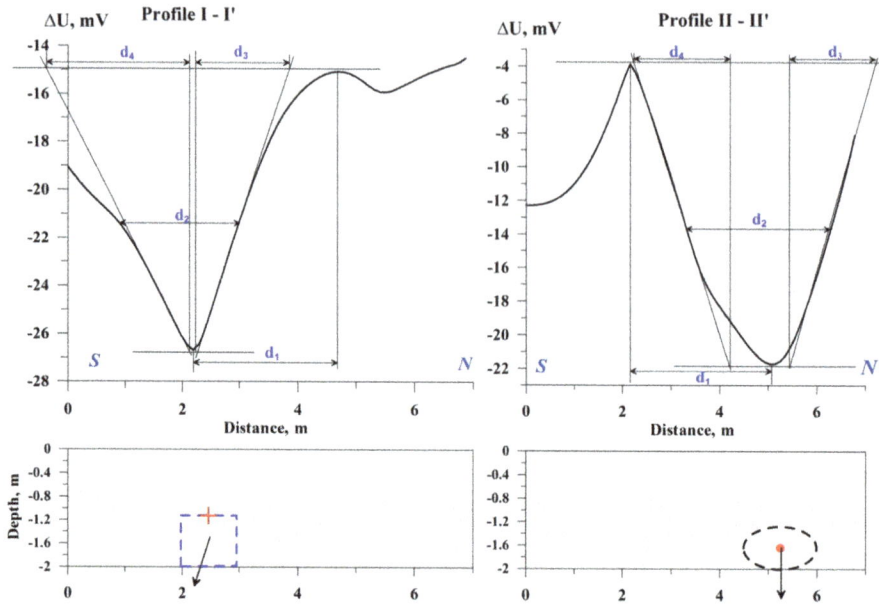

Fig. 8.16 Quantitative analysis of anomalies I–I' and II–II' (see Fig. 8.15) in the Banias site (northern Israel). Red cross indicates position of the center of upper edge, bold red point testifies position of the HCC center, and the black arrows show direction of the self-potential vector φ_p

moment is almost 1,500 times smaller than the same parameter calculated for the giant SP anomaly in the Filizchai deposit, Azerbaijan (see Fig. 8.9). For anomaly II, taking into account that parameters d_3 and d_4 are practically equal ($\theta \approx 0$), value of polarization angle φ_p is close to 90°. SP moment for anomaly II (HCC) is $M_{\Delta U} = \frac{2.79m^2 \cdot 18mV}{2.6 \cdot \cos 15°} \cong 20\,mV \cdot m^2$.

8.5.3.2 Nabatean Site of Halutza (Southern Israel)

The Halutza site is located 20 km southwest of the city of Be'er-Sheva (southern Israel). It was the central city of southern Palestine in the Roman and Byzantine periods and was founded as a way-station for Nabatean (7th–2nd centuries BC) traders traveling between Petra (Jordan) and Gaza. This site was occupied mainly throughout the Byzantine period (4th–7th centuries AD) (Kenyon 1979; Kempinski and Reich 1992).

At this site self-potential and magnetic measurements were carried out in a 20 × 10 m area with a 1 × 1 m grid (Eppelbaum et al. 2003b). The buried targets (ancient Roman limestone constructions) have produced negative anomalies in both geophysical potential fields. Results of the quantitative examination (here interpretation models of thin bed were utilized) are practically identical (Fig. 8.17a, b). Amplitude of the SP anomaly reached 40 mV; it is the largest from the anomalies

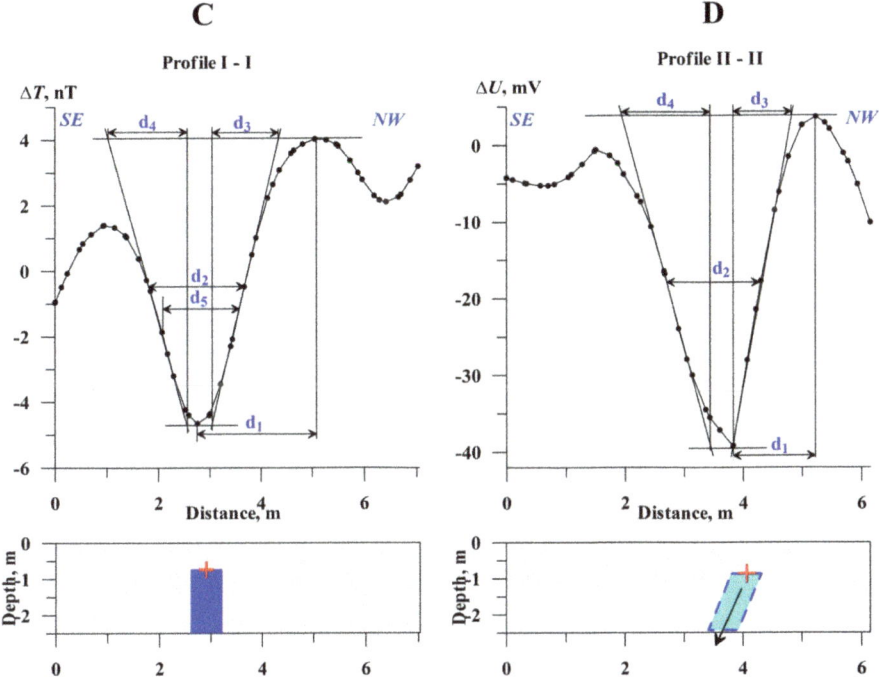

Fig. 8.17 Quantitative analysis of magnetic (A) and self-potential (B) anomalies in the site of Halutza (southern Israel). Red cross in both models indicates position of the center of the upper edge, and the red arrow shows direction of polarization vector φ_p (Fig. 8.17B)

observed in this site. The depth of both these anomalous targets is about 0.85 m. The calculated moments for the magnetic and self-potential anomalies, are following: $M_{\Delta T} = 3.61$ nT $\cdot$ m and $M_{\Delta U} = 20$ mV $\cdot$ m.φ_p angle for SP anomaly is calculated from the opposite side and consists of 70° (Fig. 8.17b). It may be also concluded that the recognized anomalous target approximated by thin bed in the SP method has not vertical dipping, but coinciding with the φ_p angle. The obtained quantitative parameters of ancient constructions agree with the results of archaeological excavations performed in the vicinity of this site.

8.5.3.3   Christian Site of Emmaus-Nikopolis (Central Israel)

The Christian archaeological site Emmaus-Nikopolis is well known in the ancient and Biblical history. The site is situated roughly halfway between Jerusalem and Tel Aviv (central Israel). The Crusaders rebuilt it on a smaller scale in the 12th century (Meyers 1996). The site of Nikopolis is displayed in almost all Christian Pilgrim texts from the 4th century onward; in majority of archaeological sources this site is named as Emmaus-Nikopolis. Many scientists note that this site is characterized by a multilayer sequence (e.g., Kempinski and Reich 1992).

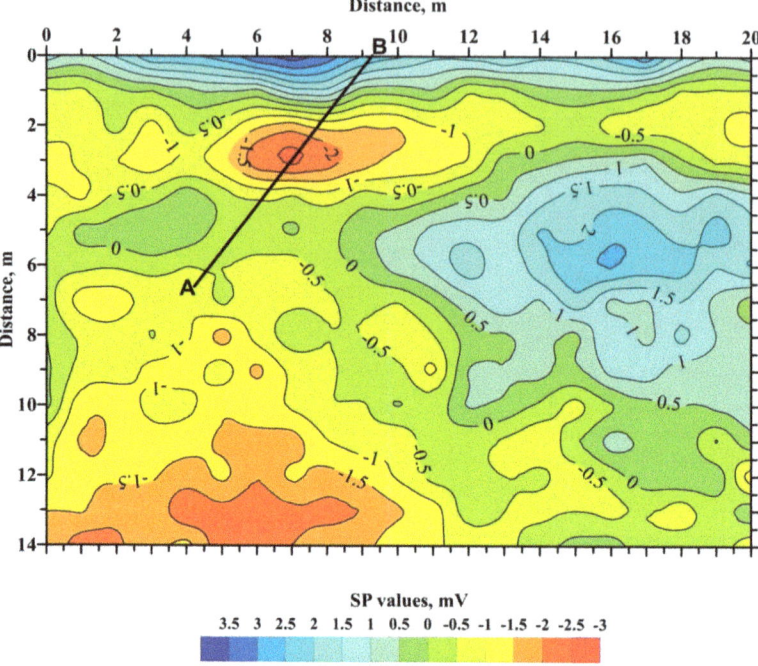

Fig. 8.18 Self-potential map observed in the site of Emmaus-Nikopolis (central Israel)

SP measurements in this site were performed by the grid of 1 × 1.5 m. In the complied SP map (Fig. 8.18), one local anomaly was selected for quantitative analysis (Profile A–B). The determined depth of the target upper edge is about 1.5 m (depth of the HCC center is about 2.1 m) (Fig. 8.19). The φ_p angle here is 85° and is calculated from the opposite side. Self-potential moment of this anomaly is 4.5 mV·m². The fragments of some glass vessels discovered in this burial cave allowed to attributing it to the Byzantine period. Interestingly to note that magnetic field examination allowed recognizing the same cave by the integrated effect from a few tens of small magnetic anomalies produced by the oil lamps (made from the fired clay). Corresponding photos of the localized buried cave and one of ancient oil lamp stored here are shown in accordingly in Fig. 8.20a, b (Eppelbaum et al. 2007).

8.5.4 Environmental Geophysics

Other SP application revealed some dangerous environmental phenomena (karst cavities, faults, rockslides) (e.g., Ogilvi and Bogoslovsky 1979; Corwin 1990, 1996; Quarto and Schiavone 1996; Gurk and Bosch 2001; Vichabian and Morgan 2002; Lapenna et al. 2003; Jardani et al. 2006a, b; Eppelbaum 2007; Jardani et al. 2007;

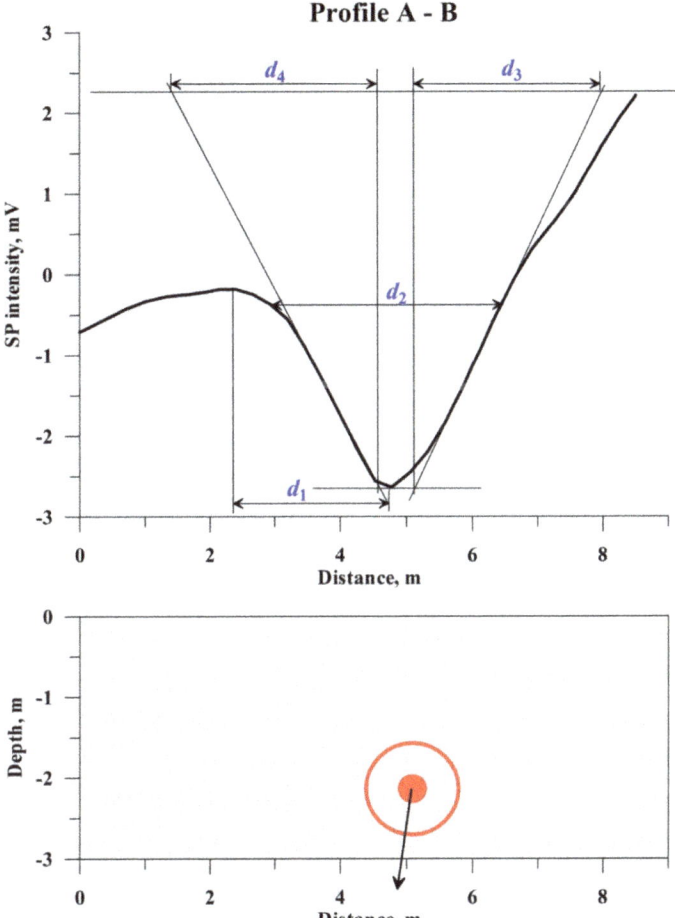

Fig. 8.19 Quantitative analysis of self-potential anomaly along profile A–B in the site of Emmaus-Nikopolis (SP map is presented in Fig. 8.18). The bold red point indicates position of the HCC center, and black arrow shows the direction of the self-potential vector φ_p

Rozycki et al. 2006; Gibert and Sailhac 2008; Srigutomo et al. 2010; Tripathi and Frayar 2016; Chen et al. 2018; Gusev et al. 2018; Oliveti and Cardarelli 2019; Eppelbaum 2021).

8.5.4.1   Buried Cavities in Dolomitic Limestone (Southern Italy)

Several impressive examples of SP application for the detection of underground cavities in southern Italy were displayed in Quarto and Schiavone (1996). Let us will consider one of these field cases, where the buried karst cavity exists in dolomitic

Fig. 8.20 The site of
Emmaus-Nikopolis (central
Israel). **a**—revealed
underground cavity (see
Figs. 8.18 and 8.19), **b**—one
of the oil lamps discovered
in the cavity

A

B

limestones (Fig. 8.21). The cavity is horizontally extended, and over it, a significant
SP anomaly (up to 100 mV) was observed. Quantitative examination along profile
A–B crossing the center of this anomaly has been performed (Fig. 8.22). For inter-
preting models of thin bed (upper edge) and center of HCC were obtained depths of
6.0 and 9.5 meters, respectively. The self-potential vector is oriented near-vertically.
Self-potential moment of anomaly from this cave (HCC model) is about 3300 mV·m^2.

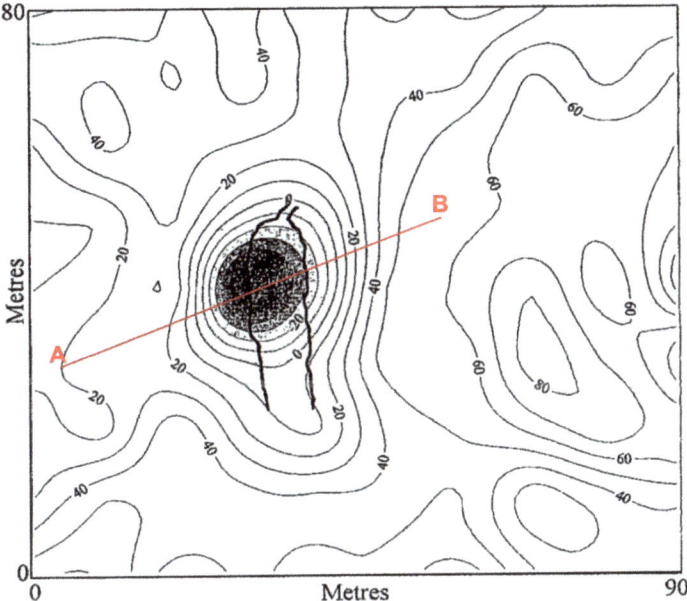

Fig. 8.21 SP map over the underground cave and position of the interpreting profile A–B (SP map after Quarto and Schiavone (1996))

8.5.4.2 Cavities in the Djuanda Forest Park (Bandung, Indonesia)

The next example displays the results of SP and electric resistivity observations carried out above cavities (built during the WW II in early 1940 s) at the Djuanda Forest Park, Bandung, Indonesia. Fascinatingly, the resistivity section (Fig. 8.23b) nicely shows two bright anomalies whereas the SP graph (Fig. 8.23a) indicates comparatively significant anomaly (amplitude is more than 20 mV) over cave II, whereas the anomaly over the cave I is only emerging (its amplitude is about 2 mV). Obviously, this fact is associated with the hydrogeological peculiarities of the subsurface section in the area under study. Quantitative analysis of SP over the cave II (here, an interpretation model of thin bed was applied) gave satisfactory results generally coinciding with the results of resistivity section. Calculated self-potential moment here is $M_{\Delta U} = 52\,\text{mV} \cdot \text{m}$.

8.5.4.3 Subvertical Fissure Zone (Russia)

Figure 8.24 displays the SP and resistivity graphs over the subvertical-fissured zone. SP quantitative examination showed significant disagreement between the results of interpretation and the available geological section. However, it is interestingly to note that the performed quantitative analysis of the resistivity curve ρa (presented in the

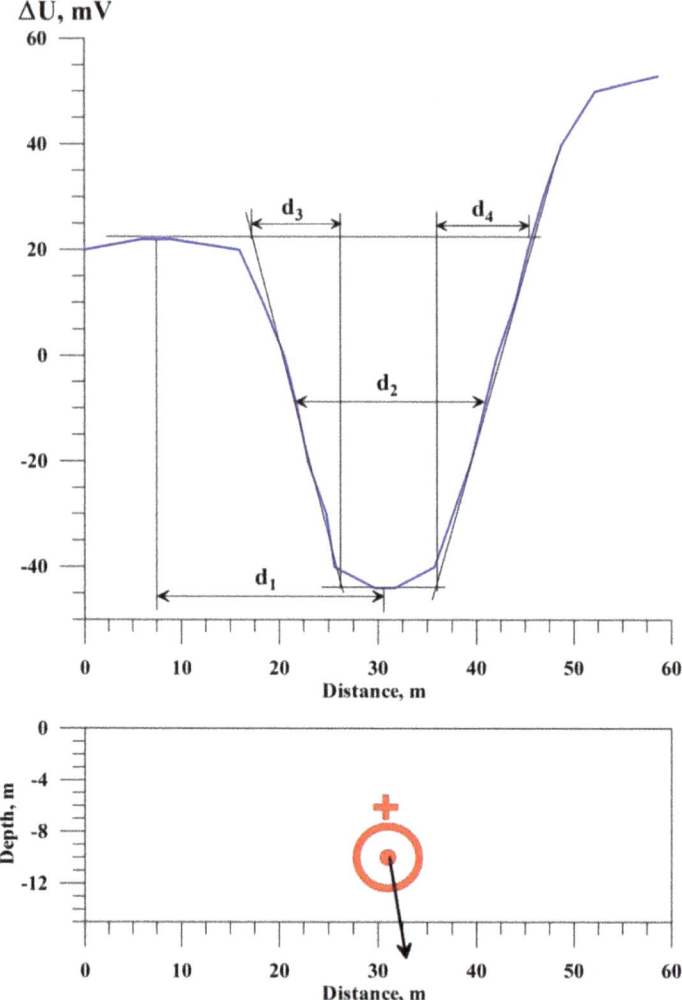

Fig. 8.22 Quantitative analysis of SP anomaly along profile A–B (location of profile is shown in Fig. 8.17). The red cross shows the position of the middle of the thin bed upper edge, and the red point shows the position of the center of the horizontal circular cylinder. The black arrow shows the orientation of the polarization vector

upper part of Fig. 8.24) by the use of the same methodology gave the similar results. Theoretical possibilities of such analysis were reported in Eppelbaum (2007) and evaluated in Eppelbaum (2019). Shevnin (2018) also indicates a good correlation between the SP and resistivity methods.

Obviously, this disagreement can be explained by some erosion of the upper part of anomalous body (fissured zone) and corresponding changes of its physical properties (possibly appearing to be close to the physical properties of the

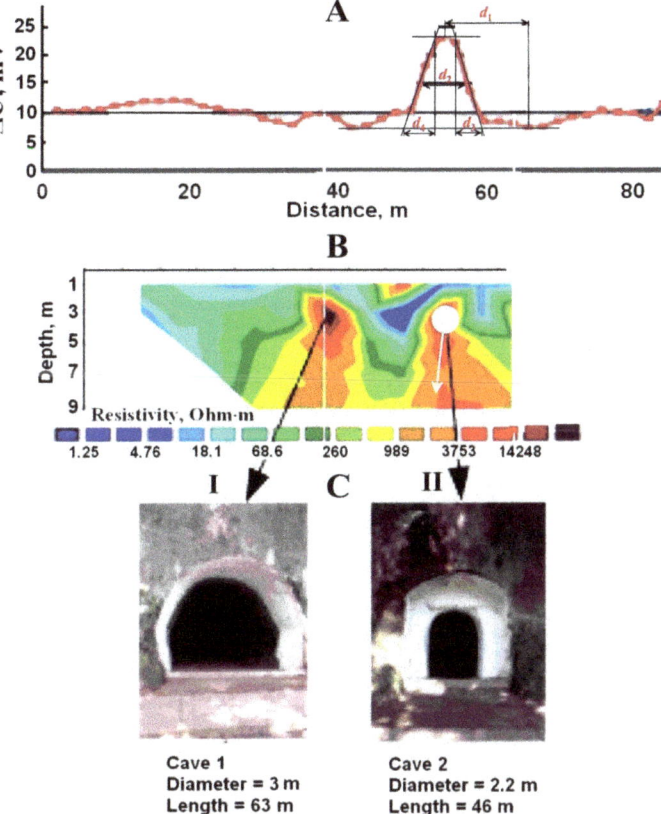

Fig. 8.23 Quantitative analysis of SP anomaly over underground cave in the Djuanda Forest Park, Bandung (Indonesia). A: SP profile with interpretation of anomaly II, B: ERT section, C: geological section. Initial data are taken from Srigumoto et al. (2010). The white circle and arrow indicate the position of upper edge of the cave II and its dipping, respectively

host media). Nevertheless, the orientation of the self-potential vector coincides with the fissured zone dipping (Fig. 8.24). Value of the self-potential moment is estimated as $M_{\Delta U} = 23\,\text{mV} \cdot \text{m}$.

8.5.5 Technogenic Geophysics

Let us designate technogenic geophysics as geophysical studies applied to the detection or determination of certain parameters of hidden modern industrial objects. The SP method fits well in such studies.

Onojasun and Takum (2015) have been successfully applied SP investigations for the localization of an underground concrete water pipeline at the Kwinana industrial

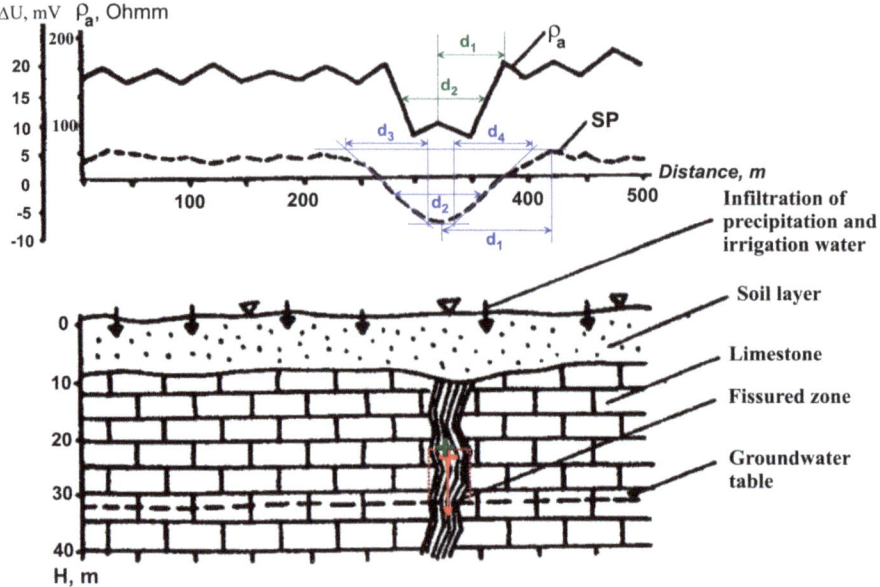

Fig. 8.24 The interpretation of SP anomaly over the fissured zone (Russia) (resistivity and SP graphs and geological section are from Ogilvy and Bogoslovsky (1979)). The red and green crosses show the positions of the upper edge of the anomalous target determined from SP and resistivity curves, respectively. The red arrow indicates position of the self-potential vector

area of Southern Perth (Western Australia). In a second example, comprehensive examination of SP imaging over a metallic contamination plume was performed by Cui et al. (2017). The authors have concluded that the SP method can be successfully used to monitor the underground metallic contaminants.

It was established that examination of the SP anomalies is significant for the localization of corrosion in the buried oil, gas and water pipes (e.g., Corwin 1996; Castermant et al. 2008; Ekine and Emujakporve 2010; Rittgers et al. 2013; Oliveti and Cardarelli 2019).

8.5.5.1 Underground Metallic Water-Pipe (Southern Russia)

Fomenko (2010) presented a case of a typical SP field distribution over the buried metallic water-pipe (Fig. 8.25). This anomaly has been interpreted by the use of tangent and characteristic point methods. The obtained position of the HCC center agrees (with some assumptions) with a center of the hidden water-pipe. The calculate self-potential moment $M_{\Delta U} = 79.8 \, \text{mV} \cdot \text{m}^2$.

Fig. 8.25 Quantitative examination of SP anomaly from the buried metallic pipe (southern Russia). Observed SP graph and environmental section are taken from Fomenko (2010). Small red circle indicates determined position of the center of HCC, and arrow shows position of the self-potential vector

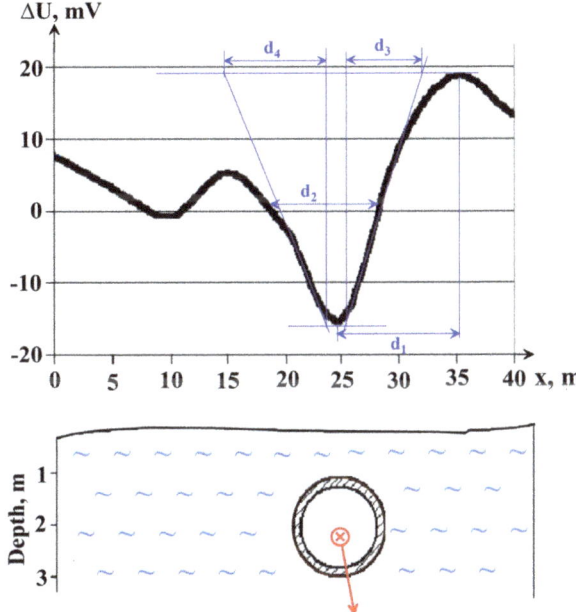

8.5.6   Generalization of the Calculated Self-Potential Moments

The calculated self-potential moments for the variety of investigated targets are compiled in Table 8.5.

The values of self-potential moments $M_{\Delta U}$ presented in Table 8.5 demonstrate a wide range of the calculated parameters. The $M_{\Delta U}$ values can be divided in three groups: (1) comparatively large values corresponding to comparatively big ore bodies, (2) middle values relating to targets studying in environmental and techno-genic geophysics, and (3) relatively small values reflecting archaeological targets. However, even the smallest $M_{\Delta U}$ value calculated, for instance, for the ancient cave in the site of Emmaus-Nikopolis has independent importance.

Table 8.5 Comparison of self-potential moments calculated from different anomalous targets (considered in Sects. 8.5.2–8.5.5)

Object	Location	Approximation model	Value of self-potential moment $M_{\Delta U}$
I. Ore geophysics			
Sariyer sulphide-pyrite deposit	Near Istanbul, Turkey	HCC	$31800\,\text{mV} \cdot \text{m}^2$
Polymetallic deposit	Russia	Thin bed	$181\,\text{mV} \cdot \text{m}$
Katsdag polymetallic deposit	Southern Greater Caucasus, Azerbaijan	Thin bed	$3450\,\text{mV} \cdot \text{m}$
Filizchai polymetallic deposit (Azerbaijan)	Southern Greater Caucasus, Azerbaijan	Thin bed	$14600\,\text{mV·m}$
Uchambo ore field	Lesser Caucasus, Georgia	HCC	$13480\,\text{mV} \cdot \text{m}^2$
Potentsialnoe polymetallic deposit	Rudnyi Altai, Russia	Thin bed, thick bed, HCC	$21890\,\text{mV} \cdot \text{m}^2$
Southern Bavarian woods, graphite body	Germany	Thin bed	$13465\,\text{mV·m}$
II. Archaeogeophysics			
Banias (anomaly I)	Northern Israel	Thin bed	$9.9\,\text{mV} \cdot \text{m}$
Banias (anomaly II)	"—"	HCC	$20\,\text{mV} \cdot \text{m}^2$
Halutza	Southern Israel	Thin bed	$20\,\text{mV} \cdot \text{m}$
Emmaus-Nikopolis	central Israel	HCC	$4.5\,\text{mV·m}^2$
III. Environmental geophysics			
Underground cave	Southern Italy	HCC	$3300\,\text{mV} \cdot \text{m}^2$
Underground cave	Bandung, Indonesia	Thin bed	$52\,\text{mV} \cdot \text{m}$
Fissured zone	Russia	Thin bed	$23\,\text{mV} \cdot \text{m}$
IV. Technogenic geophysics			
Underground metallic water-pipe	Southern Russia	HCC	$79.8\,\text{mV} \cdot \text{m}^2$

8.6   Conclusions

The self-potential method is one of the oldest and simultaneously non-expensive geophysical methods. One of its main preferences is that the presence of water in the subsurface does not limit this method capability. The various disturbances complicated the SP observations under different physical-geological environments are analyzed. The available interpretation methodologies are briefly discussed. The proved common aspects between the magnetic and self-potential fields enable to apply for interpretation of SP anomalies the modern interpreting procedures developed for complicated environments in magnetic prospecting (oblique magnetization (polarization), rugged topography and an unknown level of the normal field). These interpretation procedures applied for SP anomalies enable to obtain reliably geometric parameters of buried anomalous targets occurring in complex physical-geological environments. The suggested calculation of the direction of the electric self-polarization vector allows in many cases to estimate a dipping of anomalous objects. It is proposed to apply in SP method a calculation of self-potential moment which can be used for the classification of observed the SP anomalies. An applicability of the multimodel approach with application of the SP and other methods is demonstrated on the generalized physical-geological model of ore body of Filizchai type. Testing interpretation procedures in mining, environmental, archaeological and technogenic geophysics in various regions of the world indicates the practical effectiveness of these methodologies.

References

Abdelrahman EM, El-Araby TM, Ammar AA, Hassanein HI (1997) A least-squares approach to shape determination from selfpotential anomalies. Pure Appl Geophys 150:121–128

Abdelrahman EM, Ammar AA, Hassanein HI, Hafez MA (1998) Derivative analysis of SP anomalies. Geophysics 63:890–897

Abdelrahman EM, Sharafeldin SM (1997) A least squares approach to depth determination from residual self-potential anomalies caused by horizontal cylinders and spheres. Geophysics 62:44–48

Agarwal BNP, Srivastava S (2009) Analyses of self-potential anomalies by conventional and extended Euler deconvolution techniques. Comput Geosci 35:2231–2238

Akgün M (2001) Estimation of some bodies parameters from the self potential method using Hilbert transform. J Balkan Geophys Soc 4(2):29–44

Al-Garni MA (2017) Interpretation of spontaneous potential anomalies from some simple geometrically shaped bodies using neural network inversion. Acta Geophys 58(1):143–162

Alizadeh AM, Guliyev IS, Kadirov FA, Eppelbaum LV (2017) Geosciences in Azerbaijan. Volume II: Economic minerals and applied geophysics. Springer, Heidelberg, N.Y

Babu RHV, Rao AD (1988) Inversion of self-potential anomalies in mineral exploration. Comput Geosci 14(3):377–387

Banerjee B, Pal BP (1990) Frequency domain interpretation of Self-Potential anomalies. Gerlands Beitr Geophys 99(6):531–538

Bhattacharya BB, Shalivakhan JA, Bera A (2007) Three-dimensional probability tomography of self-potential anomalies of graphite and sulphide mineralization in Orissa and Rajasthan, India. First Break 5:223–230

Biswas A (2018) Inversion of amplitude from the 2-D analytic signal of self-potential anomalies. In: Essa KS (Ed), Minerals, IntechOpen, pp 13–45

Biswas A, Sharma SP (2016) Integrated geophysical studies to elicit the structure associated with uranium mineralization around South Purulia shear zone, India: a review. Ore Geol Rev 72:1307–1326

Bukhnikashvili AV, Kebuladze VV, Tabagua GG, Dzhashi GG, Gugunava GE, Tatishvili OV, Gogua RA (1974) Geophysical Exploration of Adjar group of copper-polymetallic deposits. Metsniereba, Tbilisi (in Russian)

Castermant J, Mendonça C, Revil A, Trolard F, Bourrie G, Linde N (2008) Redox potential distribution inferred from self-potential measurements associated with the corrosion of a burden metallic body. Geophys Prospect 56(2):269–282

Chen Y, Qin X, Huang Q, Gan F, Han K, Zheng Z, Meng Y (2018) Anomalous spontaneous electrical potential characteristics of epikarst in the Longrui Depression, Southern Guangxi Province China. Environ Earth Sci 77(659):1–9

Corry CE (1985) Spontaneous polarization associated with porphyry sulfide mineralization. Geophysics 50(6):1020–1034

Corwin RF (1990) Application of the self-potential method for engineering and environmental investigations. SAGEEP Proceedings, Tulsa, OK, USA, pp 107–121

Corwin RF (1996) The self-potential method for environmental and engineering application. In: Ward SH (Ed), Geotechnical and environmental geophysics, vol 1, Society of Exploration Geophysicists, Tulsa, pp 127–145

Cowan DR, Allchurch PD, Omnes G (1975) An integrated geo-electrical survey on the Nangaroo copper-zinc prospect, near Leonora, Western Australia. Geoexploration 13:77–98

Cui Y, Zhu X, Wei W, Liu J, Tong T (2017) Dynamic imaging of metallic contamination plume based on self-potential data. Trans Nonferrous Met Soc China 27:1822–1830

De Giorgi L, Leucci G (2017) The archaeological site of Sagalassos (Turkey): exploring the mysteries of the invisible layers using geophysical methods. Explor Geophys 49:751–761

De Giorgi L, Leucci G (2019) Passive and active electric methods: new frontiers of application. In: Persico R, Piro S, Linford N (eds) Innovation in near-surface geophysics, pp 1–21

Di Maio R, Fedi M, Lamanna M, Grimaldi M, Pappaladro U (2010) The contribution of geophysical prospecting in the reconstruction of the buried ancient environments of the house of Marcus Fabius Rufus (Pompeii, Italy). Archaeol Prospection 17:259–269

Di Maio R, Piegari E, Rani P, Avella A (2016) Self-Potential data inversion through the integration of spectral analysis and tomographic approaches. Geophys J Int 206:1204–1220

Dmitriev AN (2012) Direct and inverse SP modeling on the basis of exact model of self-potential field nature. Geology Geophys 53(6):797–812

Drahor MG (2004) Application of the self-potential method to archaeological prospection: some case histories. Archaeol Prospect 11:77–105

Drahor MG, Akyol AL, Dilaver N (2006) An application of the self-potential (SP) method in archaeogeophysical prospection. Archaeol Prospect 3(3):141–158

El-Araby HM (2004) A new method for complete quantitative interpretation of self-potential anomalies. J Appl Geophys 55:211–224

Ekine AS, Emujakporue GO (2010) Investigation of corrosion of buried oil pipeline by the electrical geophysical methods. J Appl Sci Environ Manage 14(1):63–65

Eppelbaum LV (1987) Multimodel approach to the study of geophysical targets. eposited by VINITI, USSR Academy of Sciences, №. 7842–87, 1-10 (in Russian)

Eppelbaum LV (2007) Revealing of subterranean karst using modern analysis of potential and quasi-potential fields. Proceedings of the 2007 SAGEEP conference, 20, Denver, USA, pp 797–810

Eppelbaum LV (2015) Quantitative interpretation of magnetic anomalies from bodies approximated by thick bed models in complex environments. Environ Earth Sci 74:5971–5988

Eppelbaum LV (2019a) Geophysical potential fields: geological and environmental applications. Elsevier. Amsterdam, NY

Eppelbaum LV (2019b) Advanced system of self-potential field analysis in ore deposits of the South Caucasus. Proceedings of the national Azerbaijan Academy of Sciences, No. 2, pp 21–35

Eppelbaum LV (2020) Quantitative analysis of self-potential anomalies in archaeological sites of Israel: an overview. Environ Earth Sci 79:1–15

Eppelbaum LV (2021) Review of processing and interpretation of self-potential anomalies: Transfer of methodologies developed in magnetic prospecting. Geosciences 11(6):1–22

Eppelbaum L, Ben-Avraham Z, Itkis S (2003a) Ancient Roman remains in Israel provide a challenge for physical-archaeological modeling techniques. First Break 21(2):51–61

Eppelbaum LV, Ben-Avraham Z, Itkis SE (2003b) Integrated geophysical investigations at the Halutza archaeological site. Proceed of the 64 EAGE Conf. Florence, Italy pp 151:1–4

Eppelbaum L, Ben-Avraham Z, Itkis S, Kouznetsov S (2001a) First results of self-potential method application at archaeological sites in Israel. Transactions of the XI EUG International Symposium, Strasbourg, France, p. 657

Eppelbaum LV, Khesin BE, Itkis SE (2001b) Prompt magnetic investigations of archaeological remains in areas of infrastructure development: Israeli experience. Archaeol Prospect 8(3):163–185

Eppelbaum LV, Itkis SE, Fleckenstein K-H, Fleckenstein L (2007) Latest results of geophysical-archaeological investigations at the Christian archaeological site Emmaus-Nicopolis (central Israel). Proceed of the 69th EAGE Conference, P118, London, Great Britain, pp 1–5

Eppelbaum LV, Itkis SE, Khesin BE (2000) Optimization of magnetic investigations in the archaeological sites in Israel. In: Special Issue of prosperzioni archeologiche "filtering, modeling and interpretation of geophysical fields at archaeological objects", pp 65–92

Eppelbaum LV, Khesin BE (2002) Some common aspects of magnetic, induced polarization and self-potential anomalies interpretation: implication for ore target localization. Collection of Selected Papers of the IV Internernational symposium on Problems of Eastern Mediterranean Geology, Isparta, Turkey, pp 279–293

Eppelbaum LV, Khesin BE (2012) Geophysical studies in the caucasus. Springer, Heidelberg, NY

Eppelbaum LV, Khesin BE, Itkis SE, Ben-Avraham Z (2004) Advanced analysis of self-potential data in ore deposits and archaeological sites. Proceed. of the 10th European meeting of environmental and engineering geophysics, Utrecht, The Netherlands, pp 1–4

Eppelbaum LV, Mishne AR (2011) Unmanned Airborne Magnetic and VLF investigations: Effective Geophysical Methodology of the Near Future. Positioning 2(3):112–133

Eppelbaum LV, Vaksman VL, Kouznetsov SV, Sazonova LM, Smirnov SA, Surkov AV, Bezlepkin B, Katz Y, Korotaeva NN, Belovitskaya G (2006) Discovering of microdiamonds and minerals-satellites in Canyon Makhtesh Ramon (Negev desert, Israel). Doklady Earth Sciences (Springer) 407(2):202–204

Ernstson K, Scherer V (1986) Self-potential variations with time and their relation to hydrogeologic and meteorological parameters. Geophysics 51(10):1967–1977

Erofeev L Ya, Orekhov AN, Erofeeva GV (2017) Natural electric fields in Siberian gold deposits: structure, origin, and relationship with gold orebodies. Geology Geophys 58:984–989

Eskola L, Hongisto H (1987) A macroscopic physical model for the self-potential of a sulphide deposit. Geoexploration 24:219–226

Essa K, Mehanee S, Smith PD (2008) A new inversion algorithm for estimating the best fitting parameters of some geometrically simple body to measured self-potential anomalies. Explor Geophys 39:155–163

Fitterman DV (1979) Calculation of self-potential anomalies near vertical contacts. Geophysics 44(2):195–205

Fomenko NE (2010) Ecological geophysics. South State University, Rostov-on-Don, Russia (in Russian)

Fox RW (1830) On the electromagnetic properties of metallicferous veins in the mines of Cornwall. Royal Society, London, Philosophical Transactions, pp 399–414

Fedi M, Abbas M (2013) A fast interpretation of self-potential data using the depth from extreme points method. Geophysics 78:E107–E116

Giannakis I, Tsourlos P, Papazachos C, Vargemezis G, Giannopoulos A, Papadopoulos N, Tosti F, Alani A (2019) A hybrid optimization scheme for self-potential measurements due to multiple sheet-like bodies in arbitrary 2D resistivity distributions. Geophys Prospect 67:1948–1964

Gibert D, Sailhac P (2008) Comment on "self-potential signals associated with preferential ground-water flow pathways in sinkholes" by Jardani A, Dupont JP, Revil A, J Geophys Res 113, B03210. J Geoph Res 113, B03210, 1–4

Gibert D, Pessel M (2001) Identification of sources of potential fields with the continuous wavelet transform: application to SP profiles. Geophys Res Lett 28:1863–1866

Gobashy M, Abdelazeem M, Abdrabou M, Khalil MH (2019) Estimating model parameters from self-potential anomaly of 2D inclined sheet using whale optimization algorithm: applications to mineral exploration and tracing shear zones. Natural Resour Res https://doi.org/10.1007/s11053-019-09526-o, 1–21

Göktürkler G, Balkaya Ç (2012) Inversion of self-potential anomalies caused by simple-geometry bodies using global optimization algorithms. J Geophys Eng 10:498–507

Goldie M (2002) Self-potentials associated with the Yanacocha high-sulfidation gold deposit in Peru. Geophysics 67:684–689

Gurk M, Bosch F (2001) Cave detection using the Self-Potential-Surface (SPS) technique on a karstic terrain in the Jura mountains (Switzerland), In: Hördt A, Stoll JB (eds) Trans. of the 'Kolloquium Elektromagnetische Tiefenforschung', Burg Ludwigstein, pp 283–291

Gusev AP, Kaleichik PA, Fedorsky MS, Shavrin IA (2018) Dynamics of self-potential field as indicator of dangerous rock-slide areas in industrial landscape (on example of Belarus). Bull Perm Univ (Russia), 17(2):120–127 (in Russian)

Hartal M (1997) Banias, The Aqeduct. Excavations and Surveys in Israel 16:5–8

Hristenko LA, Stepanov YuI (2012) Transformation of SP observations for solving engineering-geological problems. Mining Inform Analyt Bull 5:169–173 (in Russian)

Jardani A, Dupont J, Revil A (2006a) Self-potential signals associated with preferential groundwater flow pathways in sinkholes. J Geophys Res 111(B09204):1–13

Jardani A, Revil A, Dupont J (2006b) Self-potential tomography applied to the determination of cavities. Geophys Res Lett 23(L13401):1–4

Jardani A, Revil A (2009) Stochastic joint inversion of temperature and self-potential data. Geophys J Int 179:640–654

Jardani A, Revil A, Bolève A, Dupont JP (2008) Three-dimensional inversion of self-potential data used to constrain the pattern of groundwater flow in geothermal fields. J Geophys Res 113:B09204

Jardani A, Revil A, Santos F, Fauchard C, Dupont J (2007) Detection of preferential infiltration pathways in sinkholes using joint inversion of self-potential and EM-34 conductivity data. Geophys Prospect 55(5):749–760

Kempinski A, Reich R (eds) (1992) The architecture of ancient Israel. Israel Exploration Society, Jerusalem, Israel

Kenyon KM (1979) Archaeology in the Holy Land. Norton, USA

Khesin BE, Alexeyev VV, Eppelbaum LV (1996) Interpretation of geophysical fields in complicated environments. Kluwer Academic Publishers (Springer), Ser.: Modern Approaches in Geophysics, Boston, Dordrecht, London

Kilty KT (1984) On the origin and interpretation of self-potential anomalies. Geophys Prospect 32(1):51–62

Lapenna V, Lorenzo P, Perrone A, Piscitelli S, Sdao F, Rizzo E (2003) High-resolution geoelectrical tomographies in the study of Giarrossa landslide (southern Italy). Bull Eng Geol Environ 62:259–268

Lile OB (1996) Self potential anomaly over a sulphide conductor tested for use as a current source. J Appl Geophys 36(2–3):97–104

Logn O, Bolviken B (1974) Self potentials at the Joma pyrite deposit, Norway. Geoexploration 12:11–28

Mendonça CA (2008) Forward and inverse self-potential modeling in mineral exploration. Geophysics 73(1):F33–F43

Mauriello P, Monna D, Patella D (1998) 3D geoelectric tomography and archaeological applications. Geophys Prospect 46:543–570

Meiser P (1962) A method of quantitative interpretation of self-potential measurements. Geophys Prospect 10:203–218

Meyers EM (Ed) (1996) The Oxford encyclopedia of archaeology in the Near East. 5 Vols., Oxford University Press, Oxford

Murty BV, Haricharan P (1984) A simple approach toward interpretation SP anomaly due to 2-D sheet model of short dipole length. Geophys Res Bull 22(4):213–218

Nayak PN (1981) Electromechanical potential in surveys for sulphide. Geoexploration 18:311–320
Ogilvy AA, Bogoslovsky VA (1979) The possibilities of geophysical methods applied for investigating the impact of man on the geological medium. Geophys Prospect 27:775–789
Oliveti I, Cardarelli E (2019) Self-Potential data inversion for environmental and hydrogeological investigations. Pure Appl Geophys 176(8):3607–3628
Onojasun OE, Takum E (2015) Geophysical investigation using self-potential techniques: a case of locating underground water pipeline at Kwinana industrial area. Perth African J Geo-Sci Res 3(4):19–23
Parasnis DS (1986) Principles of applied geophysics, 4th ed., revised and supplemented. Chapman & Hall, London
Patella D (1997) Introduction to ground surface self-potential tomography. Geophys Prospect 45:653–681
Perrier F, Morat P (2000) Characterization of electrical daily variations induced by capillary flow in the non-saturated zone. Pure Appl Geophys 157:785–810
Perrier F, Pant SR (2005) Noise reduction in long-term self-potential monitoring with travelling electrode referencing. Pure Appl Geophys 162:165–179
Petrovsky A (1928) The problem of a hidden polarized sphere. Philos Magaz 5. Series 7:914–933
Quarto R, Schiavone D (1996) Detection of cavities by the self-potential method. First Break 14(11):419–430
Rao DA, Babu HV (1983) Quantitative interpretation of self-potential anomalies due to two-dimensional inclined sheets. Geophysics 48(2):1659–1664
Rao DA, Babu HV, Sivakumar GDJ (1982) A Fourier transform method for the interpretation of SP anomalies due to two-dimensional inclined sheets of finite depth extent. Pure Appl Geophys 120:365–374
Rao K, Jain S, Biswas A (2020) Global optimization for delineation of self-potential anomaly of a 2D inclined plate. Natural Resour Res, 1–15, https://doi.org/10.1007/s11053-020-09713-4
Rao SVS, Mohan NL (1984) Spectral interpretation of self-potential anomaly due to an inclined sheet. Curr Sci 53(9):474–477
Reich R (1992) Architecture of ancient Israel. Israel Exploration Society, Jerusalem
Revil A, Jardani A (2013) The self-potential method: Theory and applications in environmental geosciences. Cambridge University Press, Cambridge, UK
Rittgers JB, Revil A, Karaoulis M, Mooney MA, Slater LD, Atekwana EA (2013) Self-potential signals generated by the corrosion of buried metallic objects with application to contaminant plumes. Geophysics 78(5):EN65–EN82
Rozycki A, Fonticiella JMR, Cuadra A (2006) Detection and evaluation of horizontal fractures in Earth dams using self-potential method. Eng Geol 82:145–153
Safipour R, Hölz S, Halbach J, Jegen M, Petersen S, Swidinsky A (2017) A self-potential 382 investigation of submarine massive sulfides, Palinuro Seamount, Tyrrhenian Sea. Geophysics 383, 82(6):A51–A56
Sailhac P, Marquis G (2001) Analytic potentials for the forward and inverse modeling of SP anomalies caused by subsurface fluid flow. Geophys Res Lett 28(9):1851–1854
Sato M, Mooney HM (1960) The electrochemical mechanism of sulfide self-potentials. Geophysics 25:226–249
Semenov AS (1980) Electric prospecting by self-potential method. 4st ed., revised and supplemented. Nedra, Leningrad (in Russian)
Semenov MV (1975) Principles of prospecting and investigating pyrite-polymetallic Ore Fields by Geophysical Methods. Nedra, Leningrad (in Russian)
Sengupta SN, Bose RN, Mitra SK (1969) Geophysical investigations for copper ores in the Singhana-Gotpo area, Khetri copper belt, Rajasthan (India). Geoexploration 7:73–82
Shevnin VA (2018) Identification of self-potential anomalies of a diffusion-absorption origin. Moscow Univ Geology Bulletin 73(3):306–311

Shevnin VA, Bobachev AA, Ivanova SV, Baranchuk KI (2014) Joint analysis of self potential and electrical resistivity tomography data for studying Alexandrovsky settlement. Trans. of the 20th Meeting of environmental and engineering geophysics. Athens, Greece, Mo PA2 04, 1–5

Sindirgi P, Pamukç´u O, Özyalin S (2008) Application of normalized full gradient method to self potential (SP) data. Pure Appl Geophys 165:409–427

Sindrigi P, Özyalin S (2019) Estimating the location of a causative body from a self-potential anomaly using 2D and 3D normalized full gradient and Euler deconvolution. Turkish J Earth Sci 28:640–659

Skianis GA, Papadopoulos TD, Vaiopoulos DA (1991) 1-D and 2-D spatial frequency analysis of SP field anomalies produced by a polarized sphere. Pure Appl Geophys 137:251–260

Skianis G, Papadopoulos T, Vaiopoulos DA, Nikolaou S (1995) A new method of quantitative interpretation of SP anomalies produced by a polarized inclined sheet. Geophys Prospect 43:677–691

Srigutomo W, Arkanuddin MR, Pratomo PM, Novana EC, Agustina RD (2010) Application of qualitative and quantitative analyses of self potential anomaly in caves detection in Djuanda Forest Park. Bandung Amer Inst Phys Proc 1325(164):164–167

Srivastava S, Agarwal BNP (2009) Interpretation of self-potential anomalies by enhanced local wave number technique. J Appl Geophys 68:259–268

Stern W (1945) Relation between spontaneous polarization curves and depth, size, and dip of ore bodies. Trans Am Inst Min Metallurg Petr Eng 164:189–196

Sungkono S (2020) Robust interpretation of single and multiple self-potential anomalies via flower pollination algorithm. Arab J Geosci 13(100):1–15

Tarasov GA (1961) Electric field over a set of vertically polarized conductive spheres. Problems Mining Geophys 2:61–67 (in Russian)

Telford WM, Geldart LP, Sheriff RE (1990) Applied geophysics, 2nd edn. Cambridge University Press, Cambridge

Tlas M, Asfahani J (2013) An approach for interpretation of self-potential anomalies due to simple geometrical structures using Fair function minimization. Pure Appl Geophys 170:895–905

Tripathi GN, Frayar AE (2016) Integrated surface geophysical approach to locate a karst conduit: a case study from Royal Spring Basin, Kentucky, USA. J Nepal Geolog Soc 51:27–37

Tsokas GN, Tourlos PI, Kim J-H, Papazachos CB, Vargemezis G, Bogiatzis P (2014) Assessing the condition of the rock mass over the tunnel of Eupalinus in Samos (Greece) using both conventional geophysical methods and surface to tunnel electrical resistivity tomography. Archaeol Prospect 21:277–291

Vichabian Y, Morgan FD (2002) Self-potentials in cave detection. Lead Edge 21:866–871

Wang J-H, Geng Yu (2015) Terrain correction in the gradient calculation of spontaneous potential data. Chinese J Geophys 58(6):654–664

Wynn JC, Sherwood SI (1984) The self-potential (SP) method: an inexpensive reconnaissance and archaeological mapping tool. J Field Archaeol 11:195–204

Yüngül S (1954) Spontaneous-potential survey of a copper deposit at Sariyer. Turkey Geophys 19(3):455–458

Zaborovsky AI (1963) Electric prospecting. Gostoptekhizdat, Moscow (in Russian)

Zhdanov MS, Keller GV (1994) The geoelecrical methods in geophysical exploration. Elsevier, Amsterdam

Zhu Z, Shen J, Tao C, Deng X, Wu T, Nie Z, Wang W, Su Z (2021) Autonomous underwater vehicle based marine multi-component self-potential method: observation scheme and navigational correction. Geoscientific Instrum Methods Data Syst 10:35–43

Chapter 9
Preferential Water Flow Pathways Detection in Sinkholes Using Self-Potential (SP) Method. The Study Case of Anina Karst Region (Banat Mountains, Romania)

Laurențiu Artugyan and Petru Urdea

Abstract Limestone covers about 4500 km^2 in Romania, meaning about 2% from the country surface, being specific for the karst of the temperate zones due to its landforms, terrain diversity and the amplitude of the exo and endokarst particularities. Carbonate rocks are spread in the mountains and sub-mountains areas, with the most compact and the largest surface of this type of rocks in the Banat Mountains, more exactly in the Reşiţa-Moldova Nouă Synclinorium. The aim of this study to examine the use of the self-potential as geophysical method in detection of water flow and water resources in shallow karst depressions (sinkholes) in the Anina Karst Region, Banat Mountains, Romania. The self-potential (SP) method—also known as spontaneous potential—is founded on the measurements regarding the natural or spontaneous potentials that are forming in the ground. Our SP approach in several sinkholes in the Anina Karst Region shows that most of the anomalous zones with positive spontaneous electric potential are localized in the middle of the sinkholes, indicating the higher retention of water within the bottom part of the karst depressions. Also, the most of the values showing negative anomalies are situated in those parts where the karrens are developed. SP method applied in the Anina Karst Region, Banat Mountains, Romania, has shown the feasibility of this geophysical method in the study of water circulation in shallow karst environment. Since sinkholes are landforms that favor the fast circulation of the water from the surface into the underground, the SP approach could be a feasible method to study aquifers and the presence of contaminants in karst aquifers since water is a very important resource in karst areas. Moreover, SP method is reliable in characterization of shallow karst topography research and this geophysical method can be combined with other geophysical method in karst terrain analysis, methods like GPR and ERT.

Keywords Self-potential · Sinkholes · Aquifers · Contaminants

L. Artugyan (✉) · P. Urdea
Department of Geography, West University of Timisoara, V Pârvan no 4, 300223 Timisoara, Timis, Romania

P. Urdea
e-mail: petru.urdea@e-uvt.ro

© The Author(s), under exclusive license to Springer Nature Switzerland AG 2021
A. Biswas (ed.), *Self-Potential Method: Theoretical Modeling and Applications in Geosciences*, Springer Geophysics, https://doi.org/10.1007/978-3-030-79333-3_9

9.1 Introduction

Karst terrains represent a distinct geological environment and geomorphological based on its characteristics, both at surface and in the underground. Karst ecosystems are about to be studied and discovered even nowadays (De Waele et al. 2009). Karst environment is highly vulnerable as a consequence of the fast connection between the surface and the underground karst landforms. Pollution and deterioration events occurring at the surface in karst terrains have negative consequences in the underground (Parise and Pascali 2003). In order to solve these risks and to protect karst environment, geomorphological studies on karst topography should focus both on surface (exokarst) and underground landforms (endokarst), but also on the link between surface and subterranean karst features.

The consequence of the rock massifs dissolution that creates karst topography is represented by an effective underground flow (Waltham et al. 2005). Understanding karst topography implies the understanding of those factors those generate dissolution processes in karst terrains and the understanding of the nature of drainage generated by these dissolution processes (Ford and Williams 2011). The particularities of karst regions include the lack of surface water as a consequence of porous and fissured rocks, but also the water flowing into the underground. In karst regions the karst aquifers represent a highly important source of drinking water. However due to the fast circulation of surface water, the groundwater in karst aquifers is highly vulnerable to pollution and contamination (Bakalowicz 2005; Andreo et al. 2008; Shirazi et al. 2012; Jeannin et al. 2012). Soil permeability is another factor that contributes to the high vulnerability to pollution of karst groundwater. In karst terrains soil protective covers are usually very thin or missing completely, favouring the fast circulation of pollutants into the karst aquifers (Kaçaroğlu 1999).

The level of dissolution of the geological substrate in karst terrains is indicated by the sizes and the density of the sinkholes (Shofner et al. 2001). Since the karst system depends on the presence of fractures, fractures' orientation in karst topography provide valuable details regarding the drainage network (Chalikakis et al. 2011). In the recent years geomorphology has become a domain focused to study the human impact on landforms evolution and on the environment (Church 2010).

Limestone covers about 4500 km^2 in Romania, meaning about 2% from the country surface (Orghidan et al. 1972), being specific for the karst of the temperate zones due to its landforms, terrain diversity and the amplitude of the exo- and endokarst particularities (Cocean 2001). Among the surfaces covered by limestone Mesozoic deposits occupy the most extensive surfaces. Moreover, Mesozoic limestone has the largest thickness deposits and the consequence is represented by the highest degree of karstification processes among these limestone deposits (Trufaş and Sencu 1967). Carbonate rocks are spread in the mountains and sub-mountains areas, with the most compact and the largest surface of this type of rocks in the Banat Mountains, more exactly in the Reşiţa-Moldova Nouă Synclinorium (Orăşeanu and Iurkiewicz 2010).

Among the geophysical methods self-potential has been used in various studies in karst terrains (Revil et al. 2005; Rozycki et al. 2006; Jardani et al. 2006a, b; Wishart et al. 2006, 2008; Suski et al. 2008; Robert et al. 2011). Jardani et al. (2006a) have used previously self-potential method in sinkholes in order to delineate preferential groundwater flow pathways.

According to Chen et al. (2018), the spontaneous potential method measuring the natural electric field generated in the rock fissures networks in karst topography has been used with success in various approaches such as the detection of karst development and characteristics at macroscopic level, infiltration processes, recharge of stratified aquifers or the infiltration and migration of the groundwater. On the other hand, the SP method has been used very rare to analyze the karst water movement and the characteristics of the fracture networks at small scale in mountains karst areas or in the epikarst zone (Chen et al. 2018).

The aim of this study to examine the use of the self-potential (also known as spontaneous potential) as geophysical method in detection of water flow and water resources in shallow karst depressions (sinkholes) in the Anina Karst Region, Banat Mountains, Romania.

In Romania the self-potential method has been used only sporadically as geophysical method investigation in different environments and terrains. The self-potential method has been used in Semenic Mountains (Urdea and Țambriș 2014) and Anina Mountains (Artugyan and Urdea 2014a, b; Artugyan et al. 2015).

9.2   Study Area and Investigation Sites Description

The Anina karst region is part of the Anina Mountains, which is a subunit of Banat Mountains in the SW of Romania (Fig. 9.1). The study area is part of a folded region with a Jurrasian relief characterized by the alternation of parallel anticlines and synclines, oriented NNE-SSW (Mateescu 1961). This orientation has imposed the main drainage direction of the entire area, acting as a boundary of the representative karst systems (Iurkiewicz et al. 1996).

From geological perspective the study area is part of the most compact and the largest surface covered by carbonate rocks in Romania, namely the typical structural area Reșița-Moldova Nouă Synclinorium (Orășeanu and Iurkiewicz 2010). In this typical structural area the Paleozoic–Mesozoic geological formations are placed over the fundamental crystalline domain (Bucur 1997) since the Paleozoic and Mesozoic sedimentary deposits have been deposited before or during the Meso-Cretaceous phase (Oncescu 1965). Reșița-Moldova Nouă Synclinorium has functioned as a marine depression area where the sedimentary cover had a complete succession and erosion stage. Despite the fact that the sediments that covered a large part of the area were removed by the erosion, this structural zone is considered as the standard area of sedimentary domains (Mutihac and Ionesi 1974).

The core geomorphological characteristic of the study are is represented by the long parallel ridges that are separated by karst plateaus and deep valleys (Bucur

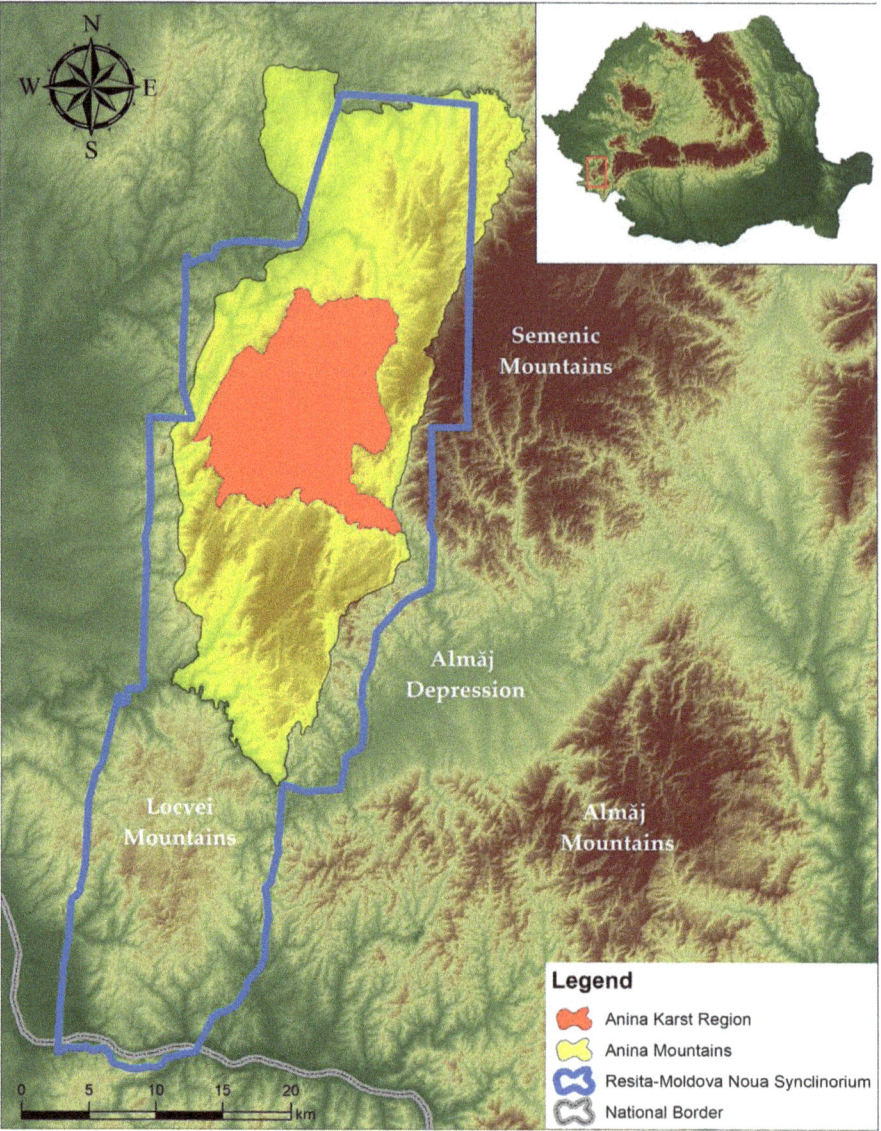

Fig. 9.1 Location of the study area within the Anina Mountains and Reşiţa-Moldova Nouă Synclinorium

1997). Due to the presence of karst plateaus that are bordered by deep valleys the Anina karst region is considered to be part of the suspended karst plateaus. These suspended karst plateaus have as particularities a high degree of karstification (Onac 2000). Due to the high level of karstification the entire region has a high diversity and very numerous landforms specific for karst terrains, both at surface and in the underground. Among exokarstic landforms that can be seen in the Anina karst region we could mention karrens, karren fields, sinkholes, sinkholes valleys, uvalas, poljes, blind valleys and dry valleys. Karst gorges sectors and karst springs are also very numerous within this karst region in the Banat Mountains. As for the underground karst landforms, caves and vertical shafts have a high density in the Anina karst region. According to Goran (1982) the caves within this karst region are part of 6 karst basins.

Within the study area the largest karst plateaus are Brădet Plateau, Mărghitaș Plateau, Colonovăț Plateau, Cârneală Plateau and Ravniștea Plateau (Fig. 9.2). Four of these karst plateaus were chosen as study sites of this research, namely Mărghitaș Plateau, Colonovăț Plateau, Cârneală Plateau and Brădet Plateau, in the Culmea Neagră (*Black Ridge*) Area. These suspendend karst plateaus (Bleahu 1982) are mostly covered with collapsed sinkholes (Sencu 1977).

All these karst plateaus have mostly a flat aspect disrupted by sinkholes and sinkhole vallyes. Most of these sinkholes are surrounded by large karrens or karren fields. Mărghitaș Plateau is covered mostly with pasture and the only clumps of trees are situated around and inside the large sinkholes present here. On the other hand, Colonovăț Plateau and Cârneală Plateau are forested karst plateaus. Brădet Plateau is the largest karst plateau in the Anina Karst Region and it is covered with both pasture and dense forests. The Culmea Neagră (*Black Ridge*) Area (where we have developed our self-potential measurements) is covered only with forest, as its names says.

There is no surface water on the Mărghitaș Plateau and Colonovăț Plateau, while Cârneală Plateau has several small creeks. Buhui Cave, Cuptoare Cave and Mărghitaș Cave are situated in Buhui Valley, this river bordering Colonovăț and Mărghitaș plateaus. Cârneală Plateau has also some blind valleys, while the most important cave in this karst plateau is Cârneală Cave. As for the Culmea Neagră Area, many of these sinkholes host the entrances of several vertical karst shafts present in this area. Among these vertical karst shafts Culmea Neagră (*Black Ridge*) Vertical Shaft has been explored after 2000 (Burdan et al. 2007) and it has shown a high potential of further exploration, pointing out the potential for further underground voids discovery. Due to this we choose to develop our measurements in three sinkholes situated in the proximity of the Culmea Neagră Vertical Shaft in order analyze the surface water drainage in order to identify possible rapid underground flow and possible other underground voids.

9.3 Self-Potential as Geophysical Method

Self-potential (SP) method has been utilized for the first time by Robert Fox in 1930 aiming to detect copper into the underground in Cornwall, England. SP is considered as the oldest geophysical method and at the same time the simplest method. However, due to the challenges and difficulties in the interpretation based on the lacuna of the information regarding the mechanisms those generated SP signals, the SP method

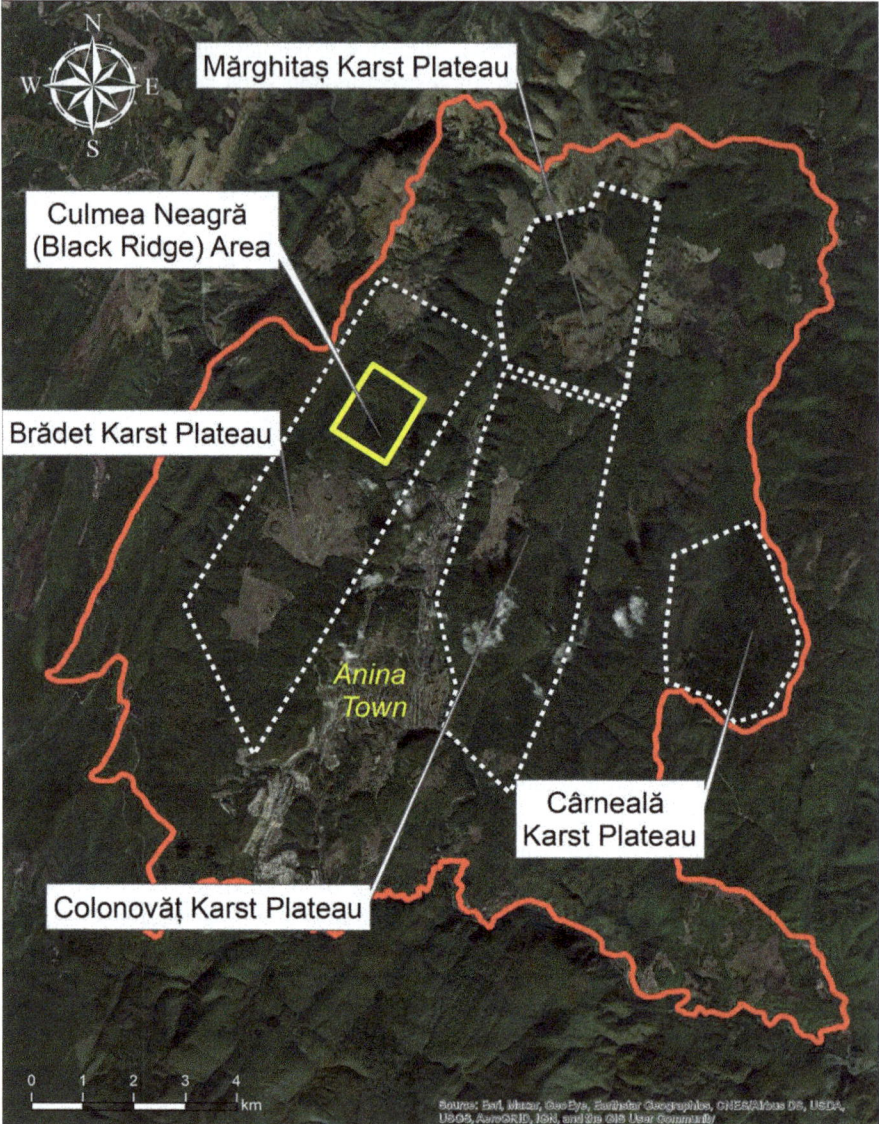

Fig. 9.2 Location of the karst plateaus representing the study areas for this research

has been a very long time an undeveloped geophysical method. Nowadays SP signals and the mechanism generating these signals are better understood (Boleve 2009). The SP method is an electrical geophysical method and it is known as a passive method, measuring the natural electrical fields occurring at the surface of the ground (Revil and Jardani 2013). Revil et al. (2013) describe the SP method as a simple and flexible method also in geophysical measurements.

The self-potential (SP) method—also known as spontaneous potential—is founded on the measurements regarding the natural or spontaneous potentials that are forming in the ground. These spontaneous potentials are a consequence of the electrochemical interactions between different minerals and subsurface fluids, or of the electro kinetic processes generated by the flow of the ionic fluids (Sharma 2002). The electro kinetic fields are generated during the infiltration of the water coming from precipitation through the porous material in the ground. In contact with precipitation water all the minerals in the soil become electrically charged. This electric field is called as the streaming potential or spontaneous potential (Sheffer 2007; Jardani et al. 2007a). In the porous medium the electrical conduction and the ionic diffusion could have a significant contribution to electrical conductivity. Therefore, it can have an important role in the resistivity of certain spontaneous electric potentials (Revil et al. 1997).

The streaming potential results in those situations when an electrolyte moves towards a stationary solid feature and this phenomenon is highly important in groundwater studies (Bérubé 2007).

According to Reynolds (1997), the processes that cause the self-potential in the earth subsurface are still not completely understood. Even if the SP method has seen important progresses, specialists consider that is a high difficulty in SP data interpretation. Also, the electrical noise that may result in SP data acquisition is seen as a challenge. On the other hand, the SP method is fast, nonintrusive and a cheap geophysical method. This fact is appreciated by specialists (Nyquist 2002).

It is important to highlight that during the measurements of the natural or spontaneous electric field method in karst areas there are several factors that may interfere. Regarding the sinkholes there are both natural and man-made features that may interfere. The natural factors include the sink points, soil holes, fault rupture zones generating the natural electric field, the gap networks, the carbonaceous limestone, while the man-made factors include different electromagnetic fields (Chen et al. 2018).

The SP method is the only one geophysical method that is directly sensitive to the flow of the underground water. The underground water flow certainly generates an electric field that can be detected on the surface (Boleve 2009).

The seepage flow of the groundwater in karst depression follows the filtration through the soil layer in the conditions of a certain thickness of the soil in the depression. The groundwater flow occurs along the fractures, by the underground river channels or by the pore infiltration in the numerous karst pipes. The consequence of these processes is represented by the natural polarization of the soil-rock layers creating conductive ions. In this process the rock contains the anions and the flowing solution transports the cations. Consequently, the cations and anions charged electrically will be diffused freely. The simultaneous existence of both rock fractures and soil generate a natural electric field in sinkholes. This natural electric field is proving that the groundwater seepage occurs. Consequently, analyzing the natural electric fields in sinkholes by their spatial and temporal distribution can locate the resources represented by the shallow karst groundwater (Chen et al. 2018).

In the recent years the SP method has an increased interest as geophysical study method. This higher interest is due to the fact that it is a non-invasive method (Jouniaux et al. 2009). Numerous studies trying to interpret and to explain self-potential data and anomalies (Abdelrahman et al. 2008, 2009; Sharma and Biswas 2013; Roudsari and Beitollahi 2013; Biswas and Sharma 2014, 2015; Asfahani and Tlas 2016; Biswas 2016, 2017; Essa 2019; Ekinci et al. 2020; Jougnot et al. 2020). Various domains like geothermal, environmental and engineering have used the SP method in different goals like locating and delineating sources related with the groundwater and thermal fluids movement (Jardani et al. 2008; Moore et al. 2011; Susilo et al. 2017; Revil et al. 2017; Chen et al. 2018; Zhu et al. 2020). Jouniaux et al. (2009) have used SP method in subsurface groundwater flow studies and also in subsurface contamination studies, Carpenter et al. (2013) used SP method together with other geophysical methods to identify contaminants in an aquifer, while salt plume monitoring is done using SP method by Martínez-Pagán et al. (2010). Mineralogical and geological conditions studies have been developed using the SP method (Chukwu et al. 2008; Chukwu 2013; Roudsari and Beitollahi 2013).

The SP method in karst terrains has been used in various studies: Stevanovic and Dragisic (1998), Lange (1999), Gurk and Bosch (2001), Fagerlund and Heinson (2003), Rozycki et al. (2006), Guichet et al. (2006), Jardani et al. (2006a, b, c,2007b), Jardani et al. (2009), Jouniaux et al. (2009), Robert et al. (2011), Susilo et al. (2017), Chen et al. (2018). Studies using SP method in detection of natural electrical potential field in epikarst have been developed by Fagerlund and Heinson (2003) and Jardani et al. (2006a, c), while Chen et al. (2018) have used SP method in order to detect water in shallow karst depressions.

9.4 Methodology

In those sinkholes where the bedrock is exposed at the surface or it presents a thin overburden the surface water infiltrates directly in the upper part of the rock fissure network. This surface water flow contributes to the recharge of the underground rivers and the karst pipes. In these conditions the natural electric field has a very weak polarization, while the dominant role is played by the polarization that is generated by the recharge in the upper rock fissure network. Accordingly, the anomaly of the natural electric field is highly developed in those karst depressions with fracture networks extending downward above the base represented by the underground river or karst pipe (Chen et al. 2018).

In order to develop self-potential field measurements we have used two Petieau nonpolarizing electrodes and a Voltcraft VC 850 digital multimeter. From those two electrodes one of them is used as a fix electrode, while the other one is used as a mobile electrode.

Field measurements using the SP method where repeated for two or even three times in order to have results from various seasons and atmospheric conditions. As SP measurements approaches we have used both linear profiles with various orientations (N-S, E-W, SE-NW and NE-SW) and grids.

Due to the predominantly forest vegetation in the study area, we had to adapt our measurements according to vegetation growth and to analyze most of the sites using profiles. Moreover, since the natural electric field values are influenced by the local soil characteristics and climatic conditions at a certain given time (Urdea and Țambriș 2014) we organized field campaigns in different seasons and various atmospheric conditions in order to do SP measurements.

The procedure of field data acquisition requires introducing both electrodes into a hole into the ground at about 10 cm deep (Fig. 9.3). For a better contact between electodes and soil, we have used humid benthonitic clay. After 1 min or even faster, when the digital voltmeter value has been stabilized, the value indicated on the display is marked. The next step is to move the mobile electrode into the next hole in the soil (Fig. 9.4). The space between stations (between the fix electrode and the first measurement point and then between the previous and the next measurement point) had various values (3 m and 5 m) depending on the study site size.

Stevanovic and Dragisic (1998) highlight that negative anomalies in the self-potential occur in those points where the percolation water met the soil surface. On the other hand, positive values in spontaneous potential measurements are specific for the issuing points.

As a consequence of the groundwater movement within the aquifer a natural electrical potential is generated (Jardani et al. 2009). The SP method can bring relevant information related with the dynamics and the geometry of the groundwater flow in real time (Jardani et al. 2007b, 2008). According to Rozycki et al. (2006),

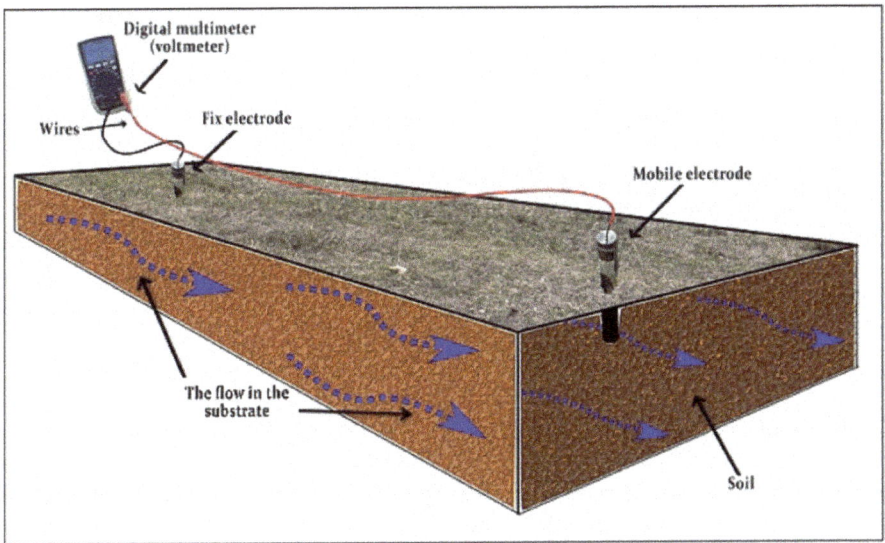

Fig. 9.3 Aquistion data using the self-potential (SP) method (after Urdea and Țambriș 2014)

Fig. 9.4 Self-potential data acquisition in the field using a digital multimeter and the fix and mobile electrodes

any anomaly in the SP obtained data may show a possible connection between the physical model and the water infiltration.

In order to model and to interpret the SP data obtained during the field campaigns the values where introduced in software able to generate profiles and grids. These profiles and grids are aimed to help us to analyze and to interpret the SP data. Accordingly, for the profiles we have used Microsoft Excel software where we have developed different graphs, while for the grids we have used ArcGIS 10 software. ArcGIS 10 software has been used to generate the gradient maps showing the SP results using Natural Neighbor interpolation method for the grid approach.

9.5 Results

The results of the SP investigations in the 4 karst plateaus in the Anina Karst Region totalized 54 measurements meaning 39 sinkholes. In this section are presented those sites with the most representative results obtained after the SP measurements for each one of the karst plateaus studied.

9.5.1 *Mărghitaş Plateau*

Site MP-S1

This sinkhole has a circular shape and it presents karrens in the NW and SE sides of the sinkhole, with thicker soil on the bottom. Even if the SP measurements were developed after a long period without precipitations, the bottom of the sinkhole presented higher moisture.

Based on the SP measurements we can observe very well the bottom of the sinkhole having the highest values of natural electric field. The edges of the sinkhole where the karrens are present indicate the highest anomalies of the SP data indicating a water flow at the surface of this shallow karst depression (Fig. 9.5).

According to the SP results the water stagnates more on the bottom of the sinkhole (Artugyan and Urdea 2014b), while the water flow is faster on the slopes of the sinkhole.

Site MP-S2

This sinkhole is located in the immediate vicinity of MP-S1 site and it has the same characteristics. For this site we have employed two SP measurements campaigns, in May 2013 and in October 2013. The measurements were done as profiles oriented N-S and W-E.

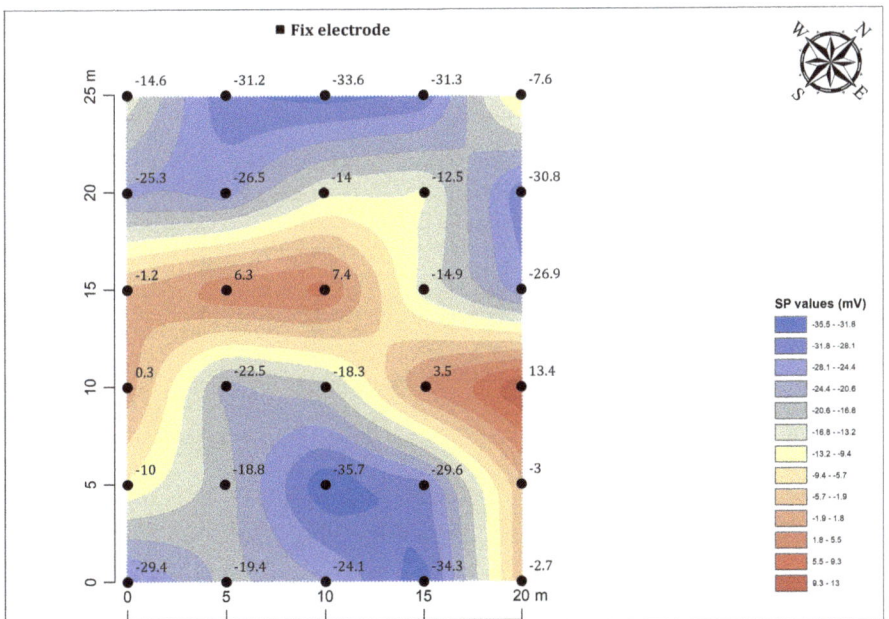

Fig. 9.5 Self-potential measurements in MP-S1 site

The obtained results indicate very similar profiles as shape, with the highest SP values in the bottom part of the sinkhole. The profiles obtained for both periods (Figs. 9.6 and 9.7) have a funnel-shaped profile. The results in this site highlight that the water circulation is faster on sinkhole edges where karrens are present and the higher retention of water happens in the middle of the sinkhole.

The SP values obtained in October 2013 are more homogeneous, providing profiles having a more linear aspect. This is due to the large periods without precipitations which preceded our field campaigns in SP measurements (Artugyan and Urdea 2014a).

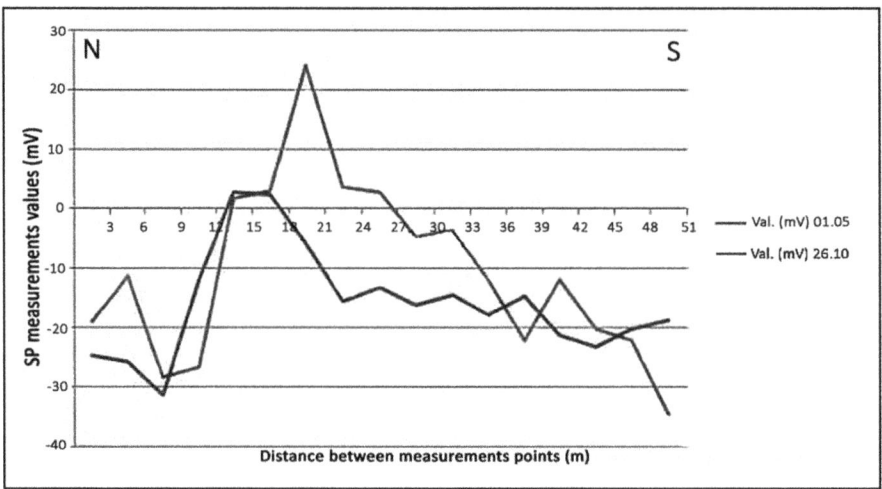

Fig. 9.6 Self-potential profiles in MP-S2 site on N-S orientation in May 2013 and October 2013

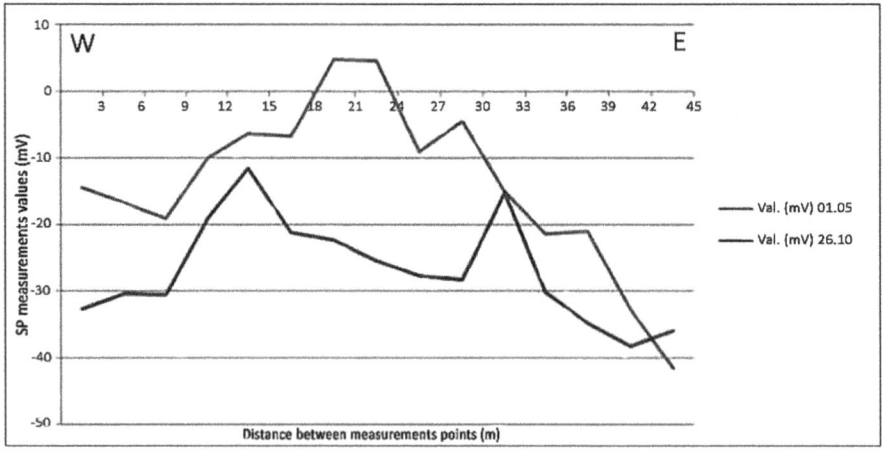

Fig. 9.7 Self-potential profiles in MP-S2 site on W-E orientation in May 2013 and October 2013

Site MP-S3

For this site we have employed 2 field campaigns, in May 2013 and November 2013. Data acquisition was done developing two grids. This sinkhole is covered with grass and karrens on the edges, being oriented NW–SE and it is more inclined in the NE. Consequently, in the southern part of the sinkhole to overburden material is thicker as a consequence of the sheet-wash transport after precipitation or after snow melting periods. This is highlighted by the SP measurements too.

The results obtained in the field campaign in May 2013 indicate mostly negative anomalies on the edges of the sinkhole, while the bottom of the sinkhole has positive anomalies (Artugyan et al. 2015).

These results (Fig. 9.8) indicate that the water circulation is faster in those areas with thinner soil cover. However, there are some negative anomalies even in the middle of the sinkhole, indicating a possible sink connected with the underground karst pipe or underground karst channel. This negative anomaly is even more obvious in the field campaign measurements developed in November 2013 when the values were almost exclusively positive, except for one negative anomaly. The highest SP values were obtained in the central part of the sinkhole, confirming the results obtained in May 2013 and showing a faster water circulation in those parts with thinner soil and with exposed rock. Moreover, in May 2013 the negative anomalies are disposed as certain alignments of negative values, being able to indicate certain fractures in the bedrock.

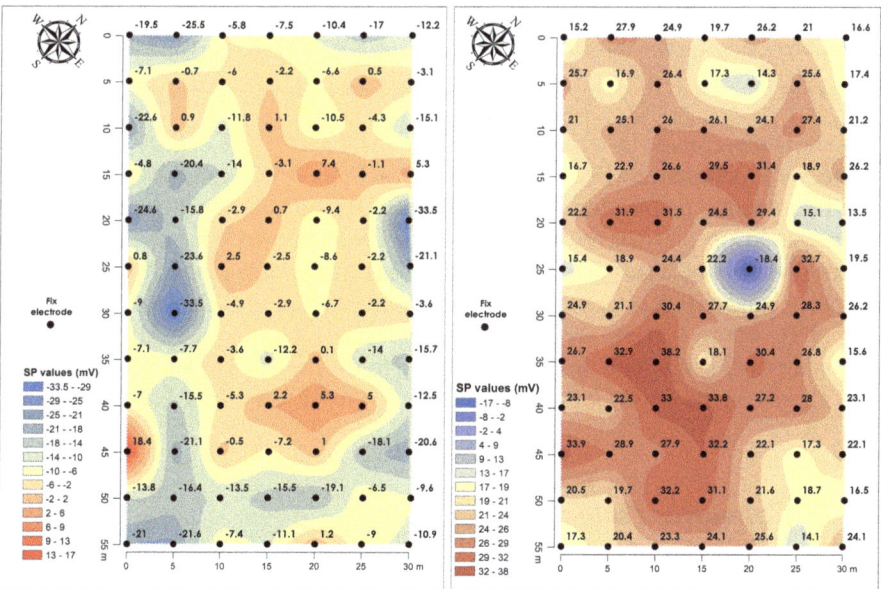

Fig. 9.8 Self-potential results in sinkhole representing MP-S3 site in May 2013 (**left**) and November 2013 (**right**)

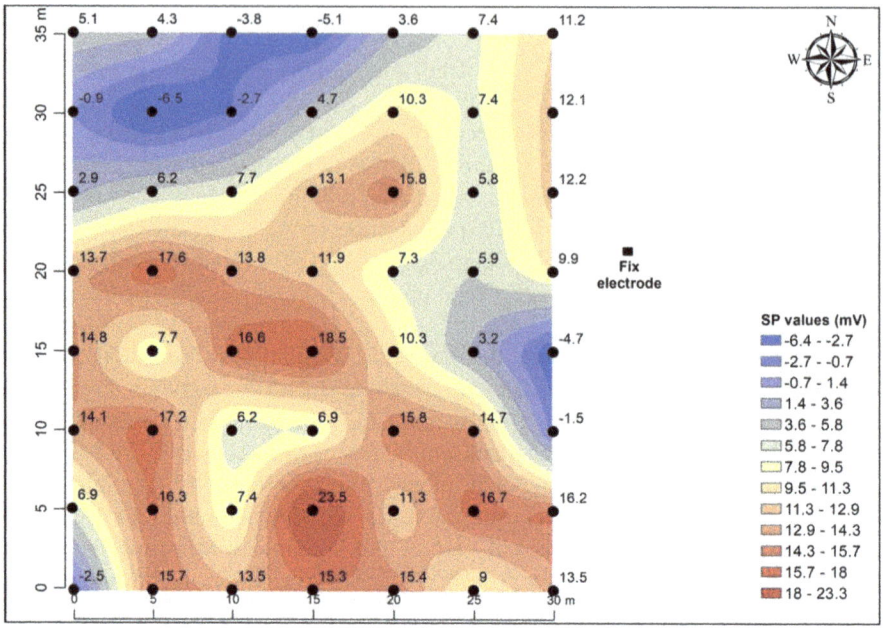

Fig. 9.9 Self-potential gradient in MP-S4 site

Based on the SP results, the water drainage has similar direction with the main tectonic orientation for this karst region, N-S or NW–SE (Artugyan et al. 2015).

Site MP-S4

This site is represented by a funnel-shaped sinkhole with a sink exactly in the middle of the karst depression. Even if that sink has formed through dissolution and it could represent a connection with underground fractures or karst channels, the SP results based on the measurements developed in May 2013 show that there is a higher retention of water indicating the fact that this sink is clogged with clay. On the other hand, SP results indicate water circulation on the edges of the sinkhole (Fig. 9.9), where karrens are present and where soil is thinner. All negative SP anomalies obtained within this site are related to the presence of karrens and thicker soil. These conditions favor faster water flow in the surface of the karst depression.

9.5.2 Colonovăț Plateau

Site CP-S1

This site is a sinkhole situated in a forested area with a large density of sinkholes. In the N and S sides of the sinkhole karrens are present. The sinkhole is not very deep

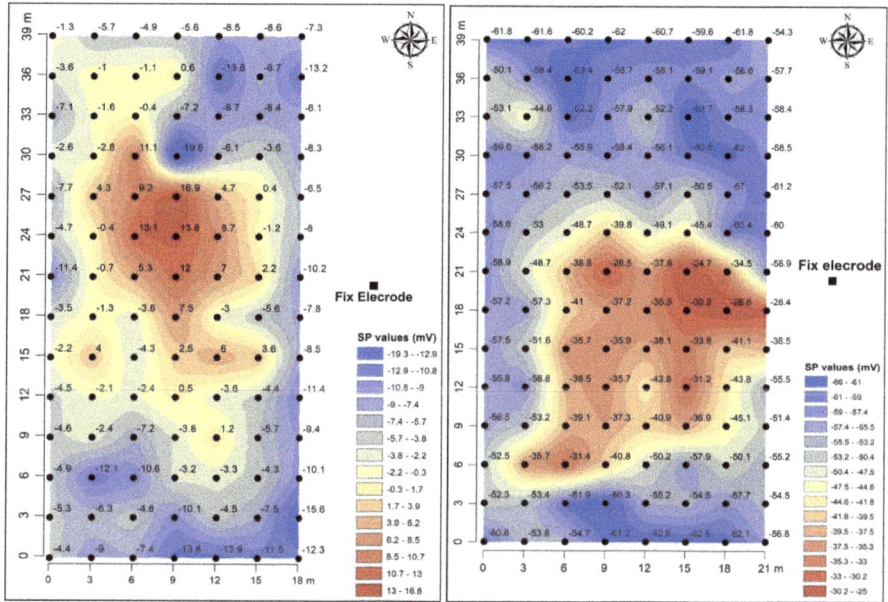

Fig. 9.10 Self-potential grids in sinkhole representing CP-M1 site in August 2013 **(left)** and October 2014 **(right)**

and it has a large flat bottom. For this site we have developed 2 field campaigns of SP measurements, in August 2013 and in October 2014. The SP measurements were done as grids. The results obtained in August 2013 (Fig. 9.10-**left**) indicate very well the bottom of the sinkhole by the positive SP values, while the negative anomalies obtained for the slopes of the sinkhole point out the water drainage. On the other hand, in October 2014 the results are described only by negative anomalies, with the higher values of SP in the middle part of the sinkhole.

Another observation regarding the SP values in October 2014 (Fig. 9.10-**right**) could highlight a possible drainage direction from E to SW indicating a possible fissure network in the underground. The higher values of SP in the middle of the karst depression in both field campaigns highlight very well the water retention proved by the presence of thicker soil and vegetation, while karrens favor water flowing in the surface of this karst depression.

Site CP-S2

This is a medium size sinkhole elongated to North and it presents steeper slopes on East side. Karrens are present in the Southern part. We have developed 3 profiles measurements for this site: September 2013, October 2013 and September 2014. Based on the SP measurements developed in October 2013 we could identify that positive values indicate acummulation tendency. However, the higher values are present in the median part of the sinkholes, while the lower SP values are overlaying those areas covered by karrens. On the steepest slope of the sinkholes the values are negative in October 2013 and September 2014 showing the faster flowing of water in the surface (Fig. 9.11).

On N-S direction the accumulation tendency based on the SP results is kept. Moreover, the SP values indicating a faster drainage are situated in the Northern part of the sinkhole, which coincide with the longer slope of this sinkhole (Fig. 9.12). From this observation arise the idea that the slope has an important role in the SP data acquisition and interpretation in the study of water flow in karst depressions.

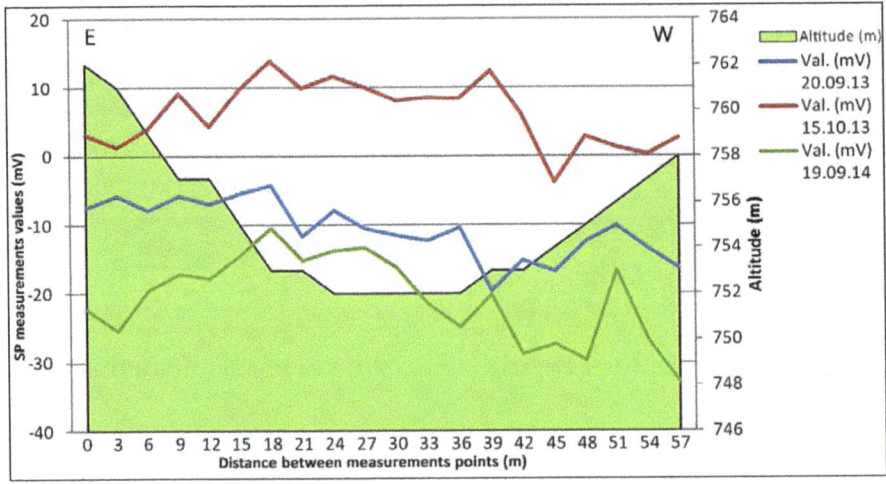

Fig. 9.11 Self-potential measurements and sinkhole topography on E-W profile in CP-S2 site

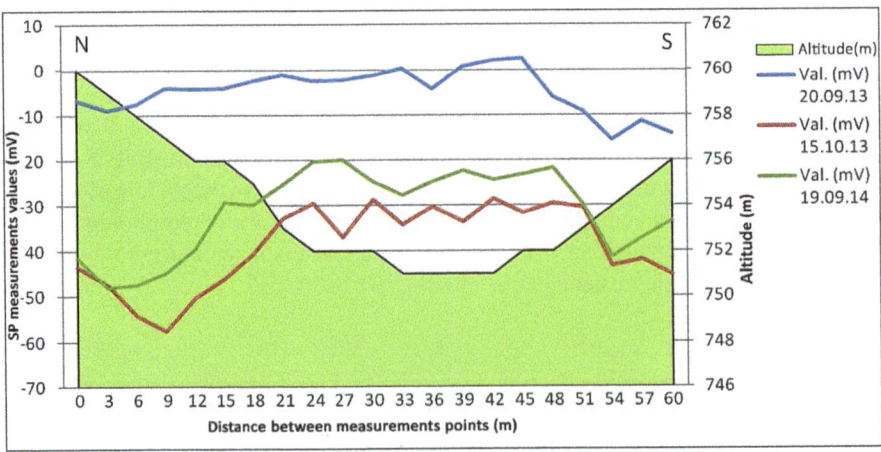

Fig. 9.12 Self-potential measurements and sinkhole topography on N-S profile in CP-S2 site

Site CP-S3

This sinkhole has a circular shape and it is located into a forested area. The sinkhole has no karrens and this is an important aspect since most of the analyzed sinkholes had karrens.

On N-S direction we had developed 3 campaigns as profiles in April 2013, October 2013 and October 2014. Most of SP values are negative indicating a fast drainage of water. Again the higher SP values are seen in the median part of the sinkhole (Fig. 9.13). However, on the profiles obtained in October 2013 and October 2014 we observe a more sinusoidal character of the profiles. This can be caused by both higher soil moisture, thicker soil cover and the missing of surface rock. Moreover, on the south side a faster drainage can be observed, especially in April 2013. Here we developed also several humidity measurements obtaining higher values on sinkhole slopes.

On E-W direction we have developed only 2 profiles. Again the values showed in October 2013 indicate a tendency of water retention comparative to the field campaign developed in October 2014. Even if the higher values are in the median part of the sinkhole, the values for both campaigns were very oscillating. However, the values of the humidity and SP in October 2014 are strongly correlated since higher values of humidity are associated to lower SP values, while lower values of humidity are associated with higher SP values (Fig. 9.14). Consequently, the SP values associated with a higher humidity are attributed to faster water flowing in this sinkhole.

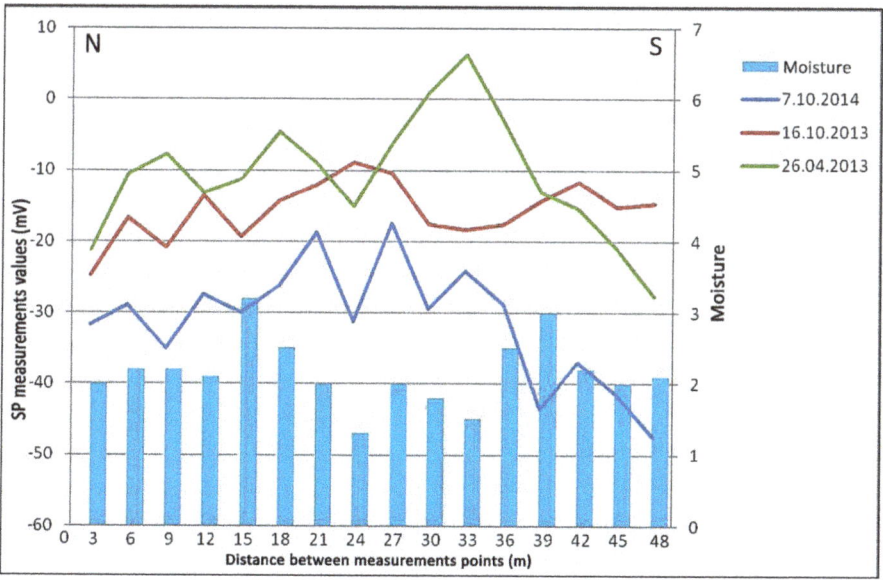

Fig. 9.13 Self-potential measurements correlated with soil moisture values on N-S profile in CP-S3 site

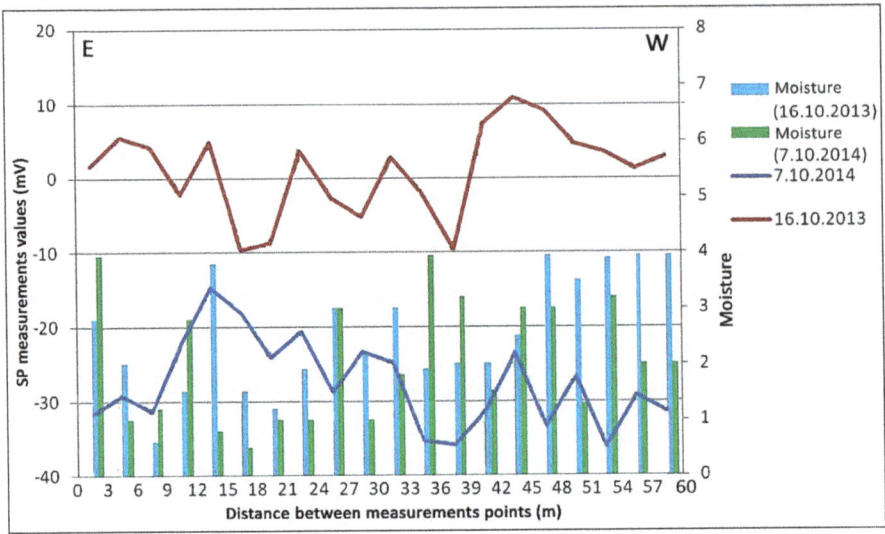

Fig. 9.14 Self-potential measurements correlated with soil moisture values on E-W profile in CP-S3 site

Site CP-S4

This site has a circular shape and it is less deep. This sinkhole is surrounded by large karrens on three sides, except the Eastern side, which makes the connection with another sinkhole. We have developed 3 SP profiles measurements, in May 2013, November 2013 and October 2014. On S–N profile all the SP values are negative indicating faster water drainage within this sinkhole. On the results obtained in October 2014 the highest values are located in the median part of the sinkhole (Fig. 9.15).

On the E-W orientation SP values are also negative. Measurements developed in May 2013 and November 2014 presents a sinusoidal aspect, while the measurements from October 2014 point out the highest values on the bottom of the sinkhole. Moreover, the sinkhole slopes are indicating a faster drainage due to the lower values of the SP (Fig. 9.16).

9.5.3 Cârneală Plateau

Sites CaP-S1, CaP-S2, CaP-S3

On Cârneală Plateau we have developed SP measurements in three sinkholes that are a chain of sinkholes. Due to this on the N-S direction the results are presented as a single profile graph, while on the W-E the results are presented as individual profiles of SP measurements.

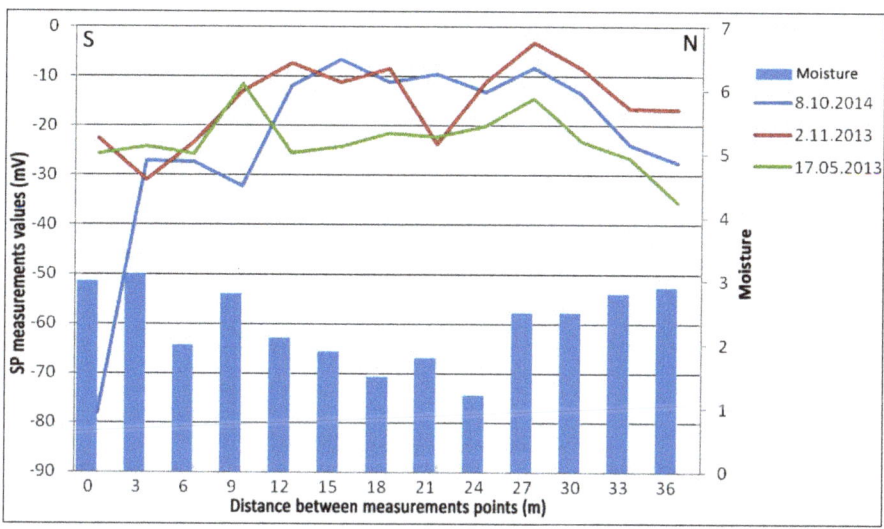

Fig. 9.15 Self-potential measurements correlated with soil moisture values on S–N profile in CP-S4 site

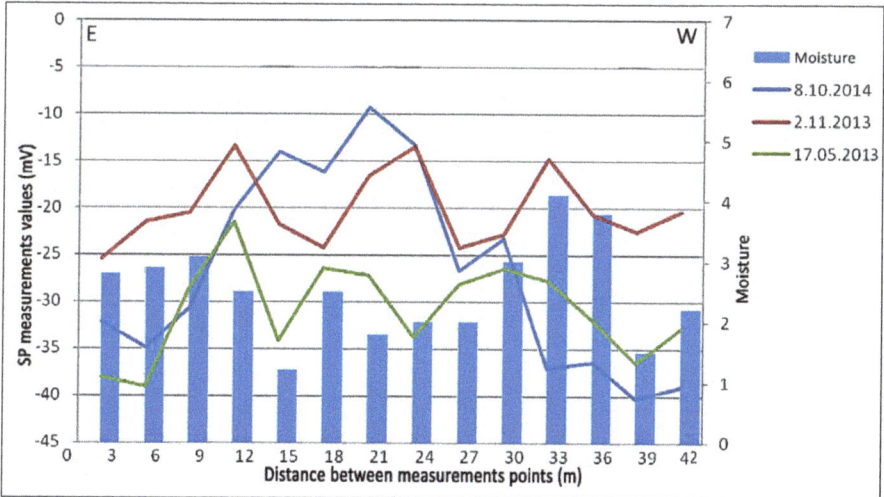

Fig. 9.16 Self-potential measurements correlated with soil moisture values on E-W profile in CP-S4 site

All sites indicate higher values in the median part of the sinkholes. Site CaP-S1 (Fig. 9.17) and site CaP-S3 have positive anomalies only on the bottom of the sinkholes, while CaP-S2 has only negative SP anomalies, but with highest of them located in the median part. These results indicate the higher retention and stagnation

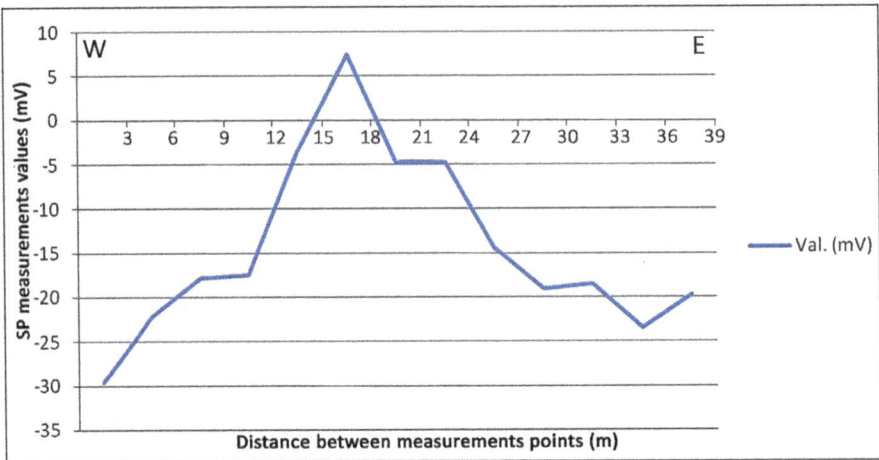

Fig. 9.17 Self-potential measurements on E-W profile in CaP-S1 site

of water on the bottom of the sinkholes, while the sinkholes' slopes favor water flowing as a consequence of the presence of karrens and thicker soil cover.

Moreover, for site CaP-S2 (Fig. 9.18) the higher slope on the Eastern side is observed also on the SP graphic with lower values indicating that slope favors the water circulation within this karst depression.

On the other hand, CaP-S3 present a negative anomaly at 12 m from starting point, indicating a possible underground karst channel or karst pipe that favors water circulation (Fig. 9.19).

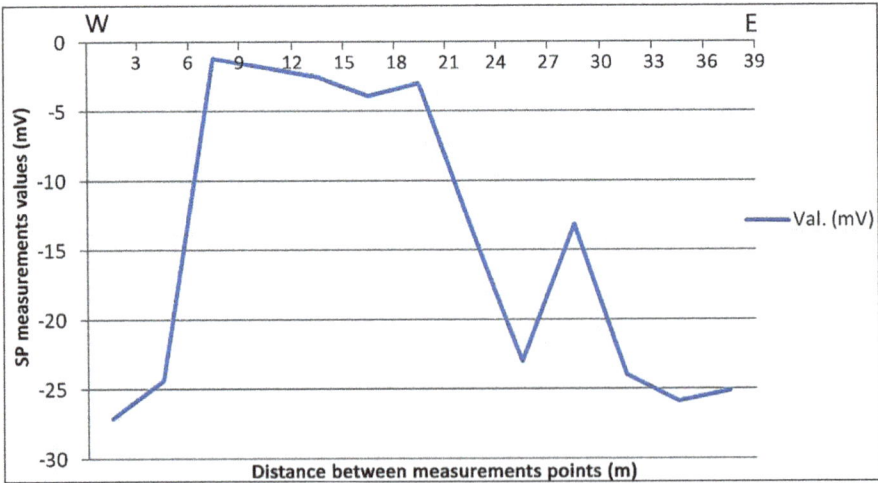

Fig. 9.18 Self-potential measurements on E-W profile in CaP-S2 site

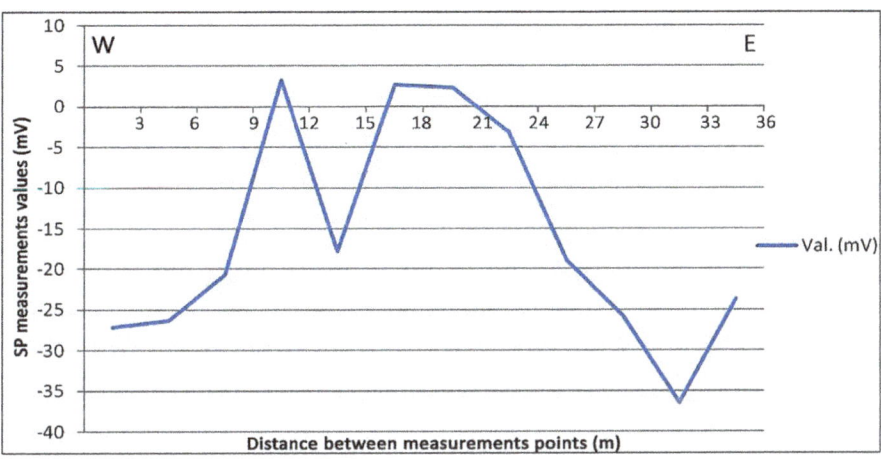

Fig. 9.19 Self-potential measurements on E-W profile in CaP-S3 site

The N-S profiles presented as a single graph (Fig. 9.20) show very well that for all the three sinkholes analyzed here the SP values are higher on the bottom of each karst depression, while the lowest values and the negative anomalies are associated with the slopes of each sinkhole.

Moreover, the SP positive anomalies are situated again only in the median part of the sinkholes, being attributed to water retention and stagnation in this part of the sinkholes. However, we could observe negative anomalies as one point in each sinkhole, considering that a karst fissure could be associated with these negative anomalies on the bottom of the sinkhole. Negative SP anomalies and the lowest values associated with sinkholes' slopes are caused by the presence of large karrens and thick soil cover, characteristics that favor water drainage.

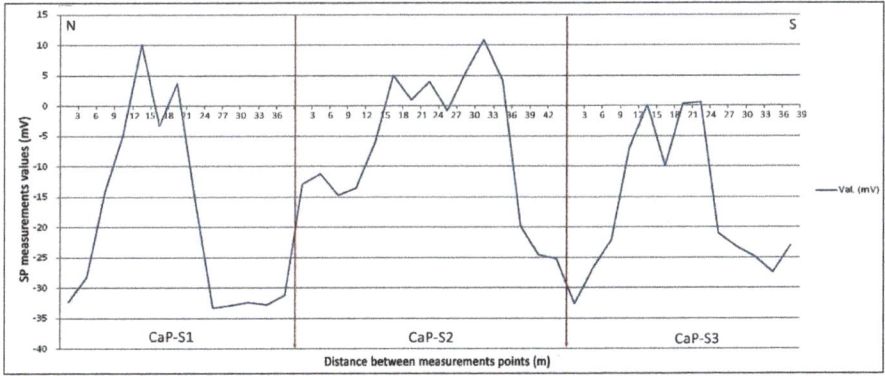

Fig. 9.20 Self-potential measurements on N-S profile in CaP-S1, CaP-S2 and CaP-S3 sites

9.5.4   Brădet Plateau—Culmea Neagră Area

Site BP-BRA-S1

This site is a less deep circular sinkhole with a diameter of about 45 m. The N-S profile is less homogeneous, but it presents the highest SP values in the median part of the sinkhole (Fig. 9.21). The S–N profile the SP values suggest a higher retention and stagnation on the bottom of the sinkhole. On the other hand, on the E-W orientation the SP values are also negative, indicating a fast water circulation (Fig. 9.22).

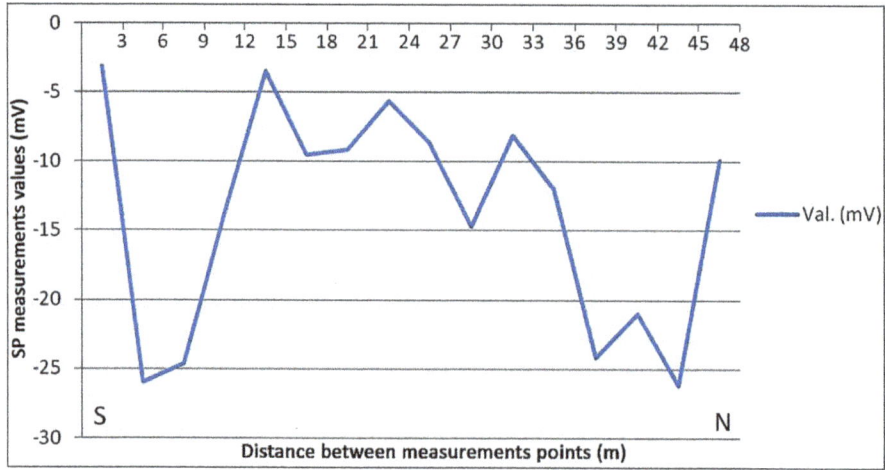

Fig. 9.21 Self-potential values measured for the BP-BRA-S1 site on S–N profile

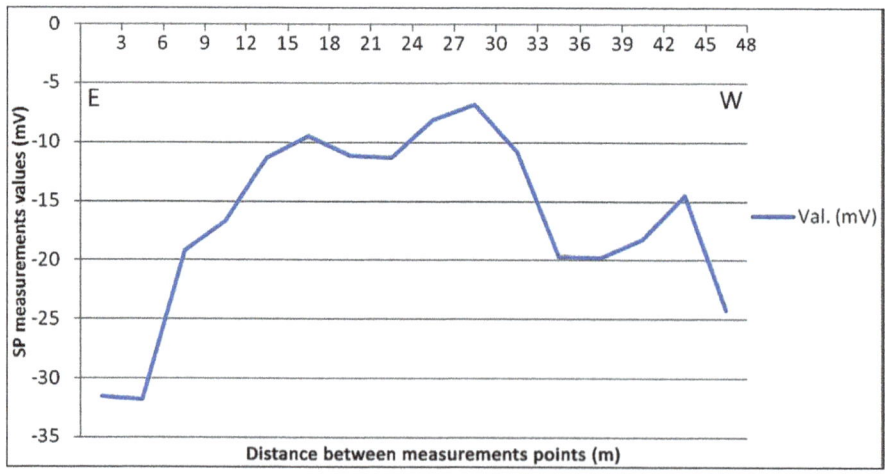

Fig. 9.22 Self-potential values measured for the BP-BRA-S1 site on E-W profile

Site BP-BRA-S2

This site is situated few meters to south to BP-BRA-S1 site. This is a deeper sinkhole elongated on N-S direction. The SP values are all negative (Fig. 9.23), an exception being present on the profile for W-E direction that presents a positive anomaly at the bottom of the sinkhole slope on the Eastern part. As in the previous sinkholes, slopes are indicating the faster drainage in water surface circulation (Fig. 9.24).

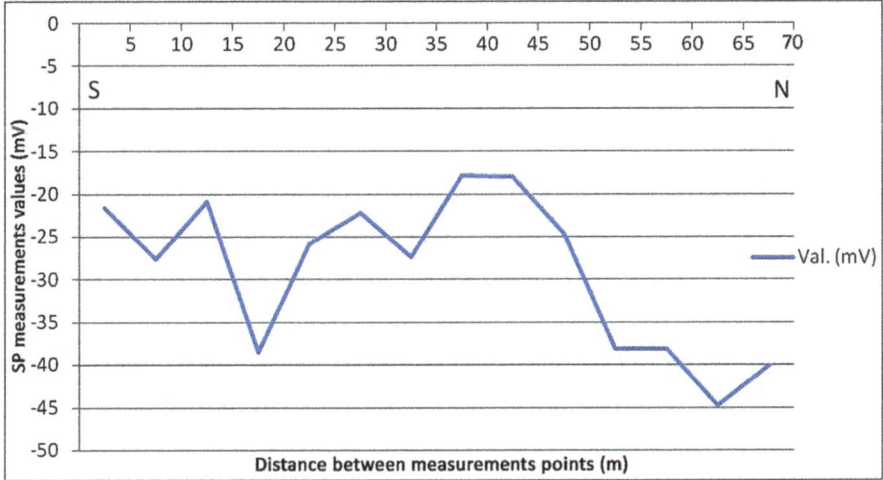

Fig. 9.23 Self-potential values measured for the BP-BRA-S2 site on S–N profile

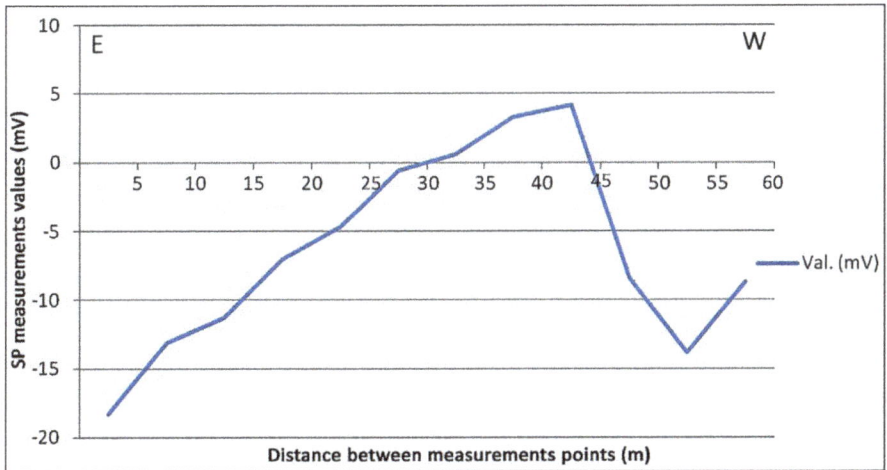

Fig. 9.24 Self-potential values measured for the BP-BRA-S1 site on W-E profile

Site BP-BRA-S3

This sinkhole is situated few meters south to BP-BRA-S2, being a less deep sinkhole
and a little bit elongated on N-S direction. As for the previous two sites in the Culmea
Neagră Area, this sinkhole indicates again SP values that show the tendency of water
retention and water stagnation in the middle part. The bottom of the sinkhole has the
highest values on both N-S and W-E profiles (Figs. 9.25 and 9.26).

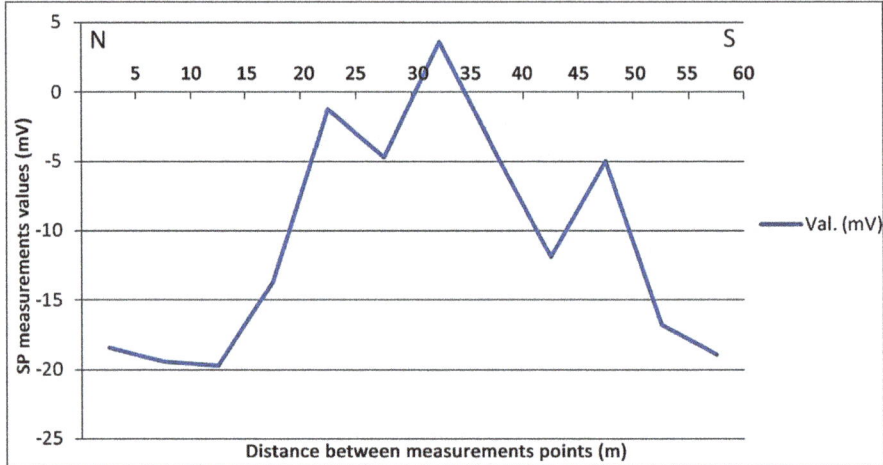

Fig. 9.25 Self-potential values measured for the BP-BRA-S3 site on N-S profile

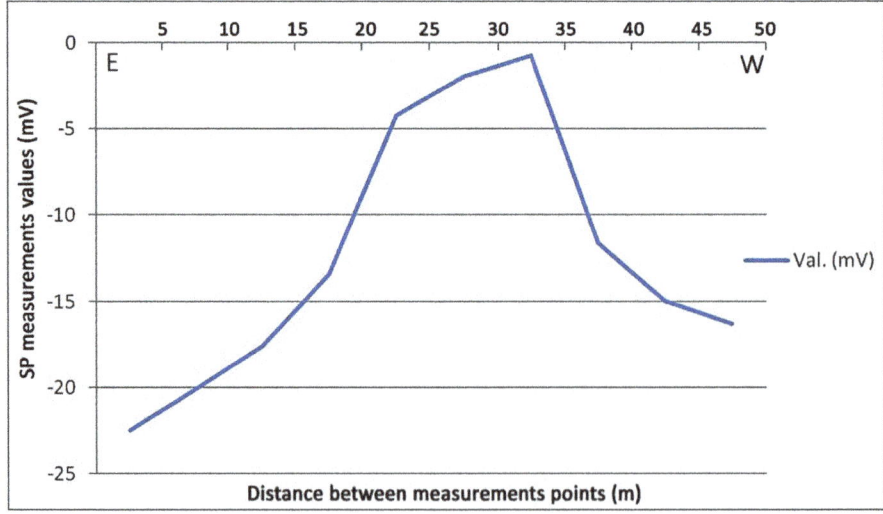

Fig. 9.26 Self-potential values measured for the BP-BRA-S3 site on E-W profile

9.6 Discussions and Conclude Remarks

The sinkholes analyzed using the SP method in the Anina Karst Region (Banat Mountains, Romania) have elongated or circular shape and various sizes. Most of the sinkholes present karrens on sinkholes' edges and flat or almost flat bottoms covered with a thick layer of clay and soil. The sinkholes situated on the Mărghitaș Plateau that are covered only with grass present smaller karrens on the edges of the sinkholes.

However, some studied sites present significant changes in SP values between measurements periods. These changes are explained by different weather conditions and soil characteristics (Urdea and Ţambriș 2014) during field measurements.

The results obtained in all the sites studied using the SP method confirm the fact that SP is a feasible method to analyze the water circulation at the surface of karst depression. Using this geophysical method we were able to point out that the bottom of the sinkhole, where the soil is thicker, retain more water. On the other hand the edges of the sinkholes, having the bedrock exposed as karrens and very thin soil, present faster water circulation.

The SP results on the karst plateaus in the Anina Karst Region are very similar with previous studies implementing the SP method in karst depressions (Chen et al. 2018), indicating that most of the positive (or higher) anomalies are located within the central part of the sinkhole. Our results indicate the positive anomalies (or higher SP values) within the center of the sinkholes proving the effectiveness of the SP method in the detection of water in epikarst zones in the karst depressions developed on the suspended karst plateaus in the Anina Karst Region, Romania. Based on these findings we consider that self-potential method has been used with success to detect water preferential pathways in shallow karst depressions.

The SP approach developed in several sinkholes in the Anina Karst Region, Romania shows that most of the anomalous zones with positive spontaneous electric potential are localized in the middle of the sinkholes, indicating the higher retention of water within the bottom part of the karst depressions. Also, the most of the values showing negative anomalies are situated in those parts where the karrens are developed. Of course, there are also some exceptions in the SP values indicating that there is a sink in the middle of the sinkhole, indicating the presence of a fast connection with a possible karst pipe or an underground channel.

The SP method in karst investigations is useful especially in narrow karst depressions since it provides better results in delineating the shallow karst formation discovered in karst depressions. Moreover, the spontaneous electric field method is easier to be implemented and provides fewer anomalous reactions (Chen et al. 2018).

In conclusion, the SP method applied in the Anina Karst Region, Banat Mountains, Romania, has shown the feasibility of this geophysical method in the study of water circulation in shallow karst environment. Based on previous studies indicating the negative or smaller values of the natural electric field attributed to faster water circulation in karst rocks (Chen et al. 2018), this study bring an important contribution in karst depressions understanding and karst terrain analysis. Furthermore, since in Romania the SP method has been used only sporadically, this study can be used as a reference study in the application of SP method in exokarst research and in water circulation in karst terrains. Moreover, SP method could be used together with other geopshyical methods in cavities detection studies (Jardani et al. 2006b) and in sinkholes and karst terrain characterization (Artugyan et al. 2015).

Since sinkholes are landforms that favor the fast circulation of the water from the surface into the underground (Jardani et al., 2006a), the SP approach could be a feasible method to study aquifers and the presence of contaminants in karst aquifers since water is a very important resource in karst areas. Moreover, SP method is reliable in characterization of shallow karst topography research and this geophysical method can be combined with other geophysical method in karst terrain analysis (like GPR and ERT).

Acknowledgements The authors are grateful and thankful to those students, colleagues and friends who helped during the self-potential field acquisition campaigns. All of them provided a great support and help during those more than 16 months of SP data field campaigns.

References

Abdelrahman EM, Essa KS, Abo-Ezz ER, Sultan M, Sauck WA, Gharieb AG (2008) New least-squares a logarithm for model parameters estimation using self-potential anomalies. Comput Geosci 34:1569–1576
Abdelrahman EM, Soliman KS, Abo-Ezz ER, Essa KS, El-Araby TM (2009) Quantitative interpretation of self-potential anomalies of some simple geometric bodies. Pure Appl Geophys 166:2021–2035
Andreo B, Vías J, Durán JJ, Jiménez P, López-Geta JA, Carrasco F (2008) Methodology for groundwater recharge assessment in carbonate aquifers: application to pilot sites in southern Spain. Hydrogeol J 16(5):911–925. https://doi.org/10.1007/s10040-008-0274-5
Artugyan L, Urdea P (2014a) Groundwater drainage monitoring and karst terrain analysis using spontaneous potential (SP) in anina mining area Banat Mountains, Romania. preliminary study. In: Proceedings of international DIKTAS Conference. Karst without boundaries, Trebinje 157–164
Artugyan L, Urdea P (2014b) Using spontaneous potential (sp) as a geophysical method for karst terrains investigation in Mărghitaş Plateau, Banat mountains, Romania. Revista De Geomorfologie 16:45–53
Artugyan L, Ardelean AC, Urdea P (2015) Characterization of Karst terrain using geophysical methods based on sinkhole analysis: a case study of anina Karstic region, Banat mountains, Romania NCKRI SYMPOSIUM 5. In: Proceedings of the 14th multidisciplinary conference on sinkholes and the engineering and environmental impacts of Karst, 387–397. https://doi.org/10.5038/9780991000951.1044
Asfahani J, Tlas M (2016) Interpretation of self-potential anomalies by developing an approach based on linear optimization. Geosci Eng 5:7–21
Bakalowicz M (2005) Karst groundwater: a challenge for new resources. Hydrogeol J 13(1):148–160. https://doi.org/10.1007/s10040-004-0402-9
Bérubé AP (2007) A graphical 3D finite element program for modelling self-potentials generated by flow through a porous medium. J Environ Eng Geophys 12(2):185–197. https://doi.org/10.2113/JEEG12.2.185

Biswas A (2016) A comparative performance of least square method and very fast simulated annealing global optimization method for interpretation of self-potential anomaly over 2-D inclined sheet type structure. J Geol Soc India 88:493–502. https://doi.org/10.1007/s12594-016-0512-8

Biswas A (2017) A review on modeling, inversion and interpretation of self-potential in mineral exploration and tracing paleo-shear zones. Ore Geol Rev 91:21–56. https://doi.org/10.1016/j.ore georev.2017.10.024

Biswas A, Sharma SP (2014) Resolution of multiple sheet-type structures in self-potential measurement. J Earth Syst Sci 123:809–825. https://doi.org/10.1007/s12040-014-0432-1

Biswas A, Sharma SP (2015) Interpretation of self-potential anomaly over idealized body and analysis of ambiguity using very fast simulated annealing global optimization. Near Surf Geophys 13:179–195. https://doi.org/10.3997/1873-0604.2015005

Bleahu M (1982) Relieful carstic. Editura Albatros, București, p 296

Boleve A (2009) Localisation et quantification des zones de fuites dans les digues et les barrages par la méthode du potentiel spontané (PhD thesis). universite de Savoie, Faculté Des Sciences: terre, univers, environnement de Grenoble, p 213

Bucur II (1997) Formaţiunile mezozoice din zona Reşiţa-Moldova Nouă (Munţii Aninei şi estul Munţilor Locvei). Ed. Presa Universitară Clujeană, Cluj-Napoca, p 214

Burdan M, Filip R, Masec F, Murvay PŞ (2007) Speleogeneza Avenului de sub Culmea Neagră şi evoluţia acviferului carstic aferent Izbucului de sub Muntele Polom. Speomond- Revista Fed. Române De Speologie 12:14–20

Carpenter PJ, Adams RF, Lenczewski Melissa, Leal-Bautista Rosa M (2013) Ground-penetrating radar, resistivity and spontaneous potential investigations of a contaminated aquifer near Cancún, Mexico. In; Proceedings of the 13th multidisciplinary conference on sinkholes and the engineering and environmental impacts of karst, Carlsbad, New Mexico, 231–237. https://doi.org/10.5038/9780979542275.1131

Chalikakis K, Plagnes V, Guerin R, Valois R, Bosch FP (2011) Contribution of geophysical methods to karst-system exploration: an overview. Hydrogeol J 19:1169–1180. https://doi.org/10.1007/s10 040-011-0746-x

Chen Y, Qin X, Huang Q, Gan F, Han K, Zheng Z, Meng Y (2018) Anomalous spontaneous electrical potential characteristics of epi-karst in the Longrui Depression, Southern Guangxi Province. China Environ Earth Sci 77:659. https://doi.org/10.1007/s12665-018-7839-y

Chukwu GU (2013) Characteristics of self-potential anomalies in Abakaliki lower benue trough of Nigeria. Int Res J Geo Min 3(7):257–269

Chukwu GU, Ekine AS, Ebeniro JO (2008) SP anomalies around Abakaliki anticlinorium of SE Nigeria. Pacific J Sci Technol 9(2):561–566

Church M (2010) The trajectory of geomorphology. Prog Phys Geogr 34(3):265–286. DOI: https://doi.org/10.1177.2F0309133310363992

Cocean P (2001) Environment threats in Romanian karst, department of geography, university of Cluj. In: 13th International congress of speleology, 4th speleological congress of Latin América and Caribbean, 26th Brazilian congress of speleology, 613–617

De Waele J, Plan L, Audra P (2009) Recent developments in surface and subsurface karst geomorphology: An introduction. Geomorphology 106:1–8. DOI: https://doi.org/10.1016/j.geomorph.2008.09.023

Ekinci YL, Balkaya Ç, Göktürkler G (2020) Global optimization of near-surface potential field anomalies through Metaheuristics. chapter 7 In: Biswas A, Sharma S (Eds) Advances in modeling and interpretation in near surface geophysics Springer, Cham, pp 155–188

Essa KS (2019) A particle swarm optimization method for interpreting self-potential anomalies. J Geophys Eng 16(2):463–477. https://doi.org/10.1093/jge/gxz024

Fagerlund F, Heinson G (2003) Detecting subsurface groundwater flow in fractured rock using self-potential (SP) methods. Environ Geol 43:782–794

Ford D, Williams P (2011) Geomorphology underground: the study of karst and karst processes. In: Gregory KJ, Goudie AS (eds): The SAGE Handbook of Geomorphology, SAGE Publications Ltd, p 648

Goran C (1982) Catalogul sistematic al peşterilor din România Inst Speol Fed Rom Turism-Alpinism Com Centr Speol Sport Bucureşti. p 496

Guichet X, Jouniaux L, Catel N (2006) Modification of streaming potential by precipitation of calcite in a sand-water system: laboratory measurements in the pH range from 4 to 12. Geophys J Int 166(1):445–460. DOI: https://doi.org/10.1111/j.1365-246X.2006.02922.x

Gurk M, Bosch F (2001) Cave detection using self-potential-surface (SPS) technique on a karstic terrain in Jura mountains Switzerland. In Proceedings of the meeting. Electrotromagnetische Tiefenforschung, 9–13 Mar

Iurkiewicz A, Dragomir G, Rotaru A, Bădescu B (1996) Karst systems in banat Mountains (Reşiţa-Nera zone). Theor Appl Karstology 9:121–140

Jardani A, Dupont JP, Revil A (2006a) Self-potential signals associated with preferential ground-water flow pathways in sinkholes. J Geophys Res Solid Earth 111(B9). https://doi.org/10.1029/2005JB004231

Jardani A, Revil A, Dupont JP (2006b) Self-potential tomography applied to the determination of cavities. Geophys Res Lett 33:L13401 https://doi.org/10.1029/2006GL026028

Jardani A, Revil A, Akoa F, Schmutz M, Florsch N, Dupont JP (2006) Least squares inversion of self-potential (SP) data and application to the shallow flow of ground water in sinkholes. Geophys Res Lett 33(19). https://doi.org/10.1029/2006GL027458

Jardani A, Revil A, Bolève A, Dupont JP, Barrash W, Malama B (2007a) Tomography of groundwater flow from self-potential (SP) data. Geophys Res Lett 34: L24403. https://doi.org/10.1029/2007GL031907

Jardani A, Revil A, Santos F, Fauchard C, Dupont JP (2007b) Detection of preferential infiltration pathways in sinkholes using joint inversion of self-potential and EM-34 conductivity data. Geophys Prospect 55:749–760. https://doi.org/10.1111/j.1365-2478.2007.00638.x

Jardani A, Revil A, Bolève A, Dupont JP (2008) Three-dimensional inversion of self-potential data used to constrain the pattern of groundwater flow in geothermal fields. J Geophys Res Solid Earth 113(B9). https://doi.org/10.1029/2007JB005302

Jardani A, Revil A, Barrash W, Crespy A, Rizzo E, Straface S, Johnson T (2009) Reconstruction of the water table from self-potential data: a Bayesian approach. Ground Water 47(2):213–227. https://doi.org/10.1111/j.1745-6584.2008.00513.x

Jeannin PY, Eichenberger U, Sinreich M, Vouillamoz J, Malard A, Weber E (2012) KARSYS: a pragmatic approach to karst hydrogeological system conceptualisation. assessment of ground-water reserves and resources in Switzerland. Environ Earth Sci 69:999–1013. https://doi.org/10.1007/s12665-012-1983-6

Jougnot D, Roubinet D, Guarracino L, Maineult A (2020) Modeling streaming potential in porous and fractured media, description and benefits of the effective excess charge density approach. chapter 4 In: Biswas A, Sharma S (Eds.) Advances in modeling and interpretation in near surface. Geophysics Springer, Cham, pp 61–96. https://doi.org/10.1007/978-3-030-28909-6_4

Jouniaux L, Maineult A, Naudet V, Pessel M, Sailhac P (2009) Review of self-potential methods in hydrogeophysics. CR Geosci 341(10–11):928–936. https://doi.org/10.1016/j.crte.2009.08.008

Kaçaroğlu F (1999) Review of groundwater pollution and protection in karst areas. Water Air Soil Pollut 113:337–356. https://doi.org/10.1023/A:1005014532330

Lange LA (1999) Geophysical studies at Kartchner Caverns State Park, Arizona. J Cave Karst Stud 61(2):68–72

Martínez-Pagán P, Jardani A, Revil A, Haas A (2010) Self-potential monitoring of a salt plume. Geophysics 75(4):WA17-WA25. https://doi.org/10.1190/1.3475533

Mateescu F (1961) Influenţe structurale în relieful Munţilor Caraşului. Probleme de geografie, vol VIII, Institutul de Geologie şi Geografie, Bucureşti, 205–219

Moore JR, Boleve A, Sanders JW, Glase SD (2011) Self-potential investigation of moraine dam seepage. J Appl Geophys 74:277–286. https://doi.org/10.1016/j.jappgeo.2011.06.014

Mutihac V, Ionesi L (1974) Geologia României. Editura Tehnică, București, 646 p
Nyquist JE, Corry CE (2002) Self-potential: The ugly duckling of environmental geophysics. Lead Edge 21(5):446–451. https://doi.org/10.1190/1.1481251
Onac B (2000) Geologia regiunilor carstice (Geology of karst terrains). Universitatea "Babes-Bolyai" Cluj-Napoca, Institutul de Speologie "Emil Racoviță" Cluj-Napoca, p 399
Oncescu N (1965) Geologia României, Editura Tehnică, București, p 534
Orășeanu I, Iurkiewicz A (2010) Karst hydrogeology of Romania, Edit. Federația Română de Speologie, Oradea, p 444
Orghidan T (1972) The fiftieth anniversary of the first speleological institute of the World. AAPG Bull Int J Speleol 4(1):1–7
Parise M, Pascali V (2003) Surface and subsurface environmental degradation in the karst of Apulia southern Italy. Environ Geol 44:247–256. DOI: https://α.org/10.1007/s00254-003-0773-6
Revil A, Jardani A (2013) The self-potential method. theory and applications in environmental geosciences, Cambridge University Press, Cambridge, p 383
Revil A, Pezard PA, Darot M (1997) Electrical conductivity, spontaneous potential and ionic diffusion in porous media. Geological Society, London, Special Publications 122:253–275. DOI: https://doi.org/10.1144/GSL.SP.1997.122.01.15
Revil A, Cary L, Fan Q, Finizola A, Trolard F (2005) Self-potential signals associated with preferential ground water flow pathways in a buried paleo-channel. Geophys Res Lett 32:L07401. DOI: https://doi.org/10.1029/2004GL022124
Revil A, Karaoulis M, Srivastava S, Byrdina S (2013) Thermoelectric self-potential and resistivity data localize the burning front of underground coal fires. Geophysics 78(5):B259–B273. https://doi.org/10.1190/geo2013-0013.1
Revil A, Ahmed AS, Jardani A (2017) Self-potential: A non-intrusive ground water flow sensor. J Environ Eng Geophys 22(3):235–247. https://doi.org/10.2113/JEEG22.3.235
Reynolds JM (1997) An introduction to applied and environmental geophysics. 1st edn, Wiley, p 806
Robert T, Dassargues A, Brouyère S, Kaufmann O, Hallet V, Nguyen F (2011) Assessing the contribution of electrical resistivity tomography (ERT) and self-potential (SP) methods for water well drilling program in fractured/karstified limestones. J Appl Geophys 75(1):42–53. https://doi.org/10.1016/j.jappgeo.2011.06.008
Roudsari MS, Beitollahi A (2013) Forward modelling and inversion of self-potential anomalies caused by 2D inclined sheets. Explor Geophys 44(3):176–184. https://doi.org/10.1071/EG12032
Rozycki A, Fonticiella Ruiz JM, Cuadra A (2006) Detection and evaluation of horizontal fractures in earth dams using the self-potential method. Eng Geol 82:145–153. https://doi.org/10.1016/j.enggeo.2005.09.013
Sencu V (1977) Carstul din Câmpul Minier Anina-St Cerc de Geol, Geofizică, Geografie 24(2):199–212
Sharma PV (2002) Environmental and engineering geophysics, Cambridge University Press, p 500
Sharma SP, Biswas A (2013) Interpretation of self-potential anomaly over 2D inclined structure using very fast simulated annealing global optimization–an insight about ambiguity. Geophysics 78:WB3–15. https://doi.org/10.1190/geo2012-0233.1
Sheffer MR (2007) Forward modelling and inversion of streaming potential for the interpretation of hydraulic conditions from self-potential data (PhD thesis). University of British Columbia, p 207
Shirazi SM, Imran HM, Akib S (2012) GIS-based DRASTIC method for groundwater vulnerability assessment: a review. J Risk Res 15(8):991–1011. https://doi.org/10.1080/13669877.2012.686053
Shofner GA, Mills HH, Duke JE (2001) A simple map index of karstification and its relationship to sinkhole and cave distribution in Tennessee. J Cave Karst Stud 63(2):67–75
Stevanovic Z, Dragisic V (1998) An example of identifying karst groundwater flow. Environ Geol 35(4):241–244. https://doi.org/10.1007/s002540050309

Susilo A, Sunaryo AT, Fitriah F, Hasan MF (2017) Identification of underground river flow in karst area using geoelectric and self-potential methods in Druju Region, southern Malang, Indonesia. Int J Appl Eng Res 12(12):10731–10738

Suski B, Ladner F, Baron L, Vuataz FD, Philippossian F, Holliger K (2008) Detection and characterization of hydraulically active fractures in a carbonate aquifer: results from self-potential, temperature and fluid electrical conductivity logging in the Combioula hydrothermal system in the southwestern Swiss alps. Hydrogeol J 16(7):1319–1328

Trufaş V, Sencu V (1967) Tipuri litologice de carst in Romania. Analele Universității București, Seria Științele Naturii, Geologie-Geografie 16(1):115–121

Urdea P, Țambriş A (2014) Spontaneous potential investigations in semenic mountains. Studia UBB-Geographia 59(2):25–46

Waltham T, Bell F, Culshaw M (2005) Sinkholes and subsidence, Springer, p 416

Wishart DN, Slater LD, Gates AE (2006) Self potential improves characterization of hydraulically-active fractures from azimuthal geoelectrical measurements. Geophys Res Lett 33(17)

Wishart DN, Slater LD, Gates AE (2008) Fracture anisotropy characterization in crystalline bedrock using field-scale azimuthal self potential gradient. J Hydrol 358(1–2):35–45

Zhu Z, Tao C, Shen J, Revil A, Deng X, Shi L, Zhou J, Wang W, Nie Z, Yu J (2020) Self-potential investigation of a deep-sea polymetallic sulfide deposit at the southwest indian ridge Indian ocean. ESSOAr, p 38 l https://doi.org/10.1002/essoar.10502023.1

Chapter 10
Interpretation of Self-Potential (SP) Log and Depositional Environment in the Upper Assam Basin, India

Dip Kumar Singha, Neha Rai, Madhvi, Mangal Maurya, Uma Shankar, and Rima Chatterjee

Abstract Analysis of self-potential (SP) log and the subsurface petrophysical parameters are important and basic study for evaluation of hydrocarbon reservoir. The mechanism of SP is the electrical potential developed in the subsurface due to some natural phenomena in the absence of any artificially applied current. The data are acquired using only two electrodes, one in a borehole and the other at a surface location that is also used for remote reference. The electric potential values recorded during the acquisition are only the relative changes in the SP voltage in the borehole environments of the formation rock. In this study, three wells (namely, KM, KT, and KJ) are used to estimate the petrophysical parameters in the several depth intervals of respective wells. The depth of the permeable zones which can act as reservoir rocks is marked in all the three wells using SP data in the formations. The wells were drilled in the sedimentary rock up to the basement in the Dhansiri valley of upper Assam basin for oil and gas exploration. The thickness of sedimentary rocks of tertiary age is found around 7000 m in the region. In the paper, the SP log data with the help of resistivity data are used to delineate the presence of hydrocarbons in the formations. The resistivity of formation water (Rw) has been determined using SP data and hence, water saturation (Sw) is estimated using Archie's equation and deep resistivity data in the reservoir and adjacent rock. Several permeable zones saturated by hydrocarbon are identified in Barail and Sylhet formations in KM well, Kopili and Sylhet formations in KT well and Sylhet and Upper gondwana formations in KJ well respectively. The values of Rw and Sw using SP data are varying 0.241–1.399 Ω-m and 20–70% in the formations. The study has further emphasized the importance of SP logs in lithofacies and depositional environment analysis. The lithofacies and environment of deposition are identified with the SP shape analysis. Two depositional features, Bell shaped and funnel shaped, are obtained in all the three wells. Bell shaped trend comprises upward fining sequences and funnel shaped trend comprises coarsening upward sequences. The shapes are identified with the

D. K. Singha (✉) · N. Rai · Madhvi · M. Maurya · U. Shankar
Department of Geophysics, Banaras Hindu University, Varanasi 221005, India
e-mail: dipkrsingha@bhu.ac.in

R. Chatterjee
Department of Applied Geophysics, IIT-ISM, Dhanbad 826004, India

© The Author(s), under exclusive license to Springer Nature Switzerland AG 2021   279
A. Biswas (ed.), *Self-Potential Method: Theoretical Modeling and Applications in Geosciences*, Springer Geophysics, https://doi.org/10.1007/978-3-030-79333-3_10

lacustrine sand, delta distributaries, turbidity channels, and proximal deep-sea fans deposit. These studies of SP data is helpful for further modeling and optimizing of reservoir saturation and other petrophysical parameters in relation with rock physics modeling.

Keywords Self-potential logging · Hydrocarbons · Upper Assam Basin

10.1 Introduction

The well log data provide almost all kind of the rock physical properties of a sedimentary basin. For oil and gas sedimentary basin, the well data are mainly used to identify the reservoir thickness and to carry out all the petrophysical parameters to characterize the reservoir fluid (Paul 2012; Li et al. 2004). The basic conventional well logs are self-potential, gamma ray, resistivity, velocity, density and neutron porosity which are used to compute porosity, volume of shale, water saturation and permeability of a reservoir (Serra 1984; Rider 2002; Singha and Chatterjee 2017). The shape of well log responses provides other geological information such as sedimentary grain size, deposition environment, compaction and many more. The interpretation of well data is greatly influenced by presence of shale in the reservoir. The shale is distributed in three ways which are laminated, structural and dispersed shale distribution (Clavaud et al. 2005; Sams and Andrea 2001). The self-potential or spontaneous (SP) well log is best suitable log to identify permeable and non-permeable zones in the formation. SP log is widely used not only in hydrocarbon expolration but for coal, mineral and ground water exploration. The objectives of the chapter are to (1) analyses the SP response of three wells in the upper Assam basin, (2) to determine formation water resistivity (Rw) from SP and temperatures data, (3) to estimate water saturation (Sw), porosity and volume of shale and (4) to study sediment grain size and depositional environments using shape of SP data.

SP log was one of the earliest measurements used in the petroleum industry, and it has continued to play a significant role in well log interpretation. Primarily, the SP log is used for determining gross lithology such as reservoir and non-reservoir through its ability to distinguish permeable zones (i.e., sandstones) from impermeable zones (i.e., shales) (Doll 1949). It is also used to correlate permeable and non-permeable zones between wells of the basin (Fig. 10.1).

The SP log is a record of direct current voltage (or potential) that develops naturally between a moveable electrode in the well bore and a fixed electrode located at the surface (Doll 1949). It is measured in millivolts (mV). Electric voltages arising primarily from electrochemical factors within the borehole and the adjacent rock create the SP log response. The SP phenomenon can be attributed to two processes which involve the moment of ions:

Fig. 10.1 The schematic diagram represents the SP currents flow in the borehole. The effect of the shale potential and the diffusion potential act together at bed boundaries and develop SP log deflections (Rider 2002)

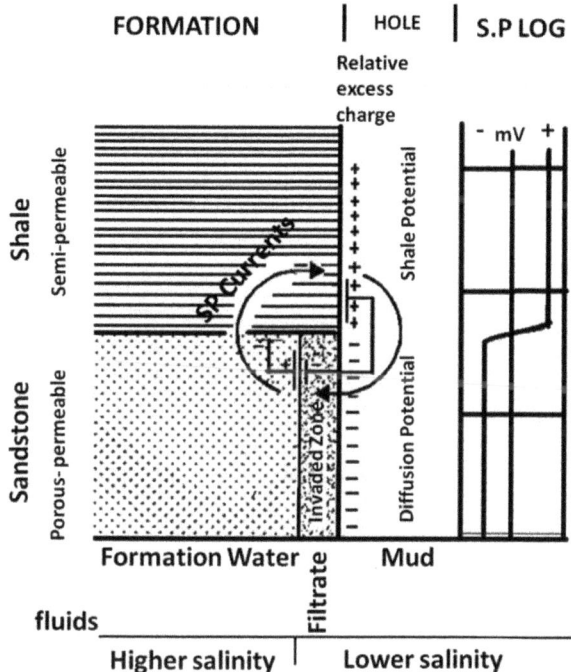

Electro kinetic potential (E_K):

Electro kinetic potential (E_K) develops when an electrolyte solution penetrates a porous, non-metalic medium under the differential pressure between the mud column and the formation. The resultant E_K is produced across the mud-cake in front of the permeable formation across the permeable formations, and across the shale beds. The equation proposed to identify E_K is shown below (Lynch 1962).

$$E_K = -\frac{\zeta \, \Delta \, \mathbf{P} \mathbf{k} \rho}{4\pi \, \eta} \qquad (10.1)$$

where

ζ: Zeta potential (adsorption),
ΔP: potential difference,
k: Solution dielectric constant,
ρ: Electrical resistivity,
η: Viscosity.

Electrochemical potential:

Electrochemical potential (E_C) develops when two fluids of different salinities are either in direct contact, or separated by a semi –permeable membrane (i.e., Shale). The electrochemical potential is the sum of two potential which is shown below:

$$E_C = E_m + E_j \tag{10.2}$$

where

E_m: Membrane potential,
E_j: Liquid junction potential or diffusion potential.

The electrochemical potential is categorized by membrane and liquid junction potential.

(a)   **Membrane potential**

The membrane potential develops when two electrolytes of different ionic concentrations such as mud and formation water are separated by shale. The value of the membrane potential develops is shown below (Serra 1984):

$$E_m = K_1 \frac{RT}{F} \ln(\frac{a_w}{a_{mf}}) \tag{10.3}$$

where

R: Ideal gas constant (8.314 J/k-mol),
T: Absolute temperature,
a_w: the ionic activity of the formation water,
a_{mf}:   the ionic activity of the mud filtrate,
K_1:   59.1 mV at 25^0C (77^0F).

(b)   **Liquid junction (diffusion) potential**

The liquid junction or diffusion potential develops at the contact of the mud filtrate and connate water in an invaded formation. The presence of different mobilities of ions causes an e.m.f., E_j The value of the diffusion potential develops is shown below (Serra 1984):

$$E_j = K_2 \frac{v - u}{v + u} \frac{RT}{F} \ln(\frac{a_w}{a_{mf}}) \tag{10.4}$$

where

v: mobility of anion,
u: mobility of cation,

a_w: the ionic activity of the formation water,
a_{mf}:   the ionic activity of the mud filtrate,
The coefficient K_2:   11.6 mV at 25^0C (77^0F).

10.2   Geology

The study area is in the Dhansiri valley of the upper Assam Basin. The upper Assam basin is south- east shelf-slope foreland basin, which is located in the north eastern part of India (Mallick et al. 1997; Naidu and Panda 1997). The study area is shielded by Mikir hills towards its west and Naga schuppen belt towards its east direction. The ~7000 m thick sedimentary rocks of tertiary period are rested upon the fractured granitic basement of Precambrian age. The successive formations such as sylhet, kopili, barail, Tipam etc. were deposited over the unconformities (Ishwar and Bhardwaj 2013; Mathur et al. 2001; Murty 1983). The sylhet formation of late paleocene to early Eocene period comprises shallow marine limestone which are interbedded with sandstone and shales. The Eocene Kopili formation comprises fluvial shales and fine grained sandstones (Kumar et al. 2018). Barail formation of late Eocene to Oligocene comprises finer to medium sandstone deposited over the grey silty shale rock and the topmost Tipam formation of early Miocene period, which consists of medium to fine grained massive sandstone having minor shales. In the post tipam formations, sands were deposited with minor layers of shales and coal (Ishwar and Bhardwaj 2013; Murty 1983).

Source Rock

The organic remains of marine life in Kopili shales (Dhansiri Valley' source rock) is mature, at the greater depth of the basin, at 3500 m depth. These rock formations include platform carbonates, shallow marine shales and the siltstones, sandstone, shales (Kumar et al. 2018; Singh et al. 2008). The production and accumulation of petroleum in Assam Arakan basin are present throughout the basin (Chandra et al. 1995; Naidu and Panda 1997).

Reservoir Rock

Reservoir rocks are present throughout most of the stratigraphic section in the Upper Assam basin. Reservoir rocks include the Eocene–Oligocene Jaintia Group Sylhet Formation limestones and Kopili Formation interbedded sandstones; Tura and Langpar (basal) marine sandstones also have reservoir potential, and Surma Group alluvial sandstone reservoirs are productive in the southwestern part of the Assam geologic province (Kumar et al. 2018; Singh et al. 2010; Mathur et al. 2001; Murty 1983). The most productive reservoirs are the Barail main pay sands and the Bokabil Group sandstones (Fig. 10.2).

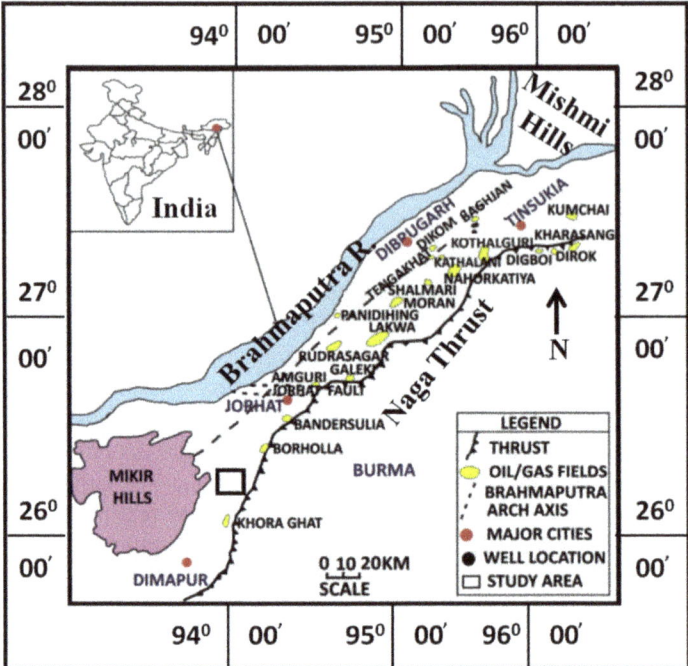

Fig. 10.2 The image illustrating the location of study area in upper Assam basin (modified after Mallick et al. 1997; Naidu and Panda 1997)

Traps and Seals

Anticlines and faulted anticlinal structures, sub-parallel to and associated with the northeast-trending Naga thrust fault, are the primary traps. Four sets of seal are observed in Dhansiri valley (Wandrey 2004; Singh et al. 2010; Gogoi and Chatterjee 2019).

- Sylhet Limestone of Early Eocene
- Grey Shales of Tipam group of mid to late Eocene
- Grey silty Shales of Bokabil formation of Middle Miocene Age &
- Bluish Grey Shales of Upper Bokabil formation of Middle Miocene Age (Fig. 10.3).

10.3 Methodology

As the basic mechanism of SP log is the potential measured between the borehole electrode and the surface electrode in the absence of any artificially applied current. The values recorded during the acquisition are only the relative changes in the SP

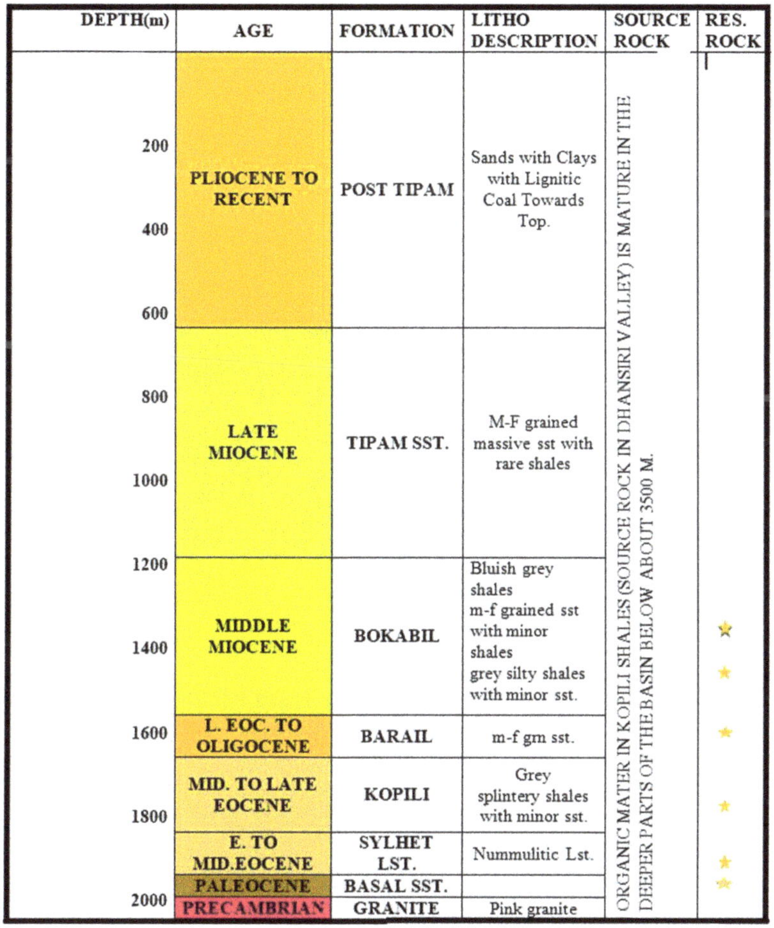

DEPTH(m)	AGE	FORMATION	LITHO DESCRIPTION	SOURCE ROCK	RES. ROCK
200 400 600	PLIOCENE TO RECENT	POST TIPAM	Sands with Clays with Lignitic Coal Towards Top.		
800 1000	LATE MIOCENE	TIPAM SST.	M-F grained massive sst with rare shales		
1200 1400	MIDDLE MIOCENE	BOKABIL	Bluish grey shales m-f grained sst with minor shales grey silty shales with minor sst.		☆ ★
1600	L. EOC. TO OLIGOCENE	BARAIL	m-f gm sst.		★
1800	MID. TO LATE EOCENE	KOPILI	Grey splintery shales with minor sst.		★
	E. TO MID.EOCENE	SYLHET LST.	Nummulitic Lst.		★
2000	PALEOCENE	BASAL SST.			☆
	PRECAMBRIAN	GRANITE	Pink granite		

(The vertical text in the SOURCE ROCK column reads: ORGANIC MATER IN KOPILI SHALES (SOURCE ROCK IN DHANSIRI VALLEY) IS MATURE IN THE DEEPER PARTS OF THE BASIN BELOW ABOUT 3500 M.)

Fig. 10.3 Generalized stratigraphy of Assam Shelf (modified after (Mathur et al. 2001; Murty 1983)

(Schlumberger 1972; Schlumberger 2000; Serra 1984; Rider 2002). We have considered three wells namely KM, KT, and KJ in the study area. The SP responses of three wells are well developed for various formations as shown in Fig. 10.4. The SP value varies in the different formations and the values vary from few tens to hundreds of millivolts, relative to the base of shale in Fig. 10.4. The correlation of Kopili, Barail and Sylhet formation are made based on the shape of SP log.

The uses of the SP log are listed below:

- The SP data is able to identify permeable beds (Fig. 10.4); where sand and shale separately indicated their presence.
- The determination of formation water resistivity (Rw).

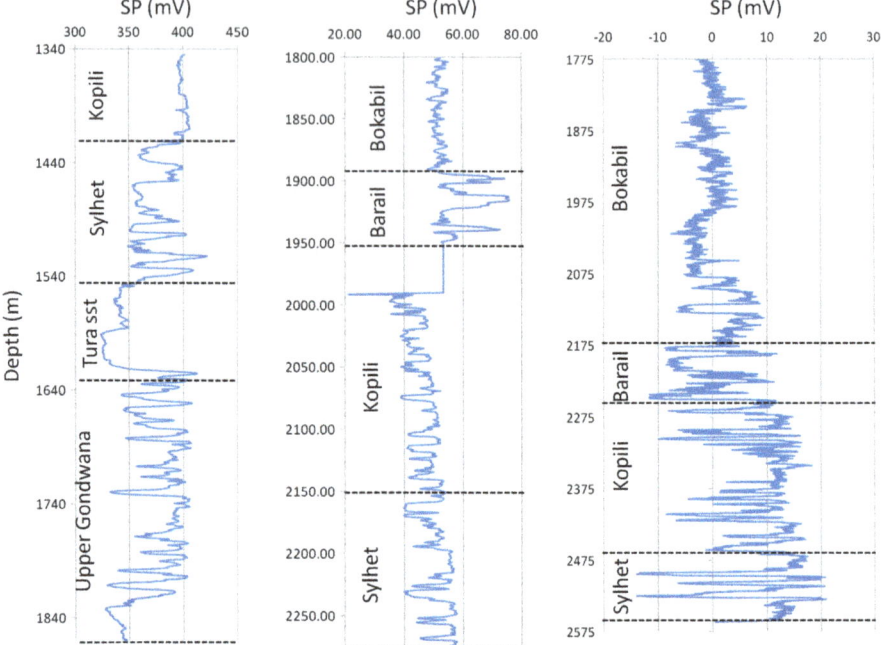

Fig. 10.4 The self-potential log data for wells KM, KT, and KJ. The relative changes of self-potential (mV) varying with subsurface depth in the formations

- The best suited method to show the quantitative availability of shale in the formations.

10.3.1 The Determination of Formation Water Resistivity (Rw)

The following procedure describes the determination of formation water (Rw) using a SP curve for all the formations in the wells KM, KT, and KJ respectively.

1. In the initiation the formation temperature (T_f) (Eq. (10.5)) of the borehole is measured, and then, the value of resistivity for the mud filtrate (Rmf) (Eq. (10.6)) and resistivity of the mud (Rm) are corrected to that formation temperature.

(a) The equation which measures the formation temperature is given below (After Western Atlas Logging Services 1985):

$$T_f = \left(\frac{BHT - AMST}{TD} * FD \right) + AM \tag{10.5}$$

where

BHT:	Bottom hole temperature
AMST:	Annual mean surface temperature
FD:	Formation depth
TD:	Total depth
Tf:	Formation temperature
AMST:	Annual mean surface temperature

(b)   The equation which calculates the resistivity of mud filtrate to the formation temperature is shown below (After Western Atlas Logging Services 1985):

$$Rmf = \frac{Rmf_{Surf}(T_{Surf} + 6.77)}{T_f + 6.77} \tag{10.6}$$

where

Rmf_{surf}:	Rmf at measured temperature T_{Surf}.
T_{Surf}:	T_{Surf} measured temperature of R_{mf}.
Rmf:	Rmf at formation temperature.

(c)   The equation which calculates the equivalent formation water resistivity (Rwe) is shown below (After Western Atlas Logging Services 1985):

$$Rwe = Rmf * 10^{SP/(61+0.133*BHT)} \tag{10.7}$$

where

BHT: Bottom hole temperature

(d)   The equation which uses for the estimation of formation water resistivity (Rw) is shown below (After Western Atlas Logging Services 1985):

$$Rw = \frac{Rwe + 0.131 * 10^{\left(\frac{1}{\log(BHT/19.9)}\right)-2}}{-0.5 * Rwe + 10^{\frac{0.0426}{\log(BHT/50.8)}}} \tag{10.8}$$

Next, the effect obtained due to different thickness in beds is minimized with the correction made to the static self-potential (SSP) level-SSP which is the maximum SP value (Fig. 10.5) obtained for the formation.

2.   Once, the SSP has been calculated, the formation water resistivity will be calculated with the following formula (Asquith 2004; Doll 1949):

$$SSP = -K^* \log\left(\frac{Rmf}{RW}\right) \tag{10.9}$$

$$K = 60 + (0.133 * T_f) \tag{10.10}$$

Fig. 10.5 The diagram
explained the definition of
SSP

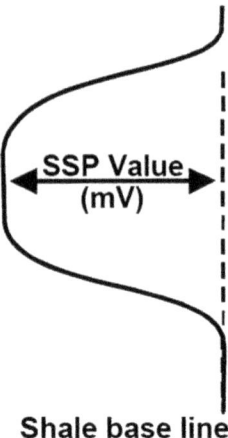

10.3.2   *The Determination of Volume of Shale (Vsh)*
in the Formations

SP method is the best suited for the quantitative estimation of shale in the formation; in the sand its availability is used to describe the shaly sand reservoirs and, it is used as a mapping parameter in the analysis of both the sandstone and the carbonate facies.

The equation, to estimate the volume of shale in the permeable zone is shown below:

$$Vsh = 1 - \frac{PSP}{SSP} \qquad (10.11)$$

where

Vshale =   Volume of shale,
PSP =   Pseudo-static spontaneous potential (maximum SP value for shale lithology),
SSP =   Static spontaneous potential of a nearby thick clean sand (Fig. 10.6).

10.3.3   *SP Log Shape Analysis*

The shape of SP curves is very important or the interpretation of depositional facies because it is directly related to the grain size of rock successions (Selley 1979). There are different curve shapes of log responses such as gamma ray log, resistivity etc. are used to interpret the depositional environment and interpretation of facies in the subsurface (Cant 1992). In the chapter, the SP log shapes have been used to identify

Fig. 10.6 The SP log varying with depth shows the shale base line in the log and with SSP and PSP the volume of shale has identified for KT well

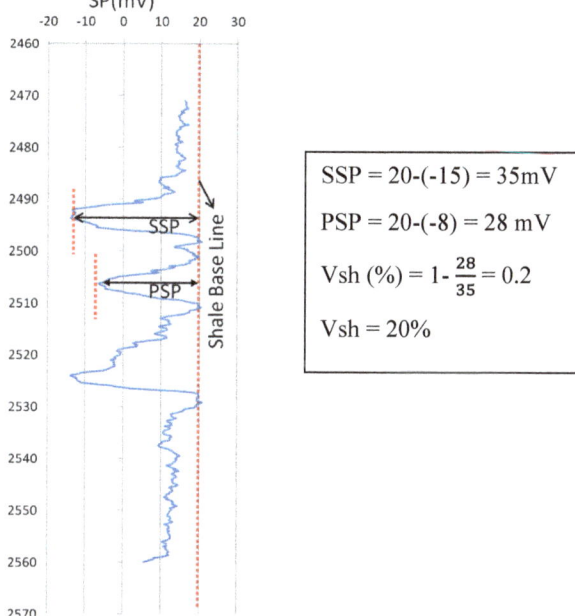

$$SSP = 20-(-15) = 35\,mV$$

$$PSP = 20-(-8) = 28 \ mV$$

$$Vsh \ (\%) = 1-\frac{28}{35} = 0.2$$

$$Vsh = 20\%$$

the sedimentary facies. SP log shapes are categorized into bell, cylinder and funnel shapes shown in Fig. 10.7 (Rider 2002).

10.3.3.1   Funnel-Shaped Successions

The funnel shape usually indicates deposition of cleaning upward sediment in onland environment or an increase in the sand content of the turbidity bodies in a deep marine setting shown by zig-zag funnel shape (Fig. 10.7). The environments of shallowing-upward and coarsening successions are sub-divided into three categories namely which are regressive barrier bars, prograding marine shelf fans and prograding delta. The first two environments are commonly deposited with glauconite, shell debris, carbonaceous detritus and mica. The prograding delta is observed comparatively large. (Nwagwu et al. 2019). The funnel shape indicates sediments size coarsening up with sharp top.

10.3.3.2   Cylindrical-Shaped Successions

Cylindrical-shaped of SP logs generally indicate aggrading sand in a slope channel and inner fan channel environments (Fig. 10.7). Cylindrical features trends with greater range of bed thickness, indicating a turbidite sands deposition. Sandstone beds are typically sharp and erosive with gradual transition to silt and mudstone. The

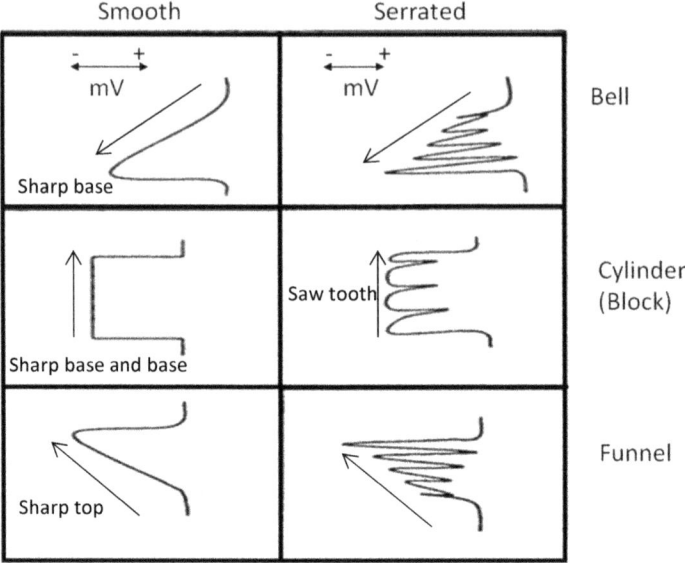

Fig. 10.7 The basic geometrical shapes which classify different depositional environments and used in the analysis of SP log (Rider 2002; Nwagwu et al. 2019)

turbidites are deposits from turbulent flow of sediment-laden turbidity current down a slope on the sea floor (Cant 1992; Nwagwu et al. 2019). This shape shows uniform grain size of sediment rock with up and top sharp.

10.3.3.3 Bell-Shaped Successions

The bell-shaped successions are usually indicative of a transgressive sand, tidal channel or deep tidal channel (Fig. 10.7). The tidal channels commonly contain glauconite and shell debris. The Bell shaped successions with carbonaceous detritus are also deposited in environments of fluvial or deltaic channels. The bell-shaped successions are thin, which indicate that the sands were deposited in environment of transgressive marine shelf (Nwagwu et al. 2019). This indicates fining up with sharp top.

10.4 Estimation of Petrophysical Parameters

The water saturation is estimated by empirical formula using the value of Rw for all the wells in the selective depth intervals in the formations. The porosity has been calculated in the permeable and non-permeable zone using sonic transit time. The following procedure are discussed below.

10.4.1 Estimation of Water Saturation in the Formations

Archie proposed an empirical model to estimate water saturation in clean sand matrix. It usually works in clean sandstones and carbonate rocks (Archie 1942). The saturation estimation does not only depend the resistivity of water but also depends upon the matrix mineral when the matrix is electrical conductive; besides the non-conductive quartz and calcite matrix grains. The empirical equation is greatly influenced by presence clastic shaly rocks with the presence of clay minerals.

$$SW = \sqrt[1/n]{\frac{a}{\Phi^m} * \frac{Rw}{Rt}} \qquad (10.12)$$

where

Sw:	saturation of water,
a:	constant; $0.5 \leq a \leq 2.5$,
m:	cementation factor; $1.3 \leq m \leq 2.5$,
n:	saturation exponent; (often the value us is ~ 2),
Rw:	formation water resistivity,
Rt:	observed bulk resistivity,
Φ:	porosity.

The Eq. (10.12) is applicable in clean sandstone formation whereas, the geology of the study area shows the availability of shale in the formations too. Hence, the above Eq. (10.12) is not enough for the evaluation of water saturation. Another model used to identify the water saturation is Poupone and Leveaux model (Poupon and Leveaux 1971); the formula used in the model is shown below:

$$Sw = \sqrt[2/n]{\left\{ \frac{\sqrt{\frac{1}{Rt}}}{\frac{Vsh^{(1-0.5Vsh)}}{\sqrt{Rsh}} + \sqrt{\frac{\Phi_e^m}{a*Rw}}} \right\}} \qquad (10.13)$$

where

Sw:	water saturation,
Rsh:	Resistivity of shale (Usually taken from the resistivity reading of nearby pure Shale),
Rt:	True formation resistivity,
Rw:	Resistivity of formation water,
Vsh:	Volume of Shale,
Φe:	porosity,
a:	Tortuosity factor,
m:	Cementation factor,
n:	Saturation exponent (Fig. 10.8).

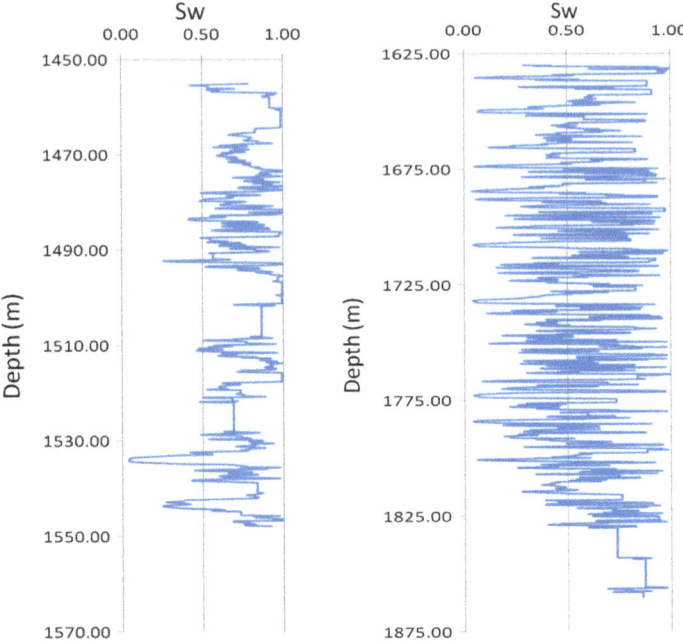

Fig. 10.8 Estimation of water saturation in well KJ for **a** Sylhet and **b** upper Gondwana formations respectively

10.4.2 Estimation of porosity Φ(s)

Wyllie time average equation (Wyllie et al. 1956) is used to estimate sonic porosity in loosely compacted sand and, the compaction factor is added to obtain the sonic porosity; the empirical correction for hydrocarbon effect is applied too in the data -when the sonic porosity value shown is too high. The Wyllie time average equation is shown below:

$$\phi(s) = \frac{\Delta t \log - \Delta tma}{\Delta tfl - \Delta tma} * \frac{1}{Cp}$$ (10.14)

$$Cp = \Delta tsh^*(C/100)$$ (10.15)

where

Cp:	Compaction factor.
C:	a constant (often it took ~ 1.0) (Hilchie 1978),
Φ(s):	sonic porosity,
Δtma:	interval transit time for matrix; which is 47 μs/ft,
Δtlog:	interval transit time recorded in log data,

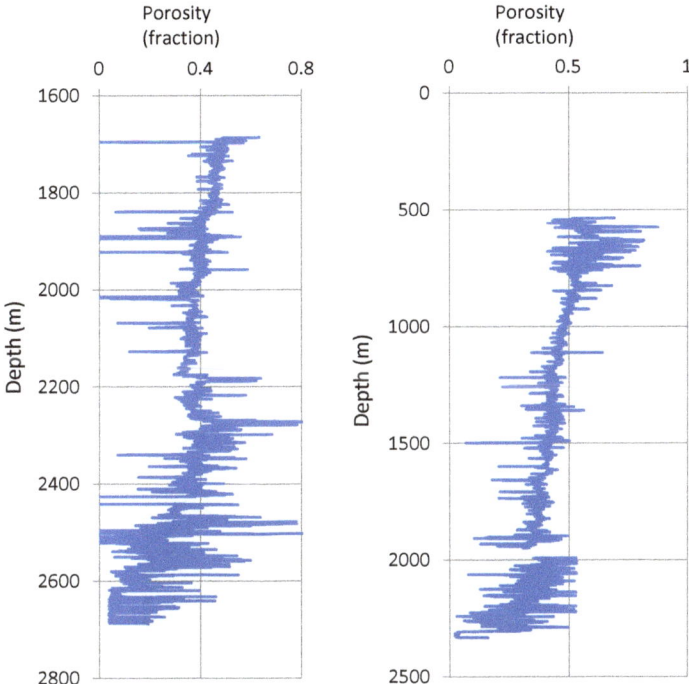

Fig. 10.9 The porosity estimated from sonic transit time using Eq. (10.14) varying with depth in KT and KM wells respectively

Δtfl: interval transit time for the presence of fluid in the rocks; which is 185 μs/ft,

Δtsh: interval transit time for the Shale adjacent zone (Fig. 10.9).

10.4.3 Estimating the Types of Shale

The SP log is a vital tool to interpret shale distribution in the formations and for identification of its different types such as laminar shale, dispersed shale, and structural shale. This distribution is identified by density and porosity cross plot in KT well shown in Fig. 10.11.

Laminar shale

In this case, the important factors are the relative thickness of the shale, the resistivity on the shape of the SP, and the permeable beds. The value of PSP (Pseudo SP or maximum SP in shaly sand formation) is used to indicate the shaliness in the formations (Serra 1984).

Dispersed shale

Dispersed shales impede the movement of Cl^- ions and the effect is strong attenuation of the SP. the attenuation occurred in the SP is due to the presence of shale in the pores. The diffusion of Cl^- ions can be reduced to zero at a certain shale percentage (Serra 1984).

Structural shale

As long as the sand grains constitute continuity in their phases the structural shales act as dispersed shales. As soon as continuity in phases break, the no deflection can be seen in the SP log and it indicates then that sand grains are surrounded by shale (Serra 1984) (Figs. 10.10 and 10.11).

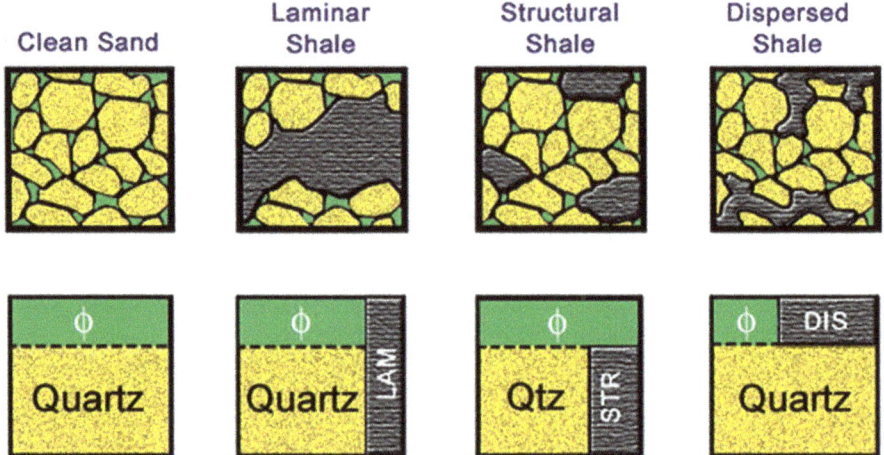

Fig. 10.10 Different shale distribution mode after Schlumberger 2000 and Serra 1984

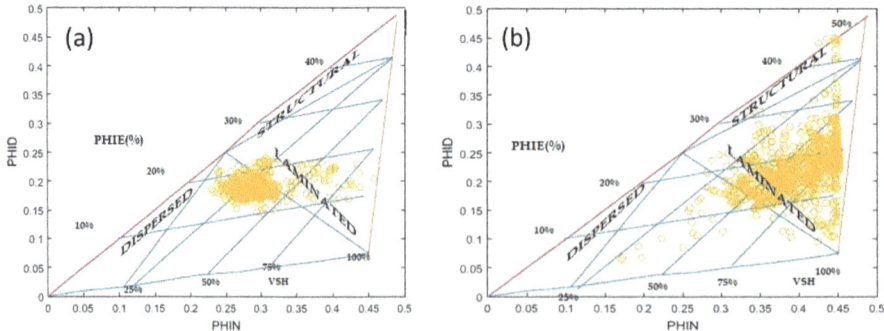

Fig. 10.11 cross plot of neutron porosity vs. density porosity for **a** KT well at depth interval and **b** KM well for depth 1950–2150 m

10.5   Results and Discussions

Two formations are encountered in each of the wells KM, KT, and KJ. The permeable zones delineated using SP log in the well KM are: Barail and Sylhet formations, similarly, for well KT: Kopili and sylhet formations, and in KJ wells: Sylhet and upper gondwana formations. First, we have identified SSP for clean sand in each formation after drawing the shale base line shown in Fig. 10.6 for KT well. Then we have estimated PSP from the shale base line for each possible permeable zone. The shape of SP is greatly influenced by presence of shaliness and as well as presence of resistive hydrocarbon in the sand zone. For example of KT well, the shale baseline is drawn at 20 mV and then, SSP and PSP are estimated 35 mV for depth interval 2488–2498 and 28 mV for depth interval 2502 to 2511 m. Similar procedure has been applied to other formations of KM and KJ well data to draw shale base line and accordingly identifying permeable zones. The preamble zones have been confirmed either by water or hydrocarbon saturated. First we have checked the higher true resistivity values shown in the tables. We have calculated saturation of the parmeable zones using Archie's equation. In the reservoirs depth interval, volume of shale calculated using SP data varies from 13 to 51% in Sylhet formation, 24–42% respectively. Minimum Sw is observed 43–59% and 40–41% for depth interval 2522–2530 in Sylhet and 1690-1695 m for upper Gondawana formations respectively. The water saturation is influenced by presence of shaliness in all the hydrocarbon reservoirs. The estimation of water saturation extremely depends on accuracy of formation water resistivity. The petrophysical parameters estimated for all the three wells KM, KT, and KJ wells are tabulated in Tables 10.1, 10.2 and 10.3.

10.6   Depositional Environment

Since shale and clay are basically fine grained than sand, a change in SP suggests a change in grain size. SP deflections can indicate depositional sequences, where either sorting, grain size and cementation change with depth and reflects a characterstic change in SP shape. The thickness of each facies or sequences of facies, and the

Table 10.1 Computed petrophysical parameters for KM Well

S.N	Depth(m)	Vsh(%)	Rt(Ω-m)	Φ(%)	Rw(Ω-m)	Sw(%)
	Barail					
1.	1902–1912	~24	~10–13	~14–25	0.243	~60–81
2.	1934–1937	~42	~7–8	~15–34	0.241	~71–87
	Sylhet					
1.	2238–2241	~17	~8–9	~21–32	0.349	~65–78
2.	2248–2265	~19	~10–16	~12–44	0.347	~63–75

Table 10.2 Computed petrophysical parameter for KT well

S.N	Depth(m)	Vsh(%)	Rt(Ω-m)	Φ (%)	Rw(Ω-m)	Sw(%)
	Kopili					
1.	2338–2341	~41	~35–40	~28	~0.30	~64–78
	Sylhet					
1.	2505–2508	~13	~40–46	~18	~0.35	~68–80
2.	2522–2530	~43	~48–52	~26	~0.35	~43–59
3.	2538–2544	~26	~80–86	~21	~0.35	~52–68

Table 10.3 Computed petrophysical parameter for KJ well

S.N	Depth(m)	Vsh(%)	Rt(Ω-m)	Φ(%)	Rw(Ω-m)	Sw(%)
	Sylhet					
1.	1483–1485	~46	~370–378	~25	~0.989	~49–67
2.	1494–1495	~51	~390–398	~33	~0.985	~67–79
3.	1520–1525	~18	~400–420	~18	~0.98	~47–65
	Upper Gondwana					
1.	1690–1695	~17	~400–406	~19	~1.399	~40–51

evolution occurring in the size of grains it accumulated are together related to its depositional environment. It has already been mentioned in the paper that the bell shaped and funnel shaped logs indicate the channel point of fluvial and delta border progradation respectively whereas, the same log shapes also indicate the transgressive marine shelf and prograding marine shelf.

In the wells KM, KT, and KJ two most profound depositional features are observed on the SP log: Bell shape and funnel shape. The bell shaped trend comprises fining upward sequences and the funnel shaped trend comprises coarsening upward sequences. The shapes are identified on the SP log are delineated in Figs. 10.12, 10.13, and 10.14.

10.7 Conclusions

The well log data are vital for the determination of petrophysical parameters for evaluation of hydrocarbon reservoirs. The shale base lines are drawn on plot of SP for individual formations in the study area for identifying permeable and non-permeable zones. Resistivity of formation water (Rw) has been calculated using SP data for all the formations and hence, water saturation is calculated using Archie's empirical formula in the upper Assam basin. Some permeable zones act as hydrocarbon reservoirs which are validated with high resistivity values and lower water saturations.

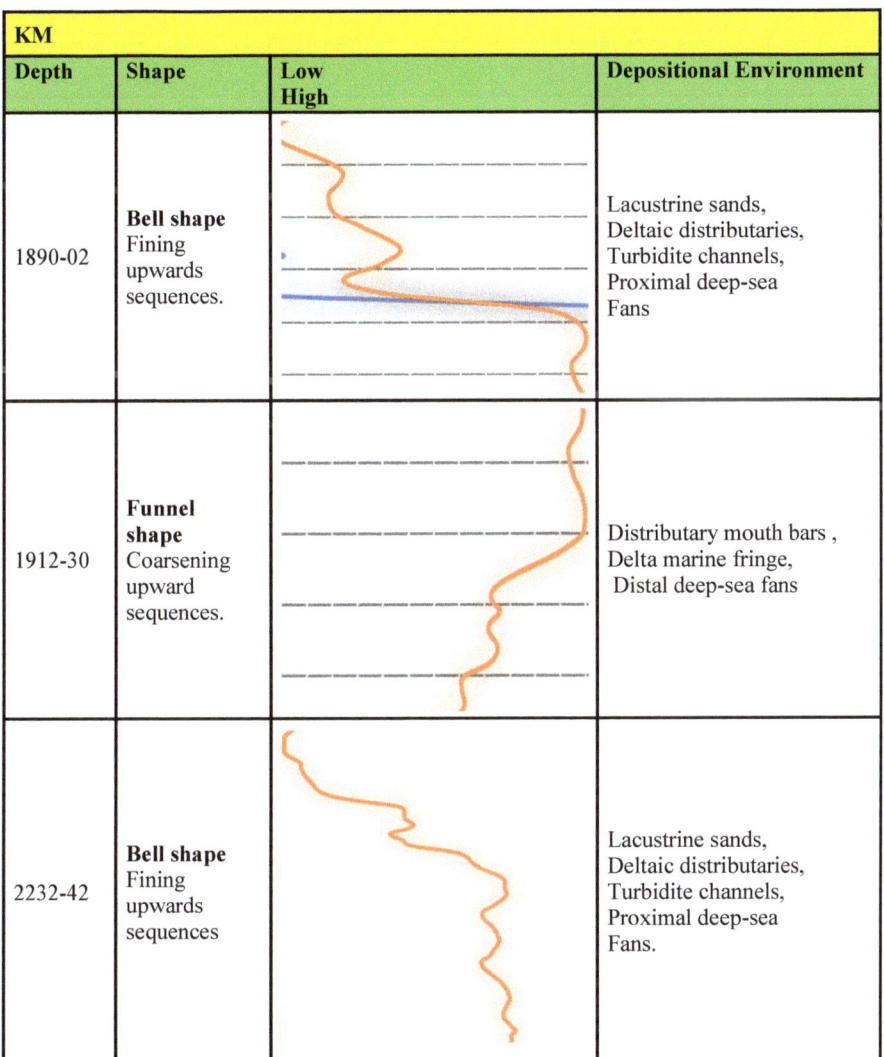

KM			
Depth	Shape	Low High	Depositional Environment
1890-02	**Bell shape** Fining upwards sequences.		Lacustrine sands, Deltaic distributaries, Turbidite channels, Proximal deep-sea Fans
1912-30	**Funnel shape** Coarsening upward sequences.		Distributary mouth bars , Delta marine fringe, Distal deep-sea fans
2232-42	**Bell shape** Fining upwards sequences		Lacustrine sands, Deltaic distributaries, Turbidite channels, Proximal deep-sea Fans.

Fig. 10.12 Depositional shape observed on KM well from self-potential log

In three wells, total twelve reservoirs have been identified from all permeable zones identified from SP log. The water saturation values are noticed higher values because presence of higher shaliness values in the reservoirs. The porosity is varying from 12 to 44% using sonic transit data. The study has further emphasized the importance of SP log in lithofacies and depositional environment analysis. The lithofacies and environment of deposition are identified with SP shape analysis using three wells KM, KT, and KJ. Two depositional features are obtained in all the three wells KM, KT, and KJ: Bell shaped and funnel shaped. Bell shaped trend comprises upward

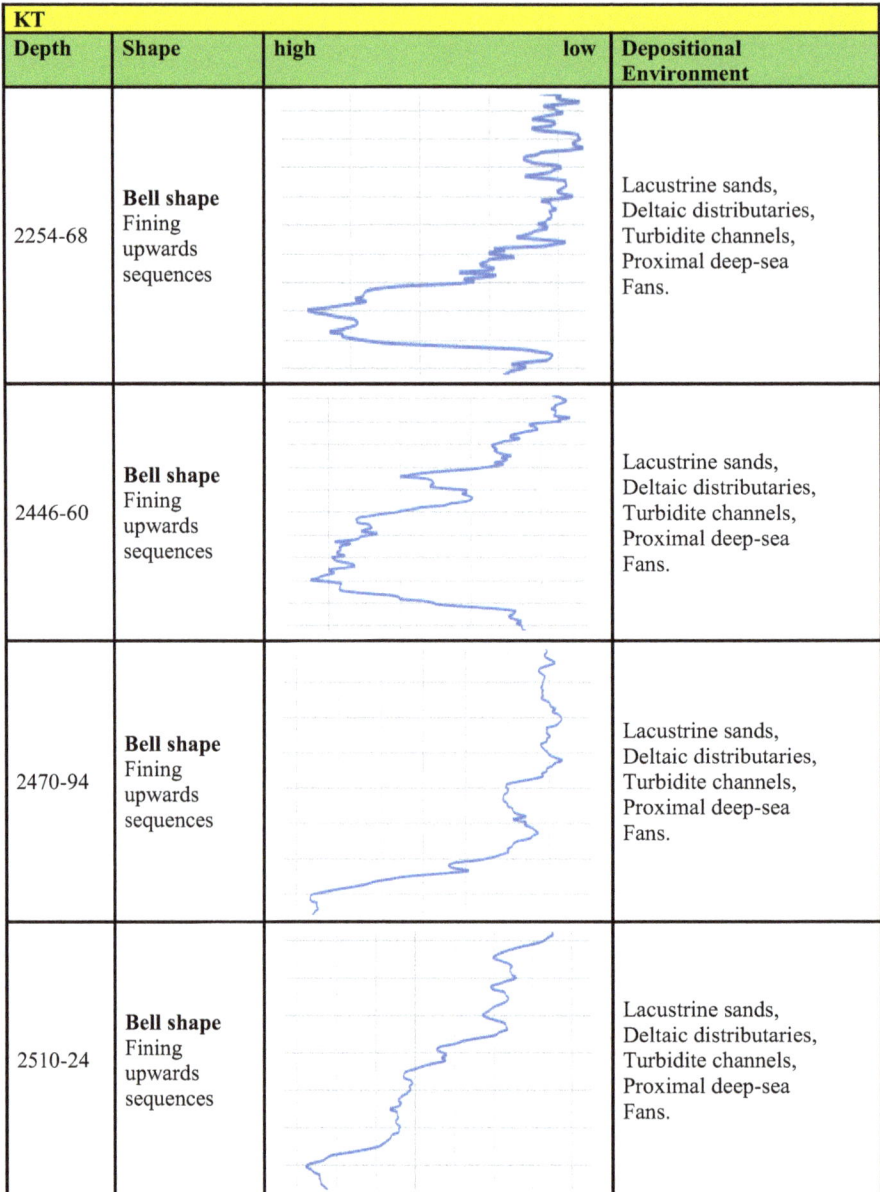

KT				
Depth	**Shape**	**high**	**low**	**Depositional Environment**
2254-68	**Bell shape** Fining upwards sequences			Lacustrine sands, Deltaic distributaries, Turbidite channels, Proximal deep-sea Fans.
2446-60	**Bell shape** Fining upwards sequences			Lacustrine sands, Deltaic distributaries, Turbidite channels, Proximal deep-sea Fans.
2470-94	**Bell shape** Fining upwards sequences			Lacustrine sands, Deltaic distributaries, Turbidite channels, Proximal deep-sea Fans.
2510-24	**Bell shape** Fining upwards sequences			Lacustrine sands, Deltaic distributaries, Turbidite channels, Proximal deep-sea Fans.

Fig. 10.13 Depositional shape observed on KT well from self-potential log

KJ			
Depth	**Shape**	**Low** **High**	**Depositional** **Environment**
1478-94	**Bell shape** Fining upwards sequences		Lacustrine sands, Deltaic distributaries, Turbidite channels, Proximal deep-sea Fans.
1502-14	**Funnel** **shape** Coarsening upward sequences.	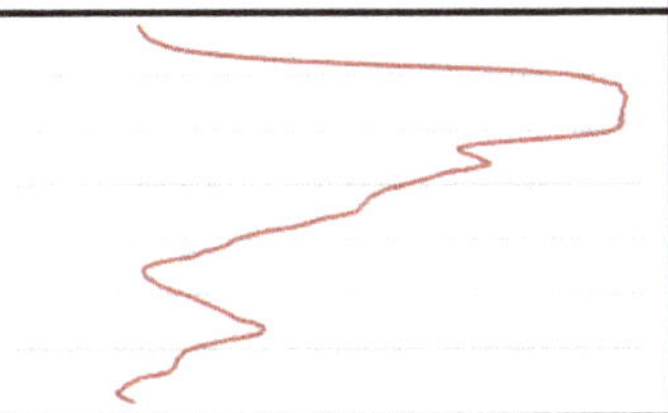	Distributary mouth Bars, Delta marine fringe, Distal deep-sea fans.
1532-46	**Funnel** **shape** Coarsening upward sequences.	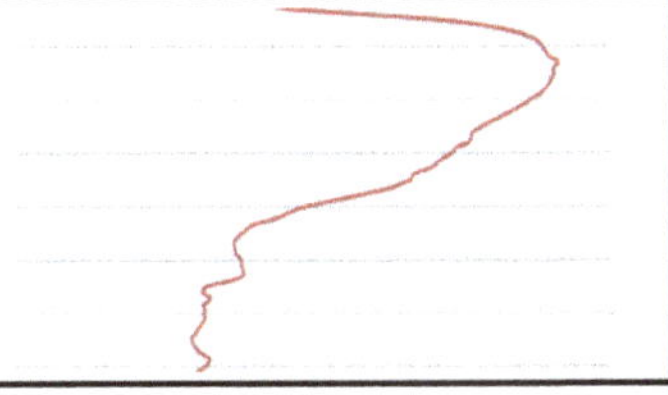	Distributary mouth Bars, Delta marine fringe, Distal deep-sea fans.
1656-82	**Bell shape** Fining upwards sequences		Lacustrine sands, Deltaic distributaries, Turbidite channels, Proximal deep-sea Fans.
1686-07	**Funnel** **shape** Coarsening upward sequences.		Distributary mouth Bars, Delta marine fringe, Distal deep-sea fans.
1818-34	**Funnel** **shape** Coarsening upward sequences		Distributary mouth Bars, Delta marine fringe, Distal deep-sea fans.

Fig. 10.14 The depositional shapes observed in KJ well from self-potential log

fining sequences and funnel shaped trend comprises coarsening upward sequences. The shapes are identified the lacustrine sand, delta distributaries, turbidite channels, and proximal deep-sea fans.

Acknowledgements Authors' sincerely thankful to the department of science and technology (DST)-INSPIRE, New Delhi, for funding the project (DST/Inspire Faculty award/2016/Inspire/04/2015/001681) dated 10-08-2015. Authors also express their gratitude to Oil India, Assam for providing us 3D seismic data, well data and geological information in the upper Assam-Arakan basin.

References

Archie GE (1942) The electrical resistivity log as an aid in determining some reservoir characteristics. Soc Petrol Eng 146

Atlas wireline services-Log interpretation charts (1985). Western Atlas International Atlas wireline services

Asquith G, Krygowski (2004) Basic well log analysis. American association of petroleum geologist, vol 16, pp 21–30

Cant DJ (1992) Subsurface facies analysis. Facies models, pp 27–45

Chandra U, Dhawan R, Mittal AK, Dwivedi P, Uniyal AK (1995) Stable isotope geochemistry of associated gases from Lakwa-Lakhmani Field, Upper Assam Basin, India. In: Proceedings first international petroleum conference & exhibition, PETROTECH-95: New Delhi, vol 3, pp 361–364

Clavaud JB, Nelson R, Guru UK, Wang H (2005) Field example of enhanced hydrocarbon estimation in thinly laminated formation with a triaxial array induction tool: a laminated sand-shale analysis with anisotropic shale. In: SPWLA 46th annual logging symposium, 26–29 June 2005, New Orleans, USA, SPWLA-2005-WW

Doll HG (1949) The SP log: theoretical analysis and principles of interpretation. Trans AIME 179(01):146–185

Gogoi T, Chatterjee R (2019) Estimation of petrophysical parameters using seismic inversion and neural network modeling in upper Assam baisn, India. Geosci Front 10:1113–1124

Hilchie DW (1978) Applied open hole log interpretation. DW Hilchie Inc, Goldon, vol 161

Ishwar NB, Bhardwaj A (2013) Petrophysical well log analysis for hydrocarbon exploration in parts of Assam Arakan Basin, India. In: 10th Biennial international conference and exposition, vol 153

Kumar M, Dasgupta R, Singha KD, Singh PN (2018) Petrophysical evaluation of well log data and rock physics modeling for characterization of Eocene reservoir in chandmari oil field of Assam Basin, India. J Petrol Explor Prod Technol 8:323–340

Li S, Henderson CM, Stewart RR (2004) Well log study and stratigraphic correlation of the Cantuar Formation, southwestern Saskatchewan. CREWES Res Rep 16:1–18

Lynch, E.J., 1962. Formation Evaluation. Harper's Geoscience series, Harper and Row, New york.

Mallick RK, Raju SV, Mathur N (1997) Geochemical characterization and genesis of Eocene cruide oils in a part of upper Assam Basin, India. Proc Second Int Petrol Conf Exhib PETROTECH 97(1):391–402

Mathur N, Raju SV, Kulkarni TG (2001) Improved identification of pay zones through integration of geochemical and log data—a case study from Upper Assam basin, India. Am Asso Petrol Geol Bull 85(2):309–323

Murty KN (1983) Geology and hydrocarbon prospects of Assam shelf. In: Proceeding of the second international petroleum conference and exhibition, petrotech, vol 97, pp 350–364

Naidu BD, Panda BK (1997) Regional source rock mapping in Upper Assam Shelf. In: Proceedings of the second international petroleum conference and exhibition (PETROTECH-97), New Delhi, vol 1, pp 350–364

Nwagwu AE, Emujajporue OG, Ugwu AS, Oghonya R (2019) Lithofacies and depositional environment from geophysical logs of EMK field, Deepwater Niger delta, Nigeria. Current Res Geosci 9:1–9

Serra O (1984) Fundamentals of well-log interpretation: 1. the acquisition of logging data. Development in petroleum science, 15A

Poupon A, Leveaux J (1971) Evaluation of water saturations in shaly formations. Log Anal 12(4):3–8

Paul WJ (2012) Petrophysics, department of geology and petroleum geology, University of Aberdeen

Rider M (2002) The geological interpretation of well logs. 2nd edn, Rider-French Consulting, Scotland

Sams MS, Andrea M (2001) The effect of clay distribution on the elastic properties of sandstones. Geophys Prospect 49(1):128–150. https://doi.org/10.1046/j.1365-2478.2001.00230.x

Schlumberger Ltd (1972) Log interpretation. Volume I principles

Schlumberger (2000) Beginnings, a brief history of Schlumberger Wireline and Testing

Selley RC (1979) Concepts and methods of subsurface facies analysis

Singh KR, Bhaumik P, Akktar SM, Siawal A, Singh JH (2008) Deeper (paleogene) hydrocarbon plays and their prospectivity in south Assam Shelf, A and A A basin, India. In: 7th Biennial international conference and exposition on petroleum geophysics, SPG, Hyderabad

Singh KR, Bhaumik P, Akhtar SM, Singh JH, Mayor S, Asthana M (2010) Major Hydrocarbon Trap Types in Dhansiri Valley, Assam and Assam Arakan Basin, India. In: 8th international conference and exhibition on petroleum geophysics at Hyderabad.

Singha DK, Chatterjee R (2017) Rock physics modeling in sand reservoir, Krishna-Godavari basin, India. Geomech Eng Int J 13:99–117

Wandrey JC (2004) Sylhet kopili/Barail-Tipam composite total petroleum system, Assam Geologic Province. India. US Geol Surv Bull 2208-D

Wyllie MRJ, Gregory AR, Gardner LW (1956) Elastic wave velocities in heterogeneous and porous media. Geophysics 21(1):41–70

Chapter 11
High Resolution Electrical Resistivity Tomography and Self-Potential Mapping for Groundwater and Mineral Exploration in Different Geological Settings of India

Dewashish Kumar

Abstract High resolution electrical resistivity tomography is the state-of-the-art geophysical technique to understand, conceptualize, delineate and demarcate the subsurface geological strata and structure(s) in terms of prospect groundwater scenario, its availability, sustainability and subsequent development. Nevertheless, exploration and prospecting for different metallic mineralization within the subsurface under various geological settings is a major concern and challenge in the field of earth science in our country. The exploration of these vital natural resources namely water and minerals from near surface layers to 200–250 m depths in a different geological environment is a major issue among the earth scientists and is of much importance and significance, which are the present and future need as well as the utmost demand of our country. This paper highlights the recent research work and their outcome on groundwater and mineral resources in the areas of exploration, prospecting and development in varied geological settings of our country.

Keywords Electrical resistivity and induced polarization tomography · Groundwater · Minerals · Exploration and prospecting · India

11.1 Introduction

Two dimensional high resolution electrical resistivity and induced polarization is one of the unique electrical tomography techniques in exploring and mapping the subsurface natural resources up to a maximum depth 250 m with a quite good precision mainly for groundwater and ploymetallic minerals deposit in different geological terrain/settings. This technique is scientifically verified and proved in a number of scientific studies conducted for groundwater and mineral resources both for the exploration, prospecting and development of the natural resources. The benefit of this technique is immense in conjunction with continuous high resolution mapping of

D. Kumar (✉)
CSIR-National Geophysical Research Institute, Hyderabad, India
e-mail: dewashishkumar@ngri.res.in

the subsurface geological formations as well as the large data density acquisition in one go and the subsequent data processing, data analysis, the requisite modeling for evaluation of the subsurface geology and the detailed interpretation of the models in terms of groundwater and mineral resources for any type of the geological settings. We acquired high resolution electrical resistivity and induced polarization full waveform sampling and recording of the dataset namely resistivity and chargeability physical parameters (ABEM 2012) with 8 windows timing set up and equal duration for current-on and current-off in the measurement cycle for a complete analysis of the output signal strength gathered from the various geological rocks. This paper highlights the important and significant results on groundwater studies on exploration, prospecting and development as well as mineral exploration in different geological settings of India.

11.2 Interpretation of 2D Inverted Resistivity and Induced Polarization Models

11.2.1 Results and Discussions

Delineation of deep-seated weathered and fractured rock for groundwater exploration: Choutuppal mandal, Nalgonda district Telangana, India

A geoscientific study was conducted at Choutuppal mandal, Nalgonda district Telangana India to tackle the heavy crisis of water. In order to understand, mapping the complex geological set up and delineated the hydrogeological strata for exploitation of groundwater resources (Kumar et al. 2016b) a systematic groundwater prospecting, exploration and development work were conducted in the study area for a long term availability and sustainability of groundwater resources within Choutuppal area. Keeping in view high resolution electrical resistivity survey was conducted in the problematic villages of the Choutuppal area where there was acute shortage of groundwater both for drinking and agriculture purpose. The 2D inverted subsurface resistivity model connecting between two villages namely Mandollagudem and Toorpugudem, which is underlain by granite hard rock is shown in Fig. 11.1. This resistivity model depicted a smooth variation of resistivity for the geological formations but with a large resistivity contrast varying from $\sim 10 - 1.6 \times 10^5$ Ω.m from south to north end of the profile (Fig. 11.1). The near surface layered formation shows low resistivity <160 Ω.m, which indicated a soil layer followed by weathered granite until ~ 30 m depth. The weathered granite is clearly mapped between 160 and 480 m lateral distance, which is very well distinguished from the massive granite (Fig. 11.1) with substantial resistivity contrast among them. At a lateral 240 m distance, the downward deepening of the geological bed indicates weathered/highly weathered granite whose resistivity vary between 450–550 Ω.m with a good resistivity contrast. The formation zone lying below 160 to 480 m lateral distance is

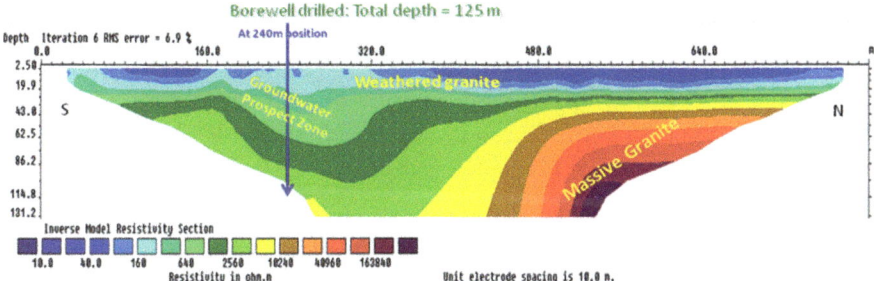

Fig. 11.1 Depicted 2D inverted resistivity model between Mandollagudem and Toorpugudem village showing a clear cut resistivity contrast between the low resistivity weathered/highly granite and the high resistivity massive granite formation in a hard rock aquifer system (*after* Kumar et al. 2016b)

delineated as the most prospect and potential area for groundwater exploitation. This is the major repository for groundwater reserve, which can be exploited to a deeper depth of 200 m or beyond for long term sustainability of the natural resources. The characteristics geological rock strata and the hydrogeological condition were highly favourable for groundwater exploration and development at this anomalous zone of the granitic rock. On the other hand, there is a clear cut indication of massive granite rock with a high resistivity of the order of ~4 × 10^4 Ω.m from ~50 m down to a deeper 131 m showing increase in resistivity with depth on northern side of the profile (Fig. 11.1). This type of high resistivity massive structure was totally devoid of water and must be avoided for groundwater exploration and exploitation strategy in a hard rock aquifer system. On overall interpretation of this resistivity model it was found that the southern side of the inverted section was more favourable in terms of groundwater prospects and exploration rather than the northern side leading to higher recharge to the groundwater table in the southern side, which leads to highly favourable hydrogeological condition for groundwater exploration and development in a granitic hard rock aquifer system (Fig. 11.1).

Delineation of a weak zone/low resistivity formation: A potential target for groundwater resources in a granite hard rock aquifer

Another study successfully conducted for mapping the potential groundwater zone in a complex granitic terrain where there is an acute shortage of drinking water in another village within Choutuppal mandal, Nalgonda district Telangana India (Kumar et al. 2017c). The 2D inverted resistivity model at Yallagiri site is quite exciting as it shows a clear cut low resistivity weak zone connected from the near surface layer and is prominent until the deeper depth up to 131 m (Fig. 11.2). This low resistivity zone was sandwiched between two high resistivity massive granite on either side of it with a significant resistivity contrast of 11,350 Ω.m among them and it lies between 480 and 640 m lateral distance (Fig. 11.2). The low resistivity zone between ~200 and <1000 Ω.m was located right below 540 m lateral distance is a major conduit of groundwater flow/recharge and is a prominent anomalous zone, which act as the

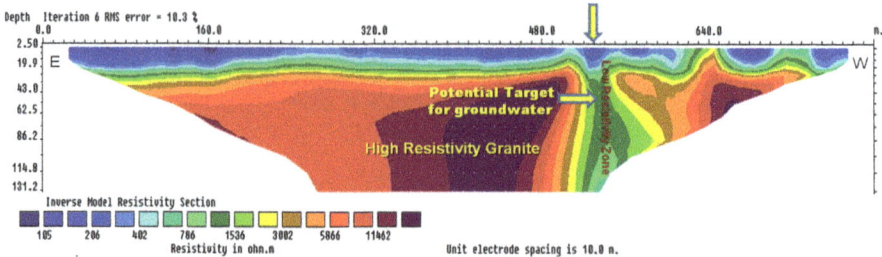

Fig. 11.2 Depicted high resolution resistivity section at Yallagiri site showing high resistivity granite towards eastern side and a low resistivity deep vertical contrasting prominent zone in west direction—is the potential target for groundwater exploration and exploitation (*after* Kumar et al. 2017c)

potential target for groundwater exploration and exploitation (Fig. 11.2) from shallow to deeper depths in this complex geological setting of Choutuppal area.

Groundwater Exploration in Granitic Terrain, Andhra Pradesh

We achieved in mapping the shallow (<100 m) as well deep (>200 m) weathered-fractured zones, which is good repository for groundwater prospecting (*after* Rao et al. 2008). At this site electrical resistivity tomography was carried out along W-E direction in a granitic terrain, Andhra Pradesh using two different array as well as Self Potential (SP) survey for a comparative study for groundwater exploration. The Wenner-Schlumberger 2D model (Fig. 11.3) depicted that up to a depth of 15 m the model shows a low resistivity layer from west to east direction, followed by increasing resistivity layers up to 40 m and then a there is a change in the geological variation at a depth of about 60 m and a sharp resistivity contrast is seen at a depth of 65 m towards westward but in the eastern direction the situation is different, a very high resistivity body is sitting at a depth of 78 m. Overall this resistivity model shows a layered geological structure (Fig. 11.3).

Whereas the pole-pole 2D resistivity model shown in Fig. 11.4 shows a layered geological structure with two high resistivity body separating almost at the center of the profile (Fig. 11.4). As it is seen and even in continuation with Wenner-Schlumberger 2D model (Fig. 11.3) above, the high resistivity is continuing upto

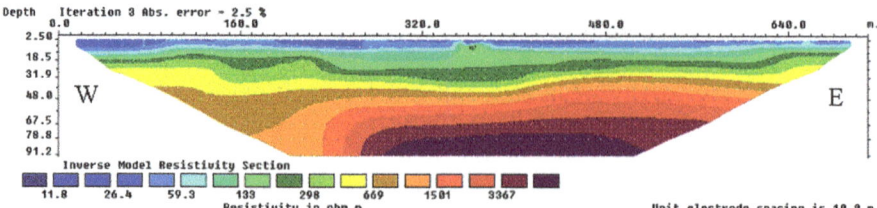

Fig. 11.3 2D Electrical resistivity model depicting a layered geological formation in a granitic terrain <100 m depth (*after* Rao et al. 2008)

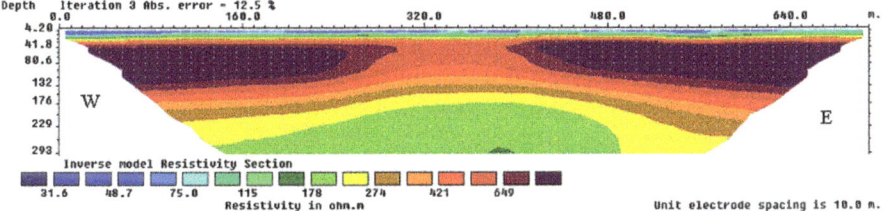

Fig. 11.4 2D Electrical Resistivity Model depicting a layered geological formation in a granitic terrain >200 m depth (*after* Rao et al. 2008)

150 m depth. Beyond 150 m depth there is a decrease in the resistivity value and at a depth of ~200 m there is a sudden drop in the resistivity value and is of the order of 200–250 Ω.m and from 210 to 293 m depth (Fig. 11.4) it revealed a further low resistivity zone, resistivity of the order < 200 Ω.m, which indicates the fissured/fractured zone saturated with water, which is the potential target for groundwater exploration.

SP Profile

The SP profile shown (Fig. 11.5) along the center of the 2D resistivity line (Figs. 11.3 and 11.4) was carried out, which indicated the anomalous zones as shown below.

As seen in the pole-pole 2D resistivity model (Fig. 11.4) there is correlation of SP variation called natural subsurface potential of the earth with the 2D resistivity model (Fig. 11.5a) along the center and it shows positive anomalies, which are the indirect indication for groundwater prospecting for exploitation of the resource.

Exploration and exploitation of deep groundwater resources in a drought prone area as well as delineation of a geological major fault structure

An integrated study was undertaken due to a major problem for groundwater resources in a drought prone Tadipatri mandal, Anantapur district in Andhra Pradesh. Both hydrogeological and high resolution geophysical surveys were carried out to

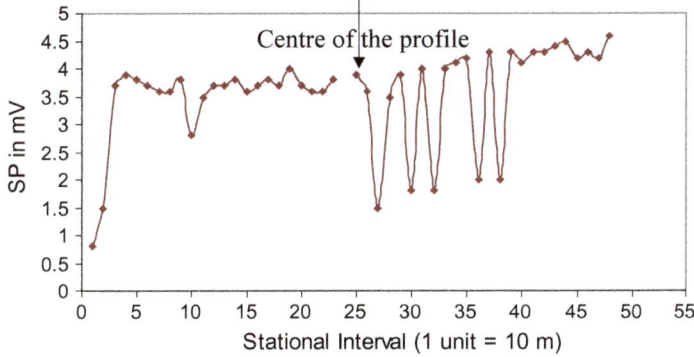

Fig. 11.5 SP Anomaly variation along the center of the 2D resistivity line

understand the complex geological setting its characteristics constituting both the hard and soft rocks (Kumar et al. 2015a, b). The high resolution electrical resistivity tomography was conducted at one of the specific site—Tummalapenta in order to evaluate and mapped the shallower to deeper geological strata and to delineate, demarcate the prospect groundwater zone for deeper groundwater exploration and exploitation of the groundwater resources (Fig. 11.6). The modeled resistivity inverted section clearly shows a large resistivity contrast of the subsurface geological scenario with a low resistivity zone <100 Ω.m towards the southern side (Fig. 11.6) as compared to a very high resistivity massive formation towards the northern side of the resistivity section (Fig. 11.6). Nevertheless, a major fault was clearly revealed and mapped, which separates the low resistivity zone and a high resistivity with a sharp resistivity contrast right below ~320 m lateral distance (Fig. 11.6). This inferred fault was very deep and was extended up to more than 250 m depth. The resistivity of the formation to the north of the fault structure was very high ranging from >17,000 Ω.m to 274 k Ω.m, which suggests no prospect groundwater zone(s) towards the northern side of the section (Fig. 11.6) where there was all the massive formation. The major potential target for groundwater is only towards the left side of the major fault structure (Fig. 11.6) showing a sharp change in resistivity all along the fault structure with a large resistivity contrast on either side of the fault (Fig. 11.6) was the key location for groundwater exploration and exploitation from shallower to deeper depths.

Exploration of deeper groundwater resources in a gneissic hard rock geological set up

A research project on groundwater was undertaken in order to delineate the shallow to deep groundwater resources in a plateau region of Chhotanagpur Gneissic Complex (CGC) near Ranchi, Jharkhand. The exploration and exploitation of the groundwater

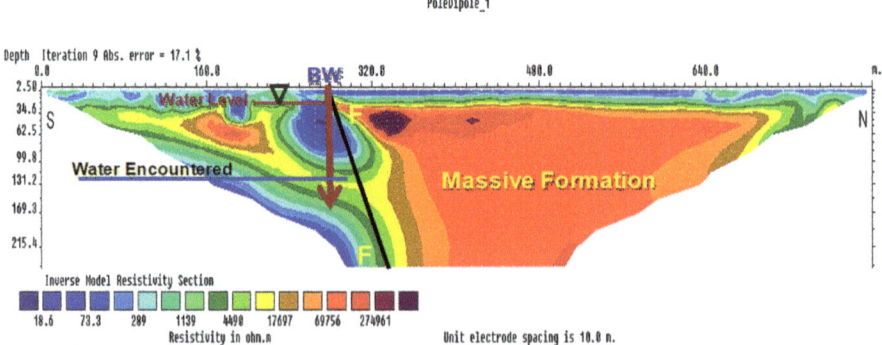

Fig. 11.6 Depicted high resolution 2D inverted resistivity model showing large resistivity contrast between the lowest and the highest resistivity values as well as delineated a major fault (**F-F**) structure and the groundwater potential zone towards the southern side of the fault plane, while on the right side of the fault structure; it is totally a very massive formation with no prospect for groundwater exploitation (*after* Kumar et al. 2015b)

resources was successfully achieved both at the shallow and deeper depths in a metamorphic East Indian shield region, which constitute the complex geological set up at Ghar Khatanga (study area) in Namkum area near Ranchi, Jharkhand. The high resolution 2D inverted resistivity section located in block-D and laid in S–N direction revealed a clear cut both the resistive and the conductive zones of geological scenario, which are of much interest and significance in terms of groundwater exploitation and development (Kumar et al. 2016a). The model resistivity section depicted a layered resistivity structure up to a depth of ~30 m (Fig. 11.7). Later down to 30 m depth the subsurface geological strata was heterogeneous in nature and it revealed a clear-cut high and low resistivity formations from 60 m to until bottom 170 m depth, which is located almost in the centre of the resistivity section (Fig. 11.7). The resistivity contrast between these two—the highest and the lowest resistivity geological formations was inferred as a fault zone (**F-F**) separating the high and low resistivity zone. It depicted a large resistivity contrast at the contact of the fault zone (Fig. 11.7). The resistivity contrast developed between the two contrasting formations is of the order of ~66,800 Ω.m, which is very high and depicts that one side of the fault is very resistive and the other side is very conductive, which is delineated between 70 and 170 m depth (Fig. 11.7). The contact of the high and low resistivity zone is the main potential target for groundwater exploration and development in this geological setting of the study area.

Delineation of conductive mineralization in Mafic–Ultramafic Intrusive Rocks

High Resolution Electrical Resistivity and Induced Polarization Tomography study was conducted in and around Betul district, Madhya Pradesh Central India for sulphide minerals prospecting and exploration. Sulphide deposits in the Betul belt define a spectrum from Zn-Cu type to Zn-Pb-Cu type of mineralization (Ghosh and Praveen 2008). The study was aimed to look and locate both expensive and inexpensive type of minerals. One of the study was conducted near Padhar area, Betul district, which is underlain by mafic–ultramafic complex of high density rocks. The

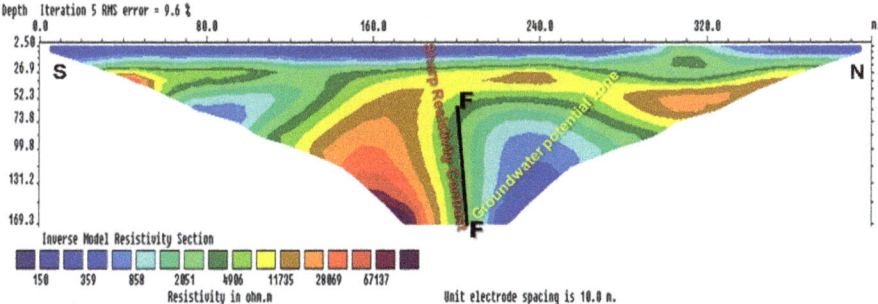

Fig. 11.7 Depicted high resolution 2D inverted resistivity model with heterogeneous subsurface formation. The model resistivity clearly delineated a high and a low resistivity geological formation, which is inferred as fault (**F-F**) structure. All along the fault plane it is the most prospect groundwater zone in the given geological setting of the area (*after* Kumar et al. 2016a)

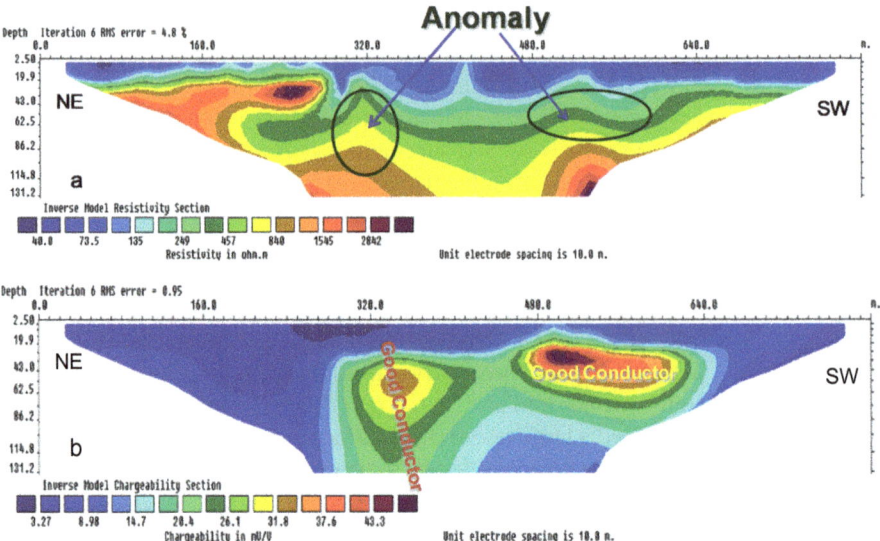

Fig. 11.8 a Depicted high resolution 2D inverted resistivity model over ultra mafic, pyroxenite gabbro host rock showing substantial resistivity contrast right over the two conductive anomalies near 320 m and around 480 m lateral distance respectively. **b**. Shows the high resolution 2D inverted chargeability model, which revealed two distinct well defined district conducting anomalies extending in vertical and horizontal directions at Padhar area, Betul Madhya Pradesh, Central India (*after* Kumar et al. 2016b)

high resolution electrical resistivity 2D inverted model depicted two sharp anomalies, which are inferred to be due to the metallic mineralization preserved within the host rock (Fig. 11.8a). The resistivity model indicated a disturbed geological structure with a large variation in resistivity between 20 Ω.m and 5,200 Ω.m up to a 131 m. Nevertheless, the 2D inverted chargeability model (Fig. 11.8b) clearly revealed two highly conducting mineralized body corresponding to two major anomalies as depicted from the resistivity model (Fig. 11.8a). The modeled data revealed a sharp contrast with a significant variation in chargeability with respect to the host rock. The 2D inverted chargeability model had evolved two prominent conducting anomalies where the magnitude of chargeability values are ~30–40 mV/V and ~25–50 mV/V lying between 300 and 600 m lateral distance (Fig. 11.8b). The nature of one conducting body is circular while the other is elongated in shape and they are lying within the depth range of 25–90 m from the surface. These anomalies indicated the strong presence of disseminated sulphide mineral deposit namely porphyry copper or the massive magnetite minerals where the body is extending both in lateral and vertical directions. The modeled data revealed the width of the main conducting anomalies are close to 90–150 m (Fig. 11.8b). The integrated study comprising the type of geophysical anomalies, detailed geological and geochemical characterization

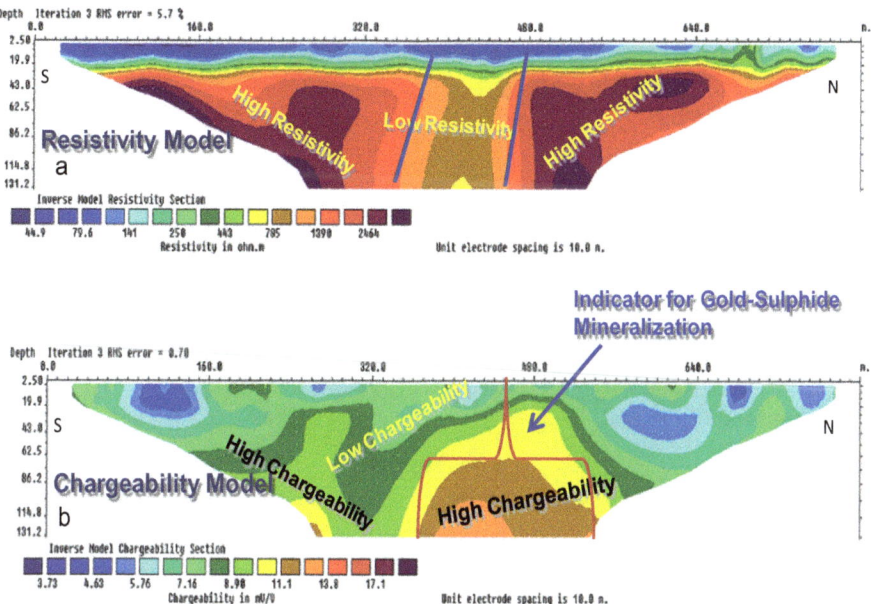

Fig. 11.9 a Depicted high resolution resistivity model with two high resistivity bodies separated by a low resistivity vertical zone with a resistivity contrast ≥1000 Ω.m. **b**. chargeability model shows a large volume of rock mass with the high chargeability ~10–14 mV/V value corresponding to the low resistivity vertical geological feature within the depth range of 40–131 m depth at the Pharsabahar area (*after* Kumar et al. 2017b)

of rocks had proved the presence of magnetite, disseminated form of pyrite and chalcopyrite with high concentration of Nickel (Ni) and Chromium (Cr) at this Padhar area (Kumar et al. 2016c).

Mapping quartz vein geological structure: Implication for Gold-Sulphide Mineralization

Yet another study deploying high resolution electrical resistivity & induced polarization tomography was conducted at Parsabahar area Jashpur district, Chhattisgarh State in Central India in the search of gold-sulphide mineralization. The aim of the research work to investigate in detail and delineate the quartz vein geological structure and their significance in terms of both precious and non-precious metallic mineralization within the host rock. The high resolution modelled data answered the main geological structure and clearly mapped the low resistivity vertical zone whose resistivity vary from 785–1390 Ω.m between ~30 m to more than 131 m depth. The low resistivity vertical geological feature was showing a sharp resistivity contrast between the low and the high resistivity geological formation (Fig. 11.9a). The resistivity model data also revealed high resistivity bodies with resistivity >400 Ω.m on either side of the low resistivity vertical zone (Fig. 11.9a) and this high resistivity

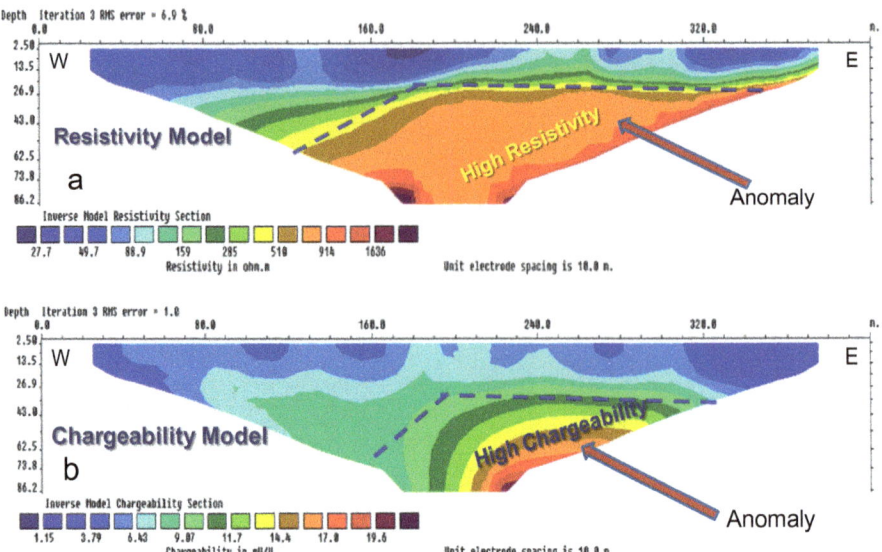

Fig. 11.10 **a** Depicted high resolution resistivity model showing high resistivity geological body with a semi-oval shaped, extending more towards eastern side and it is at a shallower depth compared to the western side of the profile. **b** chargeability model showing clear cut chargeability anomaly between ~10–22.6 mV/V was prominent from ~45 to 86 m depth, which represents metallic minerals with sulphide within the host rock (*after* Kumar et al. 2017a)

body is extending from close to 35–40 m depth to until 131 m depth and beyond. It is very clearly seen that the near surface layer shows resistivity < 100 Ω.m, which is inferred as the weathered rock materials. However, 2D inverted chargeability model describes and clearly revealed high chargeability value ~10–14 mV/V between 40 m and >131 m depth (Fig. 11.9, corresponding to low resistivity vertical zone and this zone of high chargeability is inferred as the clear cut indicator of gold-sulphide mineralization at the Parsabahar area (Fig. 11.9b). Here at the deeper depth the chargrability is the maximum and highly concentrated and it covered a substantial volume of rock mass, which is of much importance in terms of metallic mineralization in the ultramafic–mafic-granite complex of Jashpur, Bastar craton, Central India.

Another new example of mineralization in Sarni area, Betul district, Madhya Pradesh

The Sarni area, which is underlain by mafic and ultramafic intrusive rock type in Betul district, Madhya Pradesh had been explored extensively for sulphide mineralization using high resolution electrical tomography technique (Kumar et al. 2017a). The 2D inverted resistivity models at the Sarni area was quite interesting and thrilling in terms of conducting mineral resources associated with sulphides within the host rock when analyzed and viewed from the large density of good quality electrical tomography dataset. The interpretation at one of the site was presented here in order to know the magnitude of variation in the resistivity of geological structure and their corresponding chargeability developed due to sulphide mineralization inherent within the host rock, which dictates about the nature, quality and quantity of mineralization with a rock mass. One of the 2D tomography profiles revealed a clear cut high resistivity structural feature of the order of 900 Ω.m to a maximum 2900 Ω.m for the deeper geological strata (Fig. 11.10a) and this high resistivity variation appears like a semi-oval shaped. This high resistivity body is extending more towards eastern side and it is at a shallower depth compared to the western side of the profile. The modelled resistivity section depicted a weathered rock material with a resistivity <100 Ω.m from west to east direction in the near surface layer of mafic and ultramafic intrusive rock mass. The significance of this resistivity model suggests and it evaluates, there is a large range of resistivity revealed with no well defined low resistivity zones that might indicate mineral deposits within the host rock body (Fig. 11.10a). On the other hand, 2D inverted chargeability model resulted from induced polarization data shows a sharp and prominent anomaly (Fig. 11.10b) corresponding to high resistivity variation within the host rock in this area. This high chargeability anomaly ranging between ~0–22.6 mV/V is prominent from ~5 to 86 m depth (Fig. 11.10b) where the highest value of chargeability 22.6 mV/V was concentrated at the deeper depth, which suggests a strong association of metallic minerals with the sulphide within the host rock. Thus the presence of sulphide bearing mineral is clearly revealed in the present geological setting in this area.

11.3   Conclusions

High resolution electrical tomography is state-of-the-art geophysical technique in mapping the various geological terrains with different characteristics, which are inherent with different rock types. The advantage of recording the full waveform data using 4 channel Terrameter® LS system during the whole measurement cycle in data acquisition stage is immense, which helps to visualize the received measured signals in detail and is totally uncompromised. Especially during the induced polarization (IP) measurement, it records the discharge of the signals with a very high resolution, which helps in analysing the different conducting minerals lying underneath the subsurface at different depths of the earth. This technique is powerful in distinguishing different variety of rocks with the highest resolution as well as capable of acquiring large volume of data, which is utmost required for a meaningful and sound interpretation over a simple and a complex geological settings and with a maximum confidence level and accuracy in any hydrogeological, geological, structural and mineralogical interpretation. The quantitative range of resistivity and chargeability values and their contrasts evolved was very significant and helpful in demarcating the zone of interest for groundwater prospecting and exploitation of the natural resource as well as delineating the various geological structures, which controls the occurrence and movement of groundwater within the host rock. Equally it is valid for mineral exploration and prospecting where different values of modelled chargeability in association with resistivity had guided in inferring the different conductive metallic mineralisation zones in various geological settings.

Acknowledgements Author is grateful to Dr. V. M. Tiwari Director, CSIR-National Geophysical Research Institute, Hyderabad, India, for his kind permission and encouragement for publishing the current progress of research work in the field of groundwater exploration, prospecting and exploitation as well as in mineral exploration conducted in different geological terrains of the country. The financial assistance was provided by CSIR and sponsored projects are duly acknowledged. The CSIR-NGRI reference number of the manuscript is NGRI/Lib/2018/Pub-no.37.

References

ABEM (2012) ABEM instruction manual—terrameter LS, p 110

Ghosh B, Praveen MN (2008) Indicator minerals as guides to base metal sulphide mineralization in Betul Belt, central India. Jour. Earth Syst Sci 117(4):521–536

Kumar D, Mondal S, Rajesh K, Rangarajan R (2015a) Mapping shallow subsurface hard and soft rock structure in complex geological terrain for groundwater prospecting and exploitation in Anantapur district, Andhra Pradesh, presented and published in abstract volume in 52nd annual convention and meeting of IGU on near surface earth system sciences held during 3–5 Nov 2015 at NCAOR Goa, India, 10

Kumar D, Mondal S, Rajesh K, Rangarajan R (2015b) Mapping subsurface hard and soft rock with complex geological terrain for groundwater prospecting in Tadipatri Mandal, Anantapur district, Andhra Pradesh, technical report no NGRI-2015-GW-879

Kumar D, Mondal S, Rajesh K, Rangarajan R (2016a) Mapping subsurface hard rock and complex geological terrain for groundwater prospecting in the new campus at IIAB, Ranchi, technical report no NGRI-2016-GW-909

Kumar D, Mondal S, Nanda MJ, Harini P, Soma Sekhar BMV, Sen MK (2016b) Two dimensional electrical resistivity tomography (ERT) and time domain induced polarization (TDIP) study in hard rock for groundwater investigation: A case study at Choutuppal Telangana. India, Arab J Geosci Springer Publ 09(5):1–15. https://doi.org/10.1007/s12517-016-2382-1

Kumar D, Subba Rao DV, Sridhar K, Satyanarayanan M, Patro, Prasanta K (2016c) Integrated geophysical and geological studies for mineral prospecting in Betul-Chindwara belt (BCB), central India. J Geol Soc India Springer Publ 87(4): 383–396. https://doi.org/10.1007/s12594-016-0406-9

Kumar D, Mondal S, Soma Sekhar, BMV, Balaji T, Nandan MJ, Rangarajan R (2017a) Full waveform inversion of 2D resistivity and IP data for groundwater exploration in granite aquifers: implication for complex tectonical setting. J Appl Hydrol on 16 Sep 2017

Kumar D, Satyanarayanan M, Subba Rao DV (2017b) High resolution geophysical and geological study in Sarni-Mafic-ultramafic complex, Madhya Pradesh: evidence for sulphide mineralization in central India, presented as a keynote address and published as an abstract in J Geosci Res special volume no.1, 280, published by Gondwana geological society, Nagpur in national symposium on challenges and strategies in mineral exploration and mining held during 9–10, Feb 2017 at RTM nagpur university, Nagpur

Kumar D, Subba Rao DV, Mondal S, Sridhar K, Rajesh K, Satyanarayanan M (2017c) Gold-sulphide mineralization in ultramafic-mafic-granite complex of Jashpur, bastar craton, central India: evidences from geophysical studies. J Geol Soc India Springer Publ 90(2):147–153. https://doi.org/10.1007/s12594-017-0692-x

Rao VA, Kumar D, Nagaiah E, Kumar GA, Ahmed S (2008) Multi electrode resistivity imaging survey for groundwater exploration in special police 7th Batallion, Dichpally Nizamabad, Andhra Pradesh, technical report no NGRI-2008

Lightning Source UK Ltd.
Milton Keynes UK
UKHW022024270822
407877UK00001B/4